Food Engineering Series

Springer's *Food Engineering Series* is essential to the Food Engineering profession, providing exceptional texts in areas that are necessary for the understanding and development of this constantly evolving discipline. The titles are primarily reference-oriented, targeted to a wide audience including food, mechanical, chemical, and electrical engineers, as well as food scientists and technologists working in the food industry, academia, regulatory industry, or in the design of food manufacturing plants or specialized equipment.

More information about this series at http://www.springer.com/series/5996

Bosoon Park • Renfu Lu
Editors

Hyperspectral Imaging Technology in Food and Agriculture

Springer

Editors
Bosoon Park
U.S. Department of Agriculture
Agricultural Research Service
Athens, GA, USA

Renfu Lu
U.S. Department of Agriculture
Agricultural Research Service
East Lansing, MI, USA

ISSN 1571-0297
Food Engineering Series
ISBN 978-1-4939-2835-4 ISBN 978-1-4939-2836-1 (eBook)
DOI 10.1007/978-1-4939-2836-1

Library of Congress Control Number: 2015946871

Springer New York Heidelberg Dordrecht London

Printed on acid-free paper

Springer Science+Business Media LLC New York is part of Springer Science+Business Media (www.springer.com)

Preface

Over the past 15 years, we have witnessed rapid increases in R&D activities and applications of hyperspectral imaging in food and agriculture. Hyperspectral imaging integrates the main features of imaging and spectroscopic techniques so as to expand our capability of detecting minor or more sophisticated features or characteristics of an object both spatially and spectrally that would otherwise be difficult or impossible to accomplish with either imaging or spectroscopic technique. This book is intended to give a broad, comprehensive coverage of hyperspectral imaging technology and its applications in food and agriculture. It is written for both researchers who are currently engaged or interested in this area of research and advanced-level students who want to acquire special knowledge about basic concepts, principles, and applications of hyperspectral imaging. The book is organized into two parts. The first part gives readers a general introduction to the instrumentation and implementation modalities of hyperspectral imaging technology, hyperspectral image processing, and analysis techniques; it starts with basic image and spectroscopic data processing and analysis methods, followed by the specific methods and techniques for processing and analysis of hyperspectral images for quality and safety classification and prediction. The second part, consisting of ten chapters, covers a range of applications of hyperspectral imaging technology from food quality and safety inspection to plant health detection and monitoring to precision agriculture and real-time as well as microscope applications.

Athens, GA, USA — Bosoon Park
East Lansing, MI, USA — Renfu Lu

Contents

Contributors

Pepito M. Bato University of the Philippines Los Baños, Los Baños, Philippines

Deepak Bhatnagar U.S. Department of Agriculture, Agricultural Research Service, Southern Regional Research Center, New Orleans, LA, USA

Robert L. Brown U.S. Department of Agriculture, Agricultural Research Service, Southern Regional Research Center, New Orleans, LA, USA

James E. Burger (deceased)

Chris Calkins Animal Science, University of Nebraska–Lincoln, Lincoln, NE, USA

Haiyan Cen Department of Biosystems and Agricultural Engineering, Michigan State University, East Lansing, MI, USA

College of Biosystems Engineering and Food Science, Zhejiang University, Hangzhou, China

Thomas E. Cleveland U.S. Department of Agriculture, Agricultural Research Service, Southern Regional Research Center, New Orleans, LA, USA

Kim Cluff Department of Biological Systems Engineering, University of Nebraska–Lincoln, Lincoln, NE, USA

Stephen R. Delwiche U.S. Department of Agriculture, Agricultural Research Service, Food Quality Laboratory, Beltsville, MD, USA

Aoife A. Gowen School of Biosystems Engineering, University College Dublin, Dublin, Ireland

Zuzana Hruska Geosystems Research Institute, Mississippi State University, Stennis Space Center, MS, USA

Taichi Kobayashi University of Miyazaki, Miyazaki, Japan

Won Suk Lee Agricultural and Biological Engineering, University of Florida, Gainesville, FL, USA

Changying Li College of Engineering, University of Georgia, Athens, GA, USA

Renfu Lu U.S. Department of Agriculture, Agricultural Research Service, East Lansing, MI, USA

Fernando Mendoza U.S. Department of Agriculture, Agricultural Research Service, Sugarbeet and Bean Research Unit, East Lansing, MI, USA

Department of Biosystems and Agricultural Engineering, Michigan State University, East Lansing, MI, USA

Govindarajan Konda Naganathan Department of Biological Systems Engineering, University of Nebraska–Lincoln, Lincoln, NE, USA

Masateru Nagata Faculty of Agriculture, University of Miyazaki, Miyazaki, Japan

Bosoon Park U.S. Department of Agriculture, Agricultural Research Service, Athens, GA, USA

Ashok Samal Computer Science and Technology, University of Nebraska-Lincoln, Lincoln, NE, USA

Bim P. Shrestha Kathmandu University, Dhulikhel, Nepal

Jeyamkondan Subbiah Department of Biological Systems Engineering, University of Nebraska–Lincoln, Lincoln, NE, USA

Department of Food Science and Technology, University of Nebraska–Lincoln, Lincoln, NE, USA

Jasper G. Tallada Cavite State University, Indang, Philippines

U.S. Department of Agriculture, Agricultural Research Service, Manhattan, KS, USA

Weilin Wang College of Engineering, University of Georgia, Athens, GA, USA

Monsanto Company, MO, USA

Chenghai Yang Aerial Application Technology Research Unit, U.S. Department of Agriculture, Agricultural Research Service, College Station, TX, USA

Haibo Yao Geosystems Research Institute, Mississippi State University, Stennis Space Center, MS, USA

Seung Chul Yoon U.S. Department of Agriculture, Agricultural Research Service, US National Poultry Research Center, Athens, GA, USA

Part I
Image and Spectral Analysis Techniques

Chapter 1
Introduction

Renfu Lu and Bosoon Park

For the past two decades, we have witnessed the rapid developments and wide applications of imaging and spectroscopic technologies in the food and agricultural industries. Conventional imaging technology, whether monochromatic (i.e., white/black) or polychromatic (i.e., color-based), allows acquiring two- or even three-dimensional spatial information about an object. Using image processing and analysis methods and techniques, we quantify or classify the spatial characteristics or color attributes of food and agricultural products and crops or plants growing in the field. Since surface or external characteristics are important to the consumer's perception about product quality and, in many cases, are also a good indicator of product maturity and/or internal quality, imaging technology is being widely used in inspecting, monitoring, and grading a large class of agricultural and food products based on color, size/shape, and surface texture during postharvest handling, packing, and processing (Ruiz-Altisent et al. 2010; Davies 2010). Applications of imaging technology are also increasing in production agriculture such as precision chemicals application, crop yield monitoring, vision guidance for off-load vehicles, and robotic or automatic agricultural operations from seeding to weeding to harvesting (Lee et al 2010). Despite all these successful applications, conventional imaging technology generally is not suitable for detection or assessment of intrinsic properties and characteristics of products, whether they are physical and/or chemical (e.g., moisture, protein, sugar, acid, firmness or hardness, etc.).

R. Lu (✉)
U.S. Department of Agriculture, Agricultural Research Service, East Lansing, MI, USA
e-mail: renfu.lu@ars.usda.gov

B. Park
U.S. Department of Agriculture, Agricultural Research Service, Athens, GA, USA
e-mail: bosoon.park@ars.usda.gov

B. Park, R. Lu (eds.), *Hyperspectral Imaging Technology in Food and Agriculture*,
Food Engineering Series, DOI 10.1007/978-1-4939-2836-1_1

Spectroscopy, on the other hand, represents another major class of optical technology, which has been increasingly used in food and agriculture in recent years. It normally covers a portion of the electromagnetic spectrum (e.g., from x-ray to ultraviolet to visible and to infrared), and enables acquiring spectral absorption and scattering information from an object. The technology is especially suitable for quantitative or qualitative analysis of product composition and properties, because the absorption characteristics or spectral signatures of food and agricultural products are related to the chemical properties or composition as well as structural characteristics. In the early days, spectroscopy technology was primarily used as a laboratory tool due to high cost in instrumentation and slow speed or complexity in measurement. However, advances in optics, computer, and chemometric or mathematical methods for analyzing spectral data, along with dramatic reduction in instrument cost, have enabled spectroscopy technology to go well beyond the traditional domain of application nowadays. For instance, near-infrared spectroscopy is now being used for real-time, rapid online analysis, monitoring and inspection of the physical and chemical properties and compositions of many food and agricultural products (Nicolai et al. 2007). The miniaturization of spectrophotometers has further made it possible for on-site, low-cost measurement of the quality or maturity of crops in both pre- and post-harvest. In contrast to imaging technique, spectroscopic measurements normally do not provide spatially-resolved information about products.

Today, the agricultural and food industries are increasingly concerned about the sustainable production and delivery of consistent, high quality and safe food products. For many applications, there is a need for more accurate assessment and classification of food and agricultural products based on their intrinsic characteristics and properties, which may be difficult to achieve with conventional imaging or spectroscopy technology. Food and agricultural products are known for their large variations in properties and composition within and between individual product items. For instance, the quality of individual kernels, in terms of protein and moisture content, for the same lot of wheat can vary greatly (Dowell et al. 2006). Fruit growing in the same orchard or even on the same trees can have large variations in maturity and other postharvest quality attributes. Meat tenderness is greatly influenced by the type, location and direction of meat muscles (Prieto et al. 2009). The soluble solids content and textural properties like firmness vary with location and/or orientation within the same apple fruit and melon (Abbott and Lu 1996; Sugiyama 1999). In the past decade, food safety and security has received increased attention from the government as well as the general public. Prevention and early detection of food contaminants and pathogens is critical to ensure safe production and delivery of food products to the consumer. Pathogen contamination caused by animal feces is one common food safety concern, and there is zero tolerance for fecal contamination on poultry and meat products, which has been imposed in the United States and many other countries. It is difficult to achieve accurate detection of fecal matter from the products using color or monochromatic imaging technology because fecal contaminants on meat and poultry products can be indistinctive, in some cases, from the bovine animal or poultry carcasses. While visible/near-infrared spectroscopy can achieve superior detection results, it cannot pinpoint the exact location and, thus, would miss contaminants that are confined to small areas on the products. In these and many other instances, conventional imaging or spectroscopy technology has been proven insufficient to meet the food safety inspection requirements (Park et al. 2006). Thus, it is desirable or even necessary to develop and deploy a new, more effective inspection system to measure the spatial and temporal variations in the quality and condition of food products and crops growing in the field, and to detect food safety hazards that are present in harvested or processed food products.

In view of the respective merits and shortcomings of imaging and spectroscopy technologies, it is clear that great advantages can be gained if we can combine the major features of these two platforms into one single platform. The integration of imaging and spectroscopy has led to an emerging, new generation of optical technology, called *hyperspectral imaging* or *imaging spectroscopy*. Hyperspectral imaging combines the main features of imaging and spectroscopy to acquire spectral and spatial information from products simultaneously. Depending on application needs, a hyperspectral imaging system can cover a specific spectral range in the ultraviolet (UV), visible, near-infrared (NIR), or shortwave infrared (SWIR) region. The emergency of hyperspectral imaging is closely related to the advances in imaging, spectroscopy, and computer technologies in the past two decades. In the late 80s and early 90s, hyperspectral imaging technology was first used in satellite remote sensing for environmental monitoring, geological search or mineral mapping, atmosphere composition analysis and monitoring, military reconnaissance or target detection, and crop yield or growing condition monitoring or prediction (Moran et al. 1997). Development and application of hyperspectral imaging for quality and safety inspection of agricultural products has not begun until the late 1990s (Lu and Chen 1998; Martinsen and Schaare 1998). Since then, we have seen significant increases in R&D activities in hyperspectral imaging for food and agricultural product evaluation due to the advances in the high-performance digital camera and imaging spectrograph, two key optical components in the hyperspectral imaging system. Over the past 10 years, many technical symposia dedicated to hyperspectral imaging in food and agricultural applications have been held by professional societies like the American Society of Agricultural and Biological Engineers (ASABE), International Society for Optical Engineering (SPIE), and International Commission for Agricultural and Biological Engineering (CIGR). The international journal "*Sensing and Instrumentation for Food Quality and Safety*" (now renamed "*Journal of Food Measurement and Characterization*") published the first special issue "Hyperspectral and Multispectral Imaging for Food Quality and Safety" in 2008 (Lu and Park 2008). Several review articles have been written in recent years about hyperspectral imaging technology and its applications in food and agriculture (Gowen et al. 2007; Ruiz-Altisent et al. 2010). The increasing interest in hyperspectral imaging for food and agriculture applications has been further demonstrated by the exponential increase in the number of scientific publications for the past 15 years. Today, a number of manufacturers[1] (e.g., Headwall Photonics Corporation, MA, USA; Middleton Research in WA, USA; Specim, Finland) are engaged in manufacturing hyperspectral imaging instruments that are suitable for food and agriculture applications.

Earlier research in hyperspectral imaging for food and agricultural product inspection was focused on using the technique as a research tool because of the need to acquire and process a huge amount of data. As optics and computer technologies have evolved rapidly over the past 15 years, we have seen more diverse, practical applications of hyperspectral imaging in food and agriculture, including online inspection of food quality and safety and microscopic hyperspectral imaging for food safety or pathogen detection (Ariana and Lu 2008; Chao et al. 2010; Park et al. 2012; Yoon et al. 2009). The expanding use of hyperspectral imaging in food and agriculture is expected; the technology will replace imaging or spectroscopy technology in many applications.

This book is intended to give a broad, comprehensive coverage of hyperspectral imaging technology and its applications in food and agriculture. It is written for both researchers who are currently engaged or interested in this area of research and advanced-level students who want to acquire special knowledge about basic concepts, principles and applications of hyperspectral imaging. The book covers basic image and spectroscopic data processing and analysis methods, followed by the common methods and techniques for processing and analysis of hyperspectral images for quality and safety classification and prediction. It then presents a range of applications of hyperspectral imaging technology from food quality and safety inspection to plant health detection and monitoring to precision agriculture and real-time application.

References

Abbott JA, Lu R (1996) Anisotropic mechanical properties of apples. Trans ASAE 39 (4):1451–1459

Ariana DP, Lu R (2008) Quality evaluation of pickling cucumbers using hyperspectral reflectance and transmittance imaging: part I. Development of a prototype. Sensing Instrum Food Qual Safety 2(3):144–151

Chao K, Yang CC, Kim MS (2010) Spectral line-scan imaging system for high-speed non-destructive wholesomeness inspection of broilers. Trends Food Sci Technol 21(3):129–137

Davies ER (2010) The application of machine vision to food and agriculture: a review. Imaging Sci J 57(4):197–217

Dowell FE, Maghirang EB, Graybosch RA, Baenziger PS, Baltensperger DD, Hansen LE (2006) Automated single-kernel sorting to select for quality traits in wheat breeding lines. Cereal Chem 83(5):537–543

[1] Mention of commercial companies or products is solely for providing the reader with factual information, and it does not imply the endorsement of the USDA or exclusion of those not mentioned.

Gowen AA, O'Donnell CP, Cullen PJ, Downey G, Frias JM (2007) Hyperspectral imaging—an emerging process analytical tool for food quality and safety control. Trends Food Sci Technol 18(12):590–598

Lee WS, Alchanatis V, Yang C, Hirafuji M, Moshou D, Li C (2010) Sensing technologies for precision specialty crop production. Comput Electron Agric 74(1):2–33

Lu R, Chen YR (1998) Hyperspectral imaging for safety inspection of foods and agricultural products. In: Chen YR (ed) SPIE proceedings—pathogen detection and remediation for safe eating, vol 3544. SPIE, Bellingham, pp 121–133

Lu R, Park B (ed) (2008) Hyperspectral and multispectral imaging for food quality and safety—a special issue. Sensing Instrum Food Qual Safety 2(3):131–132

Martinsen P, Schaare P (1998) Measuring soluble solids distribution in kiwifruit using near-infrared imaging spectroscopy. Postharvest Biol Technol 14(3):271–281

Moran MS, Inoue Y, Barnes EM (1997) Opportunities and limitations for image-based remote sensing in precision crop management. Remote Sens Environ 61(3):319–346

Nicolaï BM, Beullens K, Bobelyn E, Peirs A, Saeys W, Theron KI (2007) Nondestructive measurement of fruit and vegetable quality by means of NIR spectroscopy: A review. Postharvest Biology and Technology 46(3):99–118

Park B, Lawrence KC, Windham WR, Smith DP (2006) Performance of hyperspectral imaging system for poultry surface fecal contaminant detection. J Food Eng 75(3):340–348

Park B, Yoon SC, Lee S, Sundaram J, Windham WR, Hinton A Jr, Lawrence KC (2012) Acousto-optic tunable filter hyperspectral microscope imaging for identifying foodborne pathogens. Trans ASABE 55(5):1997–2006

Prieto N, Roehe R, Lavín P, Batten G, Andrés S (2009) Application of near infrared reflectance spectroscopy to predict meat and meat products quality: a review. Meat Sci 83(2):175–186

Ruiz-Altisent M, Ruiz-García L, Moreda GP, Lu R, Hernández-Sanchez N, Correa, EC, Diezma B, Nicolaï BM, García-Ramos J (2010) Sensors for product characterization and quality of specialty crops - A review. Computers and Electronics in Agriculture 74(2):176–194

Sugiyama J (1999) Visualization of sugar content in the flesh of a melon by near-infrared imaging. J Agric Food Chem 47(7):2715–2718

Yoon SC, Lawrence KC, Siragusa GR, Line JE, Park B, Feldner PW (2009) Hyperspectral reflectance imaging for detecting a foodborne pathogen: campylobacter. Trans ASABE 52 (2):651–662

Chapter 2
Basics of Image Analysis

Fernando Mendoza and Renfu Lu

2.1 Introduction

Image analysis is used as a fundamental tool for recognizing, differentiating, and quantifying diverse types of images, including grayscale and color images, multi-spectral images for a few discrete spectral channels or wavebands (normally less than 10), and hyperspectral images with a sequence of contiguous wavebands covering a specific spectral region (e.g., visible and near-infrared). Earlier works on image analysis were primarily confined to the computer science community, and they mainly dealt with simple images for such applications as defect detection, segmentation and classification. Nowadays, image analysis is becoming increasingly important and widespread because it can be done more conveniently, rapidly and cost effectively (Prats-Montalbán et al. 2011). Image analysis relies heavily on machine vision technology (Aguilera and Stanley 1999). The explosive growth in both hardware platforms and software frameworks has led to significant advances in the analysis of digital images.

Image analysis has been applied to many different fields of science and technology. For example, it has been used to assess or quantify the external characteristics (i.e., color, size, shape and surface texture) and internal structures (architecture and/or connectivity of the material constituents) of food products.

F. Mendoza
U.S. Department of Agriculture, Agricultural Research Service, Sugarbeet and Bean Research Unit, 524 S. Shaw Lane, Room 105A, East Lansing, MI 48824, USA

Department of Biosystems and Agricultural Engineering, Michigan State University, East Lansing, MI, USA
e-mail: fmendoza@msu.edu

R. Lu (✉)
U.S. Department of Agriculture, Agricultural Research Service, East Lansing, MI, USA
e-mail: renfu.lu@ars.usda.gov

B. Park, R. Lu (eds.), *Hyperspectral Imaging Technology in Food and Agriculture*,
Food Engineering Series, DOI 10.1007/978-1-4939-2836-1_2

Commercial machine vision units are readily available to meet the requirements of automatic inspection for the food processing and packaging industries. As consumers are demanding better quality and safer food products, there is an increasing need for rapid and non-destructive quality evaluation of foods. In recent years, new imaging-based inspection techniques, such as multispectral and hyperspectral imaging, have been developed for quality assessment of a variety of foods, which have overcome some of the drawbacks of traditional human and instrumental inspection techniques (Du and Sun 2007). These methods, which are based on the automatic detection of various image features, correlate well with quality attributes of foods that are related to the sensorial, chemical, and physical properties (Valous et al. 2009a).

It is important to note that image analysis is part of a wider field known as image processing, where the main underlying idea is to improve the visual quality of an image and/or to extract useful information or features. The analysis is based on different image properties such as color, gloss, morphology of the objects, and texture. Image processing actions can be grouped into three sub-areas (Prats-Montalbán et al. 2011):

(a) *Image compression*, which reduces the memory requirements by removing the redundancy present in the image, that is, the image information which is not perceptible to the human eye.
(b) *Image preprocessing*, which consists of improving the visual quality of the image by reducing noise, pixel calibration and standardization, enhancing the edge detection, and making the image analysis step more reliable based on objective and well established criteria. The term image preprocessing, in general, is referred to all manipulations on an image, each of which produces a new image.
(c) *Image analysis*, which usually returns numeric values and/or graphical information about the image characteristics that are suited for classification, defect detection, or prediction of some of the quality properties of the imaged object. The term image analysis is used when the output is a number or decision, not an image.

These processing actions are related and may have a different effect or output for each application. The following sections describe basic concepts and characteristics of a digital image and how they are processed and transformed (improved), give an overview of typical image analysis methods and techniques, and show some application examples in food quality assessment and control.

2.2 The Digital Image

A *digital image* is a numerical 2D or 3D representation of a real physical object or scene, from which we can obtain an accurate spatial (geometric) and/or spectral (for the case of a hyperspectral image) representation with sufficient detail (resolution) for processing, compression, storage, printing and display. A *digital image* may be of *vector* or *raster* type, depending on whether or not the image resolution is fixed (Wikipedia 2012). Raster images are electronic files that consist of discrete picture elements, called *pixels* (short for picture elements). Associated with each pixel is a number that is the average radiance (or *brightness*) of a relatively small area within a scene, representing the color or gray-level at a single point in the image.

Vector graphics formats are complementary to raster graphics, which are the representation of images based on mathematical expressions, as are typically used in computer graphics for images made up of vectors (arrows of direction, points, and lines) that define shapes, as compared to the individual pixels used to represent a raster image. A vector image is resolution independent, which means the image can be enlarged or shrunken without affecting the output quality. The term *digital images* usually refers to raster images, which are also called bitmap images. Moreover, it is also applied to data associated to points scattered over a three-dimensional region, such as produced by tomographic equipment. In that case, each datum is called a *voxel*.

The most commonly used method for the creation of raster images (i.e., digital imaging or image acquisition) is digital photography with a CCD camera in a process called *digitization*. The digitization process requires the mapping of the image on a grid and a quantization of the intensity level. Digitization is the process of converting an analog signal into a digital signal, known as an A/D (analog to digital) conversion. For raster images, an analog voltage signal (from any of several types of imaging sensors), proportional to the amount of light reflected or transmitted by an item being digitized, is divided into discrete numeric values (Castleman 1979). In other words, the process converts an image into a series of small picture square elements or *pixels* that are either black or white (binary), a specific shade of gray (grayscale) or color. Each pixel is represented by a single or series of binary digits, either 1 s or 0 s. By measuring the color (or gray-level for black and white photos) of an image at a large number of points, we can create a digital approximation of the image from which a copy of the original can be reconstructed.

Figure 2.1 depicts an example of the process for creation of a digital image. This figure shows the energy from an illumination source being reflected from a

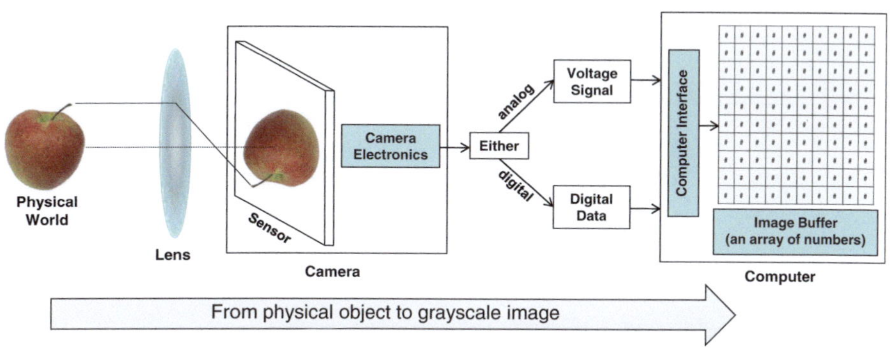

Fig. 2.1 An example of the process for creation of a digital image

scene element (apple). The imaging system collects the incoming energy and focuses it onto an image plane. The front end of the imaging system is an optical lens that projects the viewed scene onto the lens focal plane. The camera sensor is an array of photo cells which measure light at discrete points. The sensor does not directly recognize the color of the incoming light; instead, in a color camera (with 3 CCD sensors) a prism separates the light into three components or color channels i.e., *R* (for red), *G* (for green) and *B* (for blue). The response of each sensor, which is coincident with the focal plane, is proportional to the integral of the light energy projected onto the surface of the sensor. Once a digital image has been created and stored in any media, there is a corresponding digital to analog conversion that allows the computer to present the image in a human readable form on either a display or printer. Displaying an image on a computer monitor or printing the image on a printer are both examples of an analog representation of a digital image. In order to process and analyze images on a computer we first have to digitize them.

Figure 2.2 emphasizes the characteristics of a digital color image. The displayed color image is a two-dimensional matrix of thousands or millions of pixels, each of

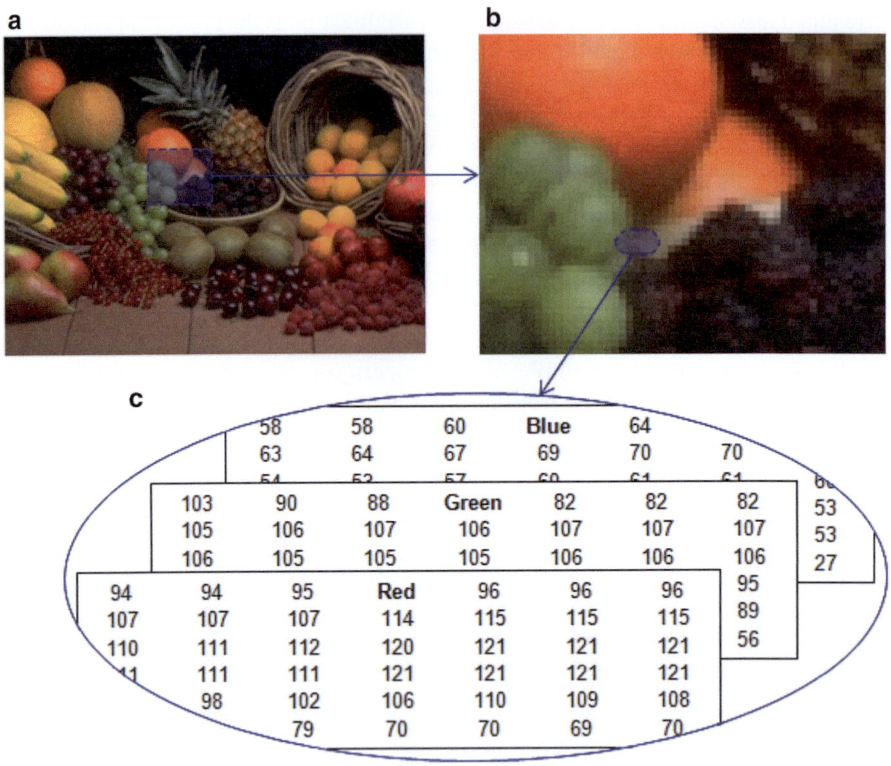

Fig. 2.2 Characteristics of a digital color image: (**a**) original image (340 × 256 pixels), (**b**) magnified color image (×1,600) to visualize the pixels, (**c**) the numerical *R*, *G*, and *B* color intensity values for a selected region

which has its own address, size, and grayscale or color channel representation. By zooming in on this digital image, it is noticed that the image is composed of a series of rows and columns of square pixels. Each pixel represents the brightness or intensity value of a given color channel (or gray-level) at any single specific point in the image. Typically, the pixel information is stored in computer memory as a raster image or raster map, a 2D array of integers. These values are often transmitted or stored in a compressed form.

Constraints of the digitized representation are that it contains much less information of the original scene since a 3D scene is reduced to a 2D representation. The sensors in any acquisition device (still or video cameras, scanners among other sensors) in general are not capable to capture and reproduce exactly, although not sensitive to the human vision, all color information from the real scene. Moreover, the size and location of the objects on the image are now estimates, whose precision and accuracy are dependent on the sampling resolution. The advantages of digital images are that they can be processed and analyzed, in many ways, by computers.

2.2.1 Basic Image Measurements

There are three basic measures of every static digital image: *spatial resolution*, *pixel bit depth*, and *color* (Puglia 2000). The specifications selected for each measure determine the amount of electronic information captured to represent the original photograph or scene. Generally, the higher the values are within these measures, the more data will be captured representing a greater amount of photographic detail from the original.

2.2.1.1 Resolution

Spatial resolution is defined as the rate, or number of times, at which an image is sampled during the acquisition or imaging process. More specifically, it is the frequency of pixels used to capture sample shades in the space of the object being digitized. *Spatial frequency* is synonymous for *spatial resolution*. Generally, more pixels per unit dimension means a higher resolution, but the overall image quality cannot be determined by spatial resolution alone. Typical array sizes in pixels (or *pixel resolution*) in many imaging sensors vary from 640×480 to $2{,}048 \times 1{,}536$ pixels. For reference human vision is >100 million pixels.

Quantitatively, *spatial resolution* (i.e., the number of pixels in an image) can be described in a number of ways, being the most common measures: *dpi* (dots per inch), *ppi* (pixels per inch), and *lpi* (line pairs per inch) (Gonzales and Woods 2008). The resolution of a digital image file is most appropriately referred to as pixels per inch or *ppi*. The *dpi* and *lpi* measures are considered printing terms and are most appropriate when referring to the resolution at which a computer

Fig. 2.3 Images saved at different *dpi* levels while keeping the same size. (**a**) typical effects on a color image; (**b**) X-ray microtomography image of 'Jonagold' apple captured with a spatial resolution of 8.5 μm/pixel, where black areas represent pores and gray areas the cellular material

printer produces a print. However, *dpi* is a more generic term and is more commonly used than *ppi* in image reproduction and photography.

In science and engineering research, however, the *dpi* term is not frequently reported when referring to the resolution of images. Generally, they are expressed by a spatial or unit distance per pixel, i.e., mm/pixel, μm/pixel, etc., and sometimes represented by a horizontal bar on the same image indicating the actual spatial length of that bar (as shown in Fig. 2.3b). Figure 2.3 presents two examples of images at different *dpi* levels while keeping their size the same. Reducing the resolution of the original image generally leads to larger pixels and less detail in the image. Images saved with the lowest *dpi* levels appear blurred and have a reduced contrast. By decreasing the image resolution of the apple tissue (Fig. 2.3b) the finest details of the tissue microstructure are lost, making it more difficult for the visualization and analytical determination of the smallest structures.

2.2.1.2 Pixel Bit Depth

This measure is sometimes called *pixel depth* or *color depth*. This defines the number of shades that can actually be represented by the amount of information saved for each pixel. Computers work on a binary system; each bit of data is either 1 or 0. Each pixel in a raster image is represented by a string of binary digits and the number of digits is known as the *bit depth*. Hence, a one-bit image can assign only one of two values to a single pixel: 0 or 1 (black or white). An 8-bit (2^8) grayscale image can assign one of 256 colors to a single pixel. A 24-bit ($2^{(3\times8)}$) *RGB* image (8-bits each for red, green and blue color channels) can assign one of 16.8 million colors to a single pixel. The *bit depth* determines the number of possible combinations of 1 s and 0 s for that number of binary digits and therefore the number of gray shades or color shades that can be represented by each pixel. This is calculated by the following formula (Puglia 2000):

$$\textit{Number of Shades} = 2^x, \text{ where } x = \text{ the bit depth.} \tag{2.1}$$

The *bit depth* influences the representation of images. The greater the *bit depth*, the finer the levels of change can be recorded. Consequently, more space is needed in the computer system to handle and store the image. Although grayscale images with bit depths of 2, 4, 6, 12, 16 and 32 exist, 8 *bpp* (i.e., byte-per-pixel) grayscale images are the most common. This is for two reasons; first, the 8 *bpp* size makes it easier to manipulate with a computer, and second, since the human eye can distinguish less than 200 shades, it can faithfully represent any grayscale image because it provides 256 distinct levels of gray. Thus, the use of 8-bit grayscale image files and corresponding 24-bit *RGB* color image files, in many cases, represent a reasonable compromise for image processing and analysis purposes. Some scientific applications requiring wider or higher dynamic ranges, as in multi- and hyper-spectral imaging, frequently use camera sensors with 14-bit (16,384 shades) or 16-bit (65,536 shades) in order to reduce the noise level for a higher signal-noise-ratio.

To achieve a desired *bit depth* without any data loss, it is necessary to digitize a photograph at a higher *bit depth* and then scale down to the desired *bit depth* after any image processing has occurred. In addition to the loss of data from small fluctuations in the acquisition system, raw digital images often require minimal processing (e.g., sharpening or minimal tonal corrections). Any processing of a digital image results in some data loss. Acquiring and processing an image at a higher *bit depth* and then reducing to the desired *bit depth* will minimize the impact of the data loss and provide a file with the desired quality.

2.2.1.3 Color Representation

Several different systems are used to represent color images. The most common are *RGB* (additive color system), *CMYK* (subtractive color system), *HSV* and the *CIELAB* color space. A color space is a specific implementation of a color model. There are many *RGB* and *CMYK* color spaces defined for specific purposes and applications (such as *sRGB* and *CMY*). The terms color space and color profile can often be used interchangeably, since a profile is a description or numerical model of a specific color space.

Operating systems and programs need to have access to an embedded profile that describes the meaning of the color values in order to interpret the color correctly. There are two types of profiles: *matrix-based* and *table-based*. Matrix-based profiles use mathematical formulas to describe the 3D color spaces. They can be relatively small and most appropriate as working spaces and as embedded profiles. Table-based profiles, as the name implies, use a large table of sample points called a

Look-Up Table or *LUT* to define the 3D color space. These profiles are more customizable, and are therefore more useful when translating color information from one space to another, or in describing the color characteristics of a particular device. Because they rely on many data points, they are much larger.

RGB is a color model that uses the three primary (red, green, blue) additive colors, which can be mixed to make all other colors. It builds its model by adding different colors of light together, where the mixture of all three colors produces white light. Grayscale values follow the line from black (the origin of the coordinate system) to white, as shown in Fig. 2.4a. Digital cameras produce *RGB* images, and monitors display *RGB* images. Mathematical conversions between different color spaces for analysis and special visualizations are also possible.

CMYK (*CMY*) is a color model based on subtracting light. The cyan (*C*), magenta (*M*), and yellow (*Y*) are the basic colors for a subtractive model, and represent the complements of the three primary colors (see Fig. 2.4a). *R*, *B*, *G*, and black (*K*) inks are used in most commercial color printing (books, magazines, etc.). Inks absorb colored light, which is why the model is called a subtractive one. *CMYK* is commonly referred to as process color, and there are many individual color spaces that use the *CMYK* color model. Conversions from *RGB* to *CMY* are done by using the following simple equation (Gonzales and Woods 2008):

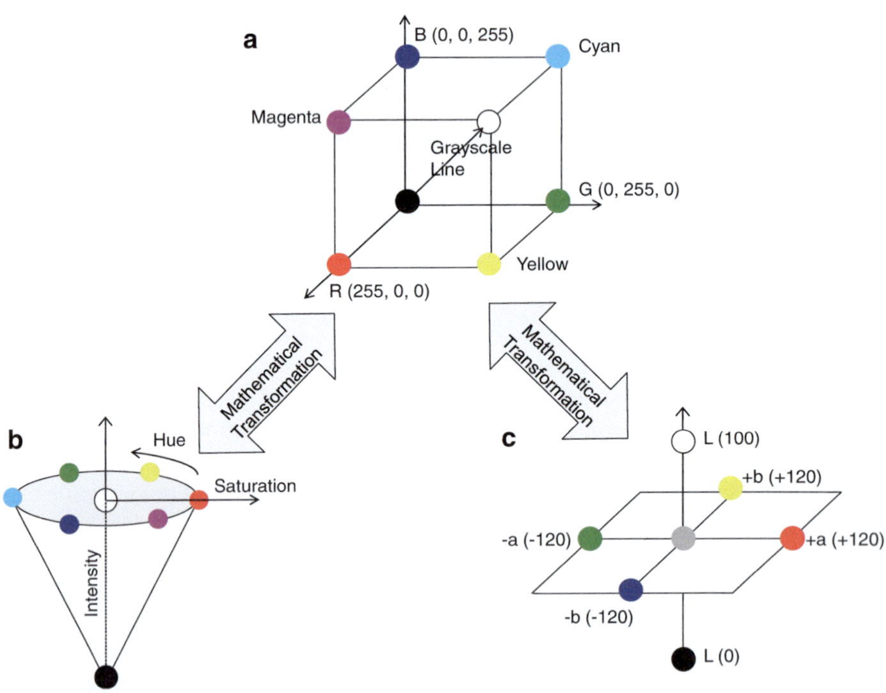

Fig. 2.4 Commonly used color spaces: (**a**) *RGB* color cube; (**b**) *HSV* color cone; and (**c**) CIELAB or $L^*a^*b^*$ color system

$$\begin{bmatrix} C \\ M \\ Y \end{bmatrix} = \begin{bmatrix} 1.0 \\ 1.0 \\ 1.0 \end{bmatrix} - \begin{bmatrix} R \\ G \\ B \end{bmatrix} \quad (2.2)$$

If a color image has to be printed, black (K) as a fourth color is added to the model to achieve a purer black than the simple combination of the other three colors, resulting in the *CMYK* model. Transformation from *CMY* to *CMYK* is done by:

$$K = min(C_{CMY}, M_{CMY}, Y_{CMY}) \quad (2.3)$$

$$C_{CMYK} = C_{CMY} - K \quad (2.4)$$

$$M_{CMYK} = M_{CMY} - K \quad (2.5)$$

$$Y_{CMYK} = Y_{CMY} - K \quad (2.6)$$

HSV is a user-oriented color model based on the artist's idea of tint, shade and tone. *HSV* expresses color into three components that vary from 0 to 1; H (hue) distinguishes among the perceived colors, such as red, yellow, green and blue, S (saturation) refers to how much of the light is concentrated at each specific wavelength of the hue; and V (value) represents the total brightness (see Fig. 2.4b). The computation of H, S, and V values from the *RGB* color space is made according to the following expressions (Du and Sun 2005):

$$V = max(nR, nG, \; nB) \quad (2.7)$$

$$S = \frac{V - min(nR, nG, nB)}{V} \quad (2.8)$$

Let

$$tR = \frac{V - nR}{V - min(nR, nG, nB)}, \quad (2.9)$$

$$tG = \frac{V - nG}{V - min(nR, nG, nB)}, \quad (2.10)$$

$$tB = \frac{V - nB}{V - min(nR, nG, nB)}, \quad (2.11)$$

Then

$$6H = \begin{cases} 5 + tB & \text{if } nR = max(nR, nG, nB) \text{ and } nG = min(nR, nG, nB) \\ 1 - tG & \text{if } nR = max(nR, nG, nB) \text{ and } nG \neq min(nR, nG, nB) \\ 1 + tR & \text{if } nG = max(nR, nG, nB) \text{ and } nB = min(nR, nG, nB) \\ 3 - tB & \text{if } nG = max(nR, nG, nB) \text{ and } nB \neq min(nR, nG, nB) \\ 3 + tG & \text{if } nB = max(nR, nG, nB) \text{ and } nR = min(nR, nG, nB) \\ 5 - tR & \text{otherwise} \end{cases} \tag{2.12}$$

where $H, S, V \in [0, \ldots, 1]$.

In food research, color is frequently represented using the *CIELAB* or $L^* a^* b^*$ color space since results closely match those of the human perception. L^* is the luminance or lightness component that goes from 0 (black) to 100 (white), and parameters a^* (from green to red) and b^* (from blue to yellow) are the two chromatic components, varying from −120 to +120 (see Fig. 2.4c). The definition of $L^* a^* b^*$ is based on the intermediate system *CIE XYZ* which is derived from *RGB* (Rec. ITU-R BT.709-5, 2002). Thus, L^*, a^*, and b^* are defined as:

$$L^* = 116\left(\frac{Y}{Y_n}\right)^{1/3} - 16 \tag{2.13}$$

$$a^* = 500\left[\left(\frac{X}{X_n}\right)^{1/3} - \left(\frac{Y}{Y_n}\right)^{1/3}\right] \tag{2.14}$$

$$b^* = 200\left[\left(\frac{Y}{Y_n}\right)^{1/3} - \left(\frac{Z}{Z_n}\right)^{1/3}\right] \tag{2.15}$$

where X_n, Y_n, Z_n correspond to the *XYZ* values of a reference white chart.

2.2.2 Types of Files

File types are used to encode digital images, allowing for compression and storage. Image files can be of different sizes and formats, and larger file types mean more disk usage and slower download. Compression is a term used to describe ways of reducing the size of the file. Compression schemes can by *lossy* or *lossless* (Shapiro and Stockman 2001). A *lossless* compression algorithm does not discard information. It looks for more efficient ways to represent an image, while making no compromises in accuracy. A *lossless* algorithm might, for example, look for a recurring pattern in the file, and replace each occurrence with a short abbreviation, thereby cutting the file size.

Contrarily, a *lossy* algorithm might accept some degradation in the image in order to achieve smaller file size. For example, *JPEG* or *JPG* file format works by analyzing images and discarding information that the eye is least likely to notice. Changes in luminance are more significant by the human observer than change in hue. Thus, in *JPEG* compression, factors of more than 20 are often quite acceptable. Better graphics programs, such as *Paint Shop Pro* and *Photoshop*, allow viewing the image quality and file size as a function of compression level, so that it is possible to conveniently choose the balance between qualities and file size. Higher *JPEG* compression equals lower image quality since the color information in individual pixels is compressed into *blocks* of pixels using mathematical algorithms that methodically blend all the pixel colors in each block. Increasing the compression produces smaller computer file sizes, whereas lower compression produces better quality but larger computer file sizes.

The most common digital image file types and their main characteristics are summarized in Table 2.1. Currently, *GIF* and *JPEG* are the formats used for nearly all web images. *PNG* is supported by most of the latest generation browsers. *TIFF* is not widely supported by web browsers, and should be avoided for web use. *PNG* does everything *GIF* does, and better; so it is expected to replace *GIF* in the future. *PNG* will not replace *JPEG*, since *JPEG* is capable of much greater compression of photographic images, even when set for quite minimal loss of quality.

Table 2.1 Common digital file types

File type		Description
TIFF	Tagged Image File Format	• *Lossless* uncompressed file format with 24 or 48 bit color support. File sizes are quite big
		• Supports embedded information like EXIF[a], calibrated color space and output profile information
		• There is a lossless compression for *TIFF* called LZW. LZW works like *zipping* the image file because there is no quality loss. An LZW *TIFF* decompresses with all of the original pixel information unaltered
PNG	Portable Network Graphics	• *Lossless* compression format with up to 16-bit depth for grayscale values and up to 48-bits for color values
		• In contrast with common *TIFF* files, it looks for patterns in the image that it can use to compress file size
		• The compression is exactly reversible, so the image is recovered exactly
JPEG	Joint Photographic Experts Group	• *Lossy* compression format optimized for photographs that contain many colors which store information as 24-bit color and the degree of compression is adjustable
		• *JPEG* compression ratio and resolution are not relational. It is possible to get a high resolution JPEG image with either low or high compression
		• Supports embedded information like EXIF, calibrated color space and output profile information
		• This is the best format for photographs to be shown on the web or as email attachments

(continued)

Table 2.1 (continued)

File type		Description
GIF	Graphics Interchange Format	• *Lossless* uncompressed 8-bit file format that supports only 256 distinct colors. *GIF* creates a table of up to 256 colors from a pool of 16 million
		• Not suitable for photographs because of its limited color support. Best used with web clip art and logo type images
BMP	Bitmap Format	• *Lossless* uncompressed file format that supports 24-bit color invented by Microsoft
		• Does not support embedded information like EXIF, calibrated color space and output profiles
		• *BMP* produces approximately the same file sizes as *TIFF* without any of the advantages of *TIFF*
RAW		• *Lossless* compressed file format that is proprietary for each digital camera manufacturer and model
		• Though *lossless*, it is a factor of three or four smaller than *TIFF* files of the same image
		• It contains the full range of color information from the sensor. RAW files converted to 16-bit *TIFF* produce the absolute best quality image available from any digital camera
		• Camera *RAW* supports imbedded EXIF data
PSD, PSP	Photoshop file, Pain Shop Pro	• Proprietary formats used by graphic programs
		• Preferred working formats as you edit images in the software, because only the proprietary formats retain all the editing power of the programs
		• These packages use layers building complex images, and layer information may be lost in the nonproprietary formats such as *TIFF* and *JPG*

[a]EXIF stands for Exchangeable Image File Format. This data includes technical information about each photograph including calibrated color space, output profile device information shutter speed and aperture used, whether or not flash was used, and the date the photo was taken

2.2.3 *Types of Digital Images*

As described above, each pixel of a *raster* image is typically associated to a specific position in the 2D region, and has a value consisting of one or more quantities (samples) related to that position. Digital images can be classified according to the number and nature of those samples in: *binary*, *grayscale*, *color*, *false-color*, *multi-spectral*, *thematic*, and *picture function* (Shapiro and Stockman 2001).

For image processing and analysis, the input image is supposed to be grayscale or *RGB*. However, there are four basic types of digital images frequently used as intermediate steps in the processing of an image, which allow to identify, enhance, quantify and also represent specific characteristics or regions of interest on an image. They are: *indexed color images*, *intensity images*, *binary*, and *labeled* images.

2.2.3.1 Indexed Color Images

A digital color image is an image that includes the R, G, and B color channel information or other color representation for each pixel. An *indexed color* image consists of two arrays: an *image matrix* and *colormap*. The *colormap* (also called a *palette*) is an ordered set of values that represent the colors in the image. For each image pixel, the *image matrix* contains a value that is an index into the *colormap*. In computing, the encoding of the color image data in this way allows for saving computer memory and file storage, while speeding up refreshing and file transfers. It is a form of vector quantization compression.

2.2.3.2 Intensity Images

An *intensity image* or *digital grayscale image* is an image in which the value of each pixel is a single sample, that is, it carries only intensity information. These images are composed exclusively of various shades of gray, varying within a given range from 0 (for black) to 1 (or 255 for white), with any fractional values in between. However, it must be noted that this does not define what black or white is in terms of colorimetry. Another convention is to employ percentages, so the scale is then from 0 to 100 %. This is used for a more intuitive approach, but if only integer values are used, the range encompasses a total of only 101 intensities, which are insufficient to represent a broad gradient of grays.

Grayscale images are often the result of measuring the intensity of light at each pixel in a single band of the electromagnetic spectrum (e.g., infrared, visible light, ultraviolet, etc.), and in such cases they are monochromatic when only a given frequency is captured. Multispectral or hyperspectral images have numerous bands or a finer spectral resolution, and they are examples of this type of intensity images. But they also can be synthesized from a full color image by converting to grayscale. Grayscale images are also called monochromatic because of the absence of any chromatic variation (i.e., one color).

If a color image has to be converted into an *intensity* or *grayscale* image, the following equations can be used. One alternative is the simple average of the R, G, B color channels:

$$I = 0.333 \cdot R + 0.333 \cdot G + 0.333 \cdot B \tag{2.16}$$

Another equation, which takes into account the luminance perception of the human eye, is

$$Y = 0.2162 \cdot R + 0.7152 \cdot G + 0.0722 \cdot B \tag{2.17}$$

The weights used to compute luminance are related to the monitor's phosphors. The explanation of these weights is due to that for equal amounts of color, the eye is more sensitive to green, then red, and then blue. This means that for equal amounts of green and blue light, the green will, nevertheless, be much brighter. Thus the image obtained by the normal averaging of an image's three color channels produces a grayscale brightness that is not perceptually equivalent to the brightness of the original color image. The weighted sum that defines Y (Eq. 2.17), however, does.

2.2.3.3 Binary

A *binary image* has only two possible values (0 and 1) for each pixel. Typically the two colors used for a binary image are black and white, though any two colors can be used. The color used for the object(s) in the image is the foreground color while the rest of the image is the background color. Thus, the set of all white pixels in a binary image is a complete morphological description of the image (Gonzales and Woods 2008). Binary images are also called *one-bit*, *bi-tonal*, *bi-level* or *two-level*. This means that each pixel is stored as a single bit (0 or 1, see Sect. 2.2.1.2). In digital image processing, binary images often arise as masks or result from certain operations such as segmentation, thresholding, and dithering. A binary image is usually stored in memory as a bitmap, a packed array of bits (Wikipedia 2012).

2.2.3.4 Labeled Images

A *labeled image* is a digital image whose pixel values are symbols from a finite alphabet. The symbol value of a pixel denotes the outcome of some decision made for that pixel. An example is the labeling of objects in a *binary image*, which means that these objects are classified and numbered. Related concepts are *thematic image*, *false-color image* and *pseudo-colored image*. In a *false-color* image this close correspondence between subject color and image color is altered. The term *false-color* is typically used to describe images whose colors represent measured intensities outside the visible portion of the electromagnetic spectrum. *Pseudo-coloring* can make some details more visible, by increasing the distance in color space between successive gray levels. *Pseudo-color* images differ from *false-color* images in that they are made from only one original grayscale image, rather than two or three.

2.3 Steps for Image Processing and Analysis

Image processing involves a series of image operations to enhance the quality of a digital image so as to remove defects such as geometric distortion, improper focus, repetitive noise, non-uniform lighting and camera motion. *Image analysis* is the process of distinguishing the objects (regions of interest or ROIs) from the background and producing quantitative information, which is subsequently used for decision making. Processing and analysis can be performed on many different types of image data. These include, in an increasing order of complexity: binary images, grayscale, color, polarized-light, multi-spectral and hyper-spectral, 3D images, multi-sensor and multimedia systems, and image sequences and video.

Gunasekaran and Ding (1994) defined three levels of image processing, named, *low level processing* which includes image acquisition and pre-processing of images, *intermediate level processing* which involves image segmentation, image representation and description, and *high level processing* which involves recognition of ROIs and interpretation for quality sorting and grading. The terms *machine vision* or *computer vision* is often used for the entire subject, including *image processing* and *analysis* and *pattern recognition* techniques. Hence, the process of making a decision involves a number of steps in sequential order. Not all situations require all of these steps or operations, but all are potentially available to deal with particular problems.

Machine vision generally consists of the following five steps or operations (Fig. 2.5): (1) *image acquisition* operations to convert images into digital form, as explained in Sect. 2.2; (2) *pre-processing* operations to obtain an improved image with the same dimensions as the original image; (3) *image segmentation* operations to partition a digital image into disjoint and non-overlapping regions; (4) *object measurement* operations to measure the characteristics of objects, such as size, shape, color and texture; and (5) *classification* or *sorting* operations to identify objects by classifying them into different groups.

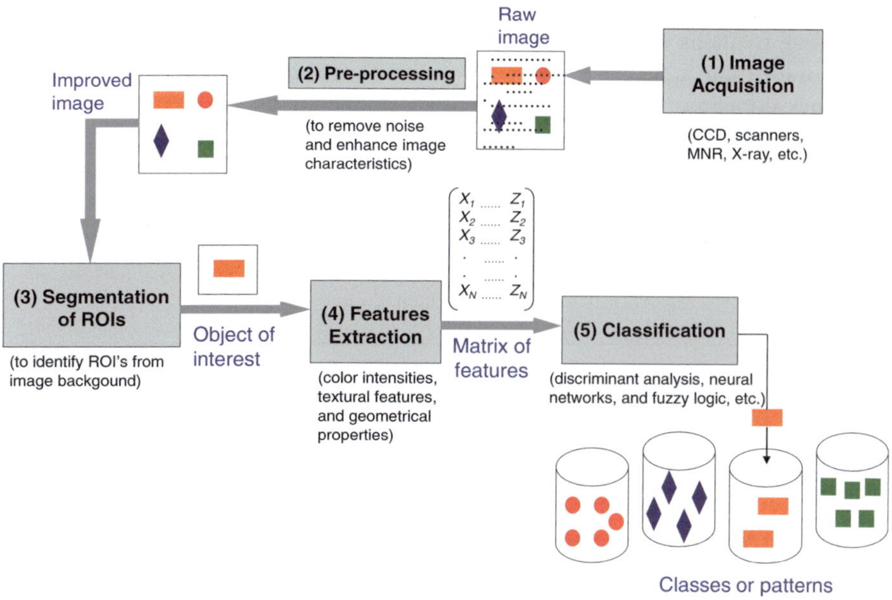

Fig. 2.5 An overview of the operational steps for a machine vision system

2.4 Image Processing

Image processing or *pre-processing* encompasses a broad range of operations, which may be treated as an end in themselves, or are intended to simplify or enhance subsequent analysis. *Pre-processing* improves the image data by removing unintended distortions or enhancing some image features that are important for further processing and creating a more suitable image than the original for a specific application. The operations that can be performed on digital images include *point*, *local* or *neighborhood*, and *global operations*.

Point operations transform pixels without regard to neighboring pixels. The gray value of the output image at a particular pixel depends only on the gray value of the same pixel in the input image. They map the pixels in one image to form another using a single mapping function. Point operations do not consider the spatial organization of the image, which forms the fundamental character of images as opposed to other types of data. Examples of these operations include contrast stretching, segmentation based on gray value, and histogram equalization (Marchant 2006). The term *contrast* refers to the amplitude of gray level variations within an image.

A *local* or *neighborhood operation* or *mask operation* generates an output pixel whose value depends on the pixel values in a neighborhood of the corresponding input point. Examples include convolution (as for image smoothing or sharpening) and spatial features detection (e.g., line, edge, and corner detection). A large and powerful class of non-linear neighborhood operations is morphological methods; they extend naturally to gray-level (and multiband) images (Soille 1999).

Finally, an operation is a *global operation* if the output value at specific coordinate is dependent on all the values in the input images. Spatial domain processing methods include all three types, but frequency domain operations, by nature of the frequency (and sequence) transforms, are global operations. Of course, frequency domain operations can become mask operations, based only on a local neighborhood, by performing the transform on small image blocks instead of the entire image. This section presents a brief description of the types of algorithms commonly utilized for digital image processing. We intentionally limit the discussion to the types of image processing algorithms that are widely used in applications for foods.

2.4.1 Grayscale Operations for Image Enhancement

Once the grayscale or color image is obtained, one of several different techniques may be used to improve the quality of the image. Image enhancement techniques are used to emphasize and sharpen image features for further analysis in order to facilitate the development of a solution to a specific application problem. Consequently, the enhancement methods are application specific and are often developed empirically.

2.4.1.1 Arithmetic Operations

All arithmetic operations performed on matrices may be performed on images. Arithmetic operations between images are array operations carried out between corresponding pixel pairs. Hence, the images normally have to be of the same size. These operators are frequently used for reducing noise and image enhancement. The four arithmetic operations are as follows:

$$s(x, y) = f(x, y) + g(x, y) \tag{2.18}$$

$$d(x, y) = f(x, y) - g(x, y) \tag{2.19}$$

$$p(x, y) = f(x, y) \times g(x, y) \tag{2.20}$$

$$v(x, y) = f(x, y) \div g(x, y) \tag{2.21}$$

where $x = 0, 1, 2, \ldots, M-1$ and $y = 0, 1, 2, \ldots, N-1$ for images with the size of M rows and N columns.

Addition is a discrete version of continuous integration, and it is an operation frequently used to create double-exposures or composites. A common variant of this operator simply allows a specified constant to be added to every pixel in an image, so as to brighten that image. *Image subtraction* is frequently used to enhance differences between images; such as finding changes between two images.

The *multiplication* or *division* of two images is performed in two main forms. The first form takes two input images and produces an output image in which the pixel values are just those of the first image, multiplied (or divided) by the values of the corresponding values in the second image. The second form takes a single input image and produces output in which each pixel value is multiplied (or divided) by a specified constant. This latter form is probably the more widely used and is generally called *scaling*.

Image averaging is an application for correcting noisy images. Averaging of multiple pictures of the same scene helps to reduce this noise. In practice, however, the images must be previously *registered* (aligned) in order to avoid the introduction of blurring and other artifacts in the output image.

Other related operators are the *logical operators* such as *And*, *Or*, *Not*, *If*. . .*Then*, *If and only If*, among others which are often used to combine (mostly binary) two images. An example is when a mask is used for selecting a region of interest from an image.

2.4.1.2 Histogram Equalization

The *histogram* of a digital image gives the numeric (quantitative) information about the distribution of the number of pixels per gray-level value. Histograms are the basis for numerous spatial domain processing techniques, and their manipulation can be used for image enhancement. In addition to providing useful image statistics, the information inherent in histograms also is quite useful in other image processing applications, such as image compression and segmentation. Mathematically, the *histogram* of a digital image is a discrete function $h(k) = n_k/n$, where $k = 0, 1, \ldots, L-1$ is the *kth* gray-level, n_k is the number of pixels in the image having gray-level k, and n is the total number of pixels in the image.

Frequently, an image is acquired in such a way that the resulting brightness values do not make full use of the available dynamic range. *Histogram equalization* is a common point operation method for *spreading* the histogram of pixel levels more evenly. The gray levels are transformed with a function such that all gray values are equally represented in the histogram. In the method each original gray-level k is mapped into new gray-level i by:

$$i = \sum\nolimits_{i=0}^{k} h(j) = \sum\nolimits_{j=0}^{k} \frac{n_j}{n} \tag{2.22}$$

where the sum counts the number of pixels in the image with gray-level equal to or less than k. Thus, the new gray-level is the cumulative distribution function of the original gray-levels, which is always monotonically increasing. The resulting image will have a histogram that is flat in a local sense, since the operation of histogram equalization spreads out the peaks of the histogram while compressing other parts of the histogram. In more complicated cases, the global histogram may not be a good representation for local statistics in two parts of the image. In such cases it is best to use an *adaptive histogram equalization* where you can divide the image into several rectangular domains, compute an equalizing histogram and modify levels so that they match across boundaries (Macaire and Postaire 1989). Figure 2.6 illustrates the image enhancement of a loaf of bread using global histogram equalization. In general, the function is non-linear. It is important to realize that whatever point operation is used, the separation into ROIs is still not possible. Thus, although the image in Fig. 2.6c appears easier to separate into components, a machine vision system will still not be able to do it on the basis of point operations only.

2.4.2 Spatial Image Filtering

With *spatial image filtering technique*, a window of finite size and shape is scanned across the entire image, transforming the local intensities in that image. The window with its weights is called the *convolution kernel* or *filter mask*. Thus, filtering creates a new pixel with coordinates equal to the coordinates of the center of the neighborhood, and whose value is the result of the filtering operation. Two main linear spatial filtering methods are *correlation* and *convolution*. Correlation is the process of moving a filter mask over the image and computing

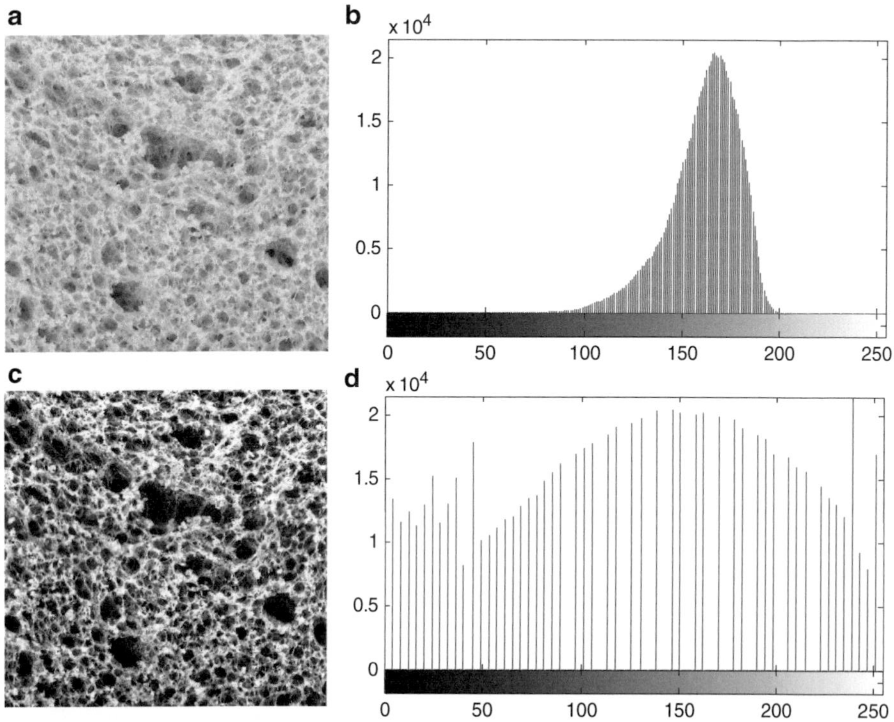

Fig. 2.6 Image quality enhancement using histogram equalization: (**a**) grayscale image of a loaf bread; (**b**) histogram of the image in (**a**); (**c**) resulting image obtained from image (**a**) by histogram equalization; (**d**) histogram of the image in (**c**)

the sum of products at each location. The mechanisms of convolution are the same, except that the filter is first rotated by 180° (Gonzales and Woods 2008). Correlation is often used to measure the similarity between images or parts of images (e.g., pattern matching). Correlation and convolution yield the same result when the filter mask is symmetric. Nonetheless, basic image processing techniques are mainly based on *convolution*.

The convolution is performed by sliding the kernel over the image, generally starting at the top left corner and moving the kernel through all the positions within the boundaries of the image. Thus, in a convolution, a *convolution kernel* is applied to every pixel to create a filtered image, $I_{out}(x, y)$:

$$I_{out}(x, y) = I_{in}(x, y)*W(x, y) = \sum_{u=-\infty}^{\infty} \sum_{u=-\infty}^{\infty} I_{in}(u, v)W(u - x, v - y) \tag{2.23}$$

where the minus signs on the right flip W (i.e., rotate it by 180°). Flipping and shifting W instead of the input image, $I_{in}(u,v)$, is done for notational simplicity and also to follow convention, which is the same as flipping and shifting the input image. It should be noticed that a special treatment is needed for calculating the border of the new image during spatial filtering. Klinger (2003) suggests four possibilities for handling the border of an image. First, if the convolution filter has a size of 3 × 3, get an image smaller by the border of one pixel. Second, keep the same gray-level values of the original image for the new border pixels. Third, use special values for the new border pixels; for example, 0, 127, or 255. Finally, use pixels from the opposite border for the calculation of the new values.

2.4.2.1 Image Smoothing and Blurring

All smoothing filters build a weighted average of the surrounding pixels, and some of them also use the center pixel itself. *Averaging* and *Gaussian* filters are linear filters often used for noise reduction with their operation causing a smoothing in the image but having the effect of blurring edges.

The *average* or *low-pass* filter is a linear filter and one of the simplest types of neighborhood operation. In the *average filtering* the new value is calculated as the average of all nine pixels (for a [3 × 3] kernel) using the same weight. The elements of the mask must be positive and the coefficients for the center pixel are either 0 or 1. With the averaging filter, the mean value of the pixels in the neighborhood is used to form a pixel in a new image at the same location as the center pixel in the original image. Like all averaging operations, this can be used to reduce some types of noise at the expense of a loss of sharpness as it is shown in Fig. 2.7. The averaging operation is represented by,

$$I_{out}(x,y) = \frac{1}{9}\sum_{u=x-1}^{x+1}\sum_{v=y-1}^{y+1} I_{in}(u,v) \tag{2.24}$$

Rather than weight all input pixels equally, it is better to reduce the weight of the input pixels with increasing distance from the center pixel. The *Gaussian filter* does this and is perhaps the most commonly used of all filters (Shapiro and Stockman 2001). The center pixel coefficient of a Gaussian kernel is always greater than 1, and thus greater than the other coefficients because it simulates the shape of a Gaussian curve according to,

Fig. 2.7 Examples of averaging filter using masks [3 × 3] and [9 × 9]

$$G_{out}(x,y) = \frac{1}{\sqrt{2\pi}\sigma} exp\left(-\frac{d^2}{2\sigma^2}\right) \tag{2.25}$$

where $d = \sqrt{(x - x_c)^2 + (y - y_c)^2}$ is the distance of the neighborhood pixel $[x,y]$ from the center pixel $[x_c,y_c]$ of the output image to which the filter is being applied. Convolution with this kernel forms a weighted average which stresses the point at the center of the convolution window, and incorporates little contribution from those at the boundary. As σ increases, more samples must be obtained to represent the Gaussian function accurately. Therefore, σ controls the amount of smoothing. A second derivative filter based on the Laplacian of the Gaussian is called a *LOG filter*. Examples of Gaussian filter using masks [3 × 3] and [9 × 9] with $\sigma = 3$ are given in Fig. 2.8.

Sometimes, non-linear operations on neighborhoods yield better results. An example is the use of a *median filter* to remove noise. *Median filtering* replaces each pixel by the median in a neighborhood around the pixel. Consider an area of a scene of reasonably uniform gray-level; the resulting image may have a single pixel with a very different gray-level due to some random noise. The output of an averaging filter would contain a proportion of this outlier pixel value. However, the *median filter* would set the output pixel to the median value (the 5th largest gray-level in a 3 × 3 neighborhood) and so it would be unaffected by the outlier. The method is very effective for removing salt and pepper noise (i.e., random occurrences or impulses of black and white pixels) (Gonzales and Woods 2008), as shown in Fig. 2.9. A disadvantage is that the median filter may change the contours of objects in the image.

Fig. 2.8 Examples of Gaussian filter using masks [3 × 3] and [9 × 9]

Fig. 2.9 Example of median filter using a kernel [3 × 3]: the input image (*left*) contains Gaussian noise, and the noise is removed in the resultant image (*right*) after 3 × 3 median filtering

Computing the median requires more computation time than computing a neighborhood average, since the neighborhood values must be partially sorted. Moreover, median filtering is not so easily implemented in special hardware that might be necessary for real-time processing. However, in many image analysis tasks, its value in image enhancement is worth the time spent.

2.4.2.2 Edge Detection and Enhancement

The detection of edges in an image is an important area of study. An *edge* is an area of an image characterized by sharp changes in gray-level or brightness. The process of edge detection attenuates high fluctuations in color, i.e., dramatic change in intensity. In the frequency domain, this process refers to the attenuation of high frequencies. Among the families of *edge detection* filters are: *gradient filters*, *Laplacian*, and *wavelet transform* (Klinger 2003). Both *gradient* and *Laplacian kernels* are of the *high-pass filter*, which operates by differencing adjacent pixels, because the sharp edges can be described by high frequencies. However, as in other areas of signal processing, *high-pass* filtering amplifies noise, if an appropriate attempt is made to find object boundaries in the image with a simple edge detector (Marchant 2006).

A *gradient filter* extracts a significant brightness change in a specific direction and is thus able to extract edges perpendicular to this direction. These filters are known as *Prewitt filter masks*. The Prewitt operator is based on convolving the image with a small, separable, and integer valued filter in horizontal and vertical direction and is therefore relatively inexpensive in terms of computations. Because the components of the gradient vector are derivatives, they are linear operators.

Another group of gradient masks are the *Sobel filters* or *Sobel kernels*. A Sobel operator gives the specified filter direction a stronger weight. In general, the gradient specifies the amount of change of a value in a certain direction. First derivatives in image processing are implemented using the magnitude of the gradient. The simplest filter kernels, used in two orthogonal directions, are:

$$G_x = \begin{bmatrix} 0 & 0 & 0 \\ 0 & -1 & 0 \\ 0 & 1 & 0 \end{bmatrix} \tag{2.26}$$

and

$$G_y = \begin{bmatrix} 0 & 0 & 0 \\ 0 & -1 & 1 \\ 0 & 0 & 0 \end{bmatrix} \tag{2.27}$$

resulting in two images, $I_x = (S_x(x, y))$ and $I_y = (S_y(x, y))$. The mask coefficients sum to zero, as expected of a derivative operator. The value and direction of the gradient are therefore calculated as follows:

$$I = \sqrt{S_x(x, y)^2 + S_y(x, y)^2} \tag{2.28}$$

and

$$\theta = arctan\left(\frac{S_x(x, y)}{S_y(x, y)}\right) \tag{2.29}$$

The Laplacian operator is an example of a second order or second derivative method of enhancement. It is particularly good at finding the fine detail in an image. All masks of the *Laplacian filter* group are omni-directional, meaning that they provide edge information in each direction. The *Laplacian operators* show two interesting effects. First, if the sum of all coefficients is equal to 0, the filter kernel shows all image areas with a significant brightness change; that means it works as an isotropic or omni-directional edge detector. In other words, isotropic filters are *rotation invariant* (Gonzales and Woods 2008), in the sense that rotating the image and then applying the filter gives the same result as applying the filter to the image first and then rotating the result. Second, if the center coefficient is greater than the sum of the absolute values of all other coefficients, the original image is superimposed over the edge information (Klinger 2003).

The simplest isotropic derivative operator is the *Laplacian*, which for a image *f*(*x*,*y*) of two variables, is defined as

$$\nabla^2 f = \frac{\partial^2 f}{\partial x^2} + \frac{\partial^2 f}{\partial y^2} \tag{2.30}$$

Laplacian as derivative operator highlights intensity discontinuities in an image and deemphasizes regions with slowly varying intensity levels. This will tend to produce images that have grayish edge lines and other discontinuities, all superimposed on a dark, featureless background. Hence, it is important to define the type of Laplacian operator used. If the filter mask has a negative center coefficient, then it subtracts, rather than adds, the Laplacian image to obtain a sharpened result. Typical Laplacian masks are:

$$L_{subtract} = \begin{bmatrix} 0 & 1 & 0 \\ 1 & -4 & 1 \\ 0 & 1 & 0 \end{bmatrix} \tag{2.31}$$

and

$$L_{add} = \begin{bmatrix} -1 & -1 & -1 \\ -1 & 8 & -1 \\ -1 & -1 & -1 \end{bmatrix} \tag{2.32}$$

The basic representation using these Laplacian masks for image sharpening is,

$$g(x, y) = f(x, y) + c\left[\nabla^2 f(x, y)\right] \tag{2.33}$$

where $f(x,y)$ and $g(x,y)$ are the input and sharpened images, respectively. The constant $c = -1$ if the Laplacian filter $L_{subtract}$ is used and $c = 1$ if L_{add} is used. Since derivative filters are very sensitive to noise, it is common to smooth the image (e.g., using a Gaussian filter) before applying the Laplacian. This two-step process is call the Laplacian of Gaussian (LoG) operation.

2.4.3 *Frequency Filtering*

Filtering in the frequency domain consists of modifying the Fourier transform of an image and then computing the inverse transform to obtain the processed result. Thus, the Fourier basis can be used to remove high frequency noise from the image of signal, to extract texture features that can be used to classify the type of objects in an image region, and also for image compression. Formally, the Fourier transform is defined by

$$F(u, v) = \int_{-\infty}^{\infty}\int_{-\infty}^{\infty} f(x, y)e^{-i2\pi(xu+yv)}dx.dy \tag{2.34}$$

with $F(u,v)$ as the frequencies and $f(x,y)$ as the pixel intensities. The letter $i = \sqrt{-1}$ denotes the *imaginary unit* of complex numbers. The exponential function of Eq. 2.34 satisfies the *Eulerian* formula

$$e^{-i2\pi\alpha} = \cos 2\pi\alpha - i\ \sin 2\pi\alpha \tag{2.35}$$

for any real α. Thus, the entire set of pixel intensities can be described by a sum of sine and cosine functions but the result is complex.

The *Fast Fourier Transform* (FFT) is more efficient than Fourier transform since it saves computational time by sharing common operations for different pairs of u, v and is usually used on square images of size $2^m \times 2^m$. It can be described by

$$F(u, v) = \frac{1}{m^2}\sum\nolimits_{x=0}^{m-1}\sum\nolimits_{y=0}^{m-1} f(x, y)e^{-i2\pi\left(\frac{xu+yv}{m}\right)} \tag{2.36}$$

where m are the numbers of pixels in x and y directions assuming square images. Despites its common use in image processing, the Fourier transform can cause unwanted degradation in local features of images because it is a global transform that uses all image pixels in every computation of $F(u,v)$ (Shapiro and Stockman 2001). An example of FFT spectrum of a potato chip image and frequency filtering is shown in Fig. 2.10.

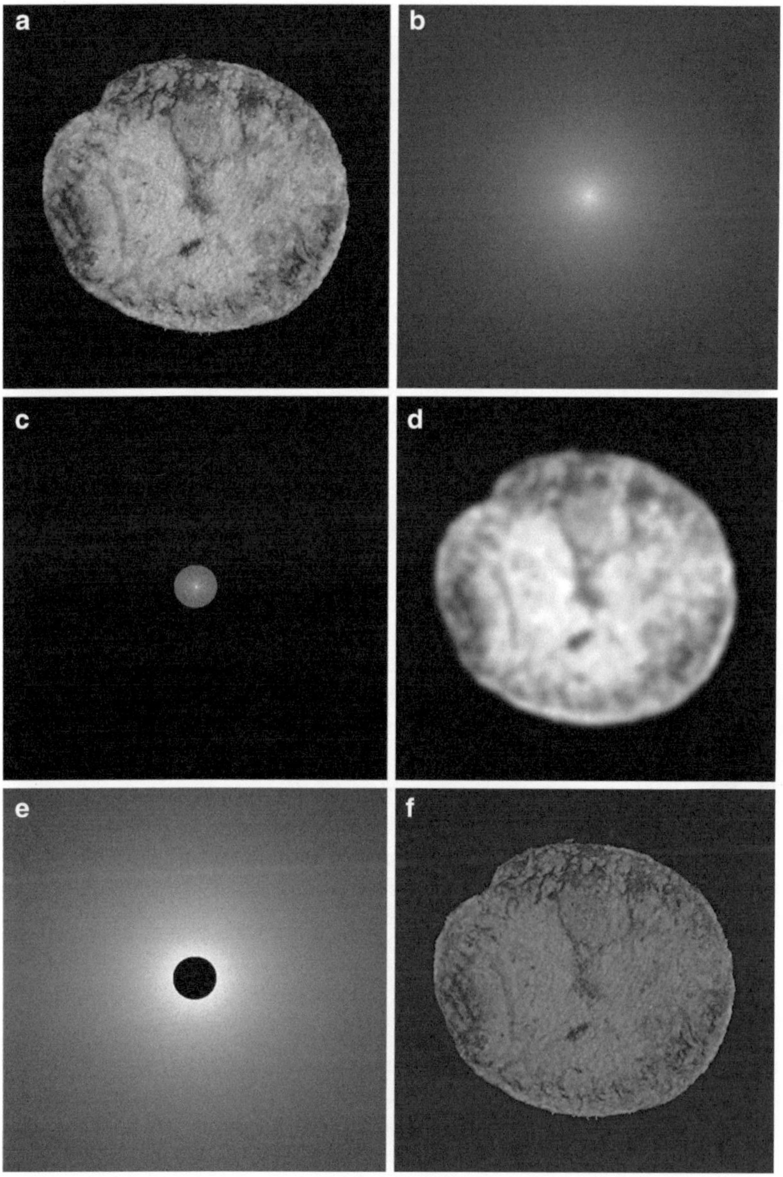

Fig. 2.10 Example of fast Fourier transform (FFT) and frequency filtering: (**a**) *grayscale* image of a potato chip; (**b**) FFT spectrum of (**a**) where low frequency components are grouped in the centre (*brighter pixels*) and high frequencies near to the corners (*darker pixels*); (**c**) low-pass filtered transform defined by a radius = 100 pixels around the central point and zero-out every point in the Fourier image that is beyond that radius; (**d**) low-pass filtered image after applying an inverse Fourier transform to smooth regions of dark and bright, but lose the sharp contours and crisp edges; (**e**) high-pass filtered transform using the same spatial frequency threshold of (**c**), where only the higher spatial frequency components are preserved; (**f**) high-pass filtered image after applying an inverse Fourier transform, which preserves all of the sharp crisp edges from the original, but loses the larger regions of dark and bright

2.4.4 Wavelet Transform

The *wavelet transform* is used primarily for smoothing, noise reduction and lossy compression. In all cases the procedure is to first perform a forward wavelet transform, and then perform some operations on the transformed image followed by an inverse wavelet transform. The reconstructed image exhibits a number of compression-related artifacts, but it is worth noting that unlike an FFT based low-pass filter, the advantage of the wavelet transform is that the image contains a fair amount of high-frequency content.

The wavelet transform is a convolution product of a signal with a scaled and translated kernel (usually a *n-th* derivative of a smoothing kernel in order to precisely detect singularities)

$$Wf(u,s) = \frac{1}{s}\int_{-\infty}^{\infty} f(x)\psi\left(\frac{x-u}{s}\right) \cdot dx \tag{2.37}$$

where s and u are real numbers (s and $u > 0$) which are discretized for computational purposes. The wavelet transform performs a transformation of a function $f(x)$ into a function defined over the scale-space plane (pair of values u and s). For implementation, typical wavelet bases are: *Daubechies* of order 4, *Symlet* of order 2 and 4, first derivative of a *Gaussian* and *Mexican Hat* wavelets.

Two simple multi-scale wavelet transforms can be used for image processing: the discrete wavelet transform (DWT) and the continuous wavelet transform (CWT). For DWT computations, a number of discrete levels can be computed from the images, and four sets of coefficients are extracted for each level: horizontal, vertical, diagonal and approximation coefficients (Mallat 1989). For CWT computations, the extension of the previous concept from 1D (Eq. 2.25) to 2D is given by Piñuela et al. (2007):

$$Mf(u,v,s) = \sqrt{\left|W^1 f(u,v,s)\right|^2 + \left|W^2 f(u,v,s)\right|^2} \tag{2.38}$$

with u and v denoting the 2D coordinates and the scale parameter being usually used as $s = 2j$. Now, two separate wavelet transforms are computed: W^1 refers to the wavelet transform performed along the horizontal dimension and W^2 refers to the vertical one.

2.4.5 Binary Morphological Operations

Morphological operations are methods for processing binary images based on shapes. Basically, morphological operations change the structure of the elements or particles in an image. In a binary image, the elements or particles are defined as

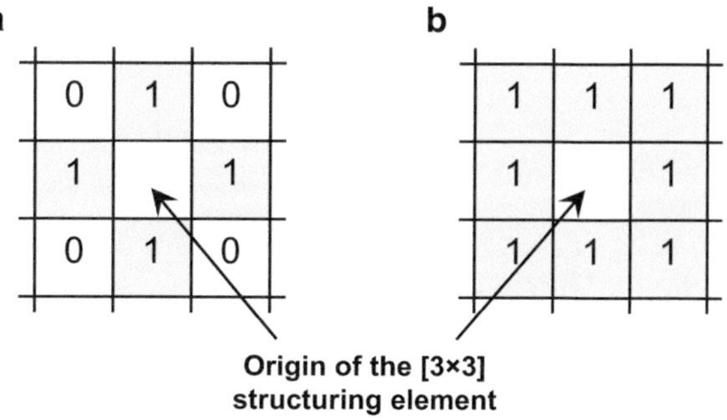

Fig. 2.11 Examples of square structuring elements (a) connectivity 4; (b) connectivity 8

the segmented regions (or ROIs) in which the pixel value is 1. The rest of the image (pixel value 0) is called background. In these operations, the value of each pixel in the output image is based on the corresponding input pixel and its neighbors. Morphological operations can be used to construct filters similar to the spatial filters discussed above. The basic operations of binary morphology are *dilation*, *erosion*, *closing*, and *opening*.

Morphological operations use a structuring element to calculate new pixels, which plays the role of a neighborhood or convolution kernel in other image processing operations (as shown in filter mask operations). Figure 2.11 shows two typical examples of structuring elements. The shape of the structuring element could be rectangular, square, diamond, octagon, disk, etc. The connectivity defines whether four or all eight surrounding pixels are used to calculate the new center pixel value (in the case of a [3×3] structuring element) (Klinger 2003).

2.4.5.1 Erosion and Dilation

These two operators are fundamental for almost all morphological operations (Gonzales and Woods 2008). Opening and closing are also duals of each other with respect to set complementation and reflection. Thus, *Erosion* is an operator that basically removes objects smaller than the structuring element and removes perimeter pixels from the border of larger image objects (sets the pixel value to 0). If I is an image and M is the structuring element (mask), the erosion (operator $\ominus$) is defined as:

$$erosion(I) = I \ominus M = \cap_{a \in M} I_{-a} \tag{2.39}$$

where I_a indicates a basic shift operation in the direction of element a of M and I_{-a} would indicate the reverse shift operation.

Contrarily, a *dilation* operation enlarges a region. A dilation adds pixels to the perimeter of each image object (sets their values to 1), filling in holes and broken areas, and connecting areas that are separated by spaces smaller than the size of the structuring element. The dilation (operator $\oplus$) is defined as:

$$dilation(I) = I \oplus M = \cup_{a \in M} I_a \tag{2.40}$$

2.4.5.2 Opening and Closing

Both operations generate a certain amount of smoothing on an object's contour. An *opening* operation (erosion then dilation) can separate objects that are connected in a binary image. Opening generally smoothes the contour of an object, breaks narrow isthmuses, and eliminates thin protrusions. Mathematically, the opening function can be described by

$$opening(I) = dilation(erosion(I)) \tag{2.41}$$

or, using the operator $\circ$,

$$I \circ M = (I \ominus M) \oplus M \tag{2.42}$$

The closing operation is defined as dilation followed by an erosion using the same structuring element. A *closing* operation can close up internal holes and gaps in a region and eliminate bays along the boundary.

$$closing(I) = erosion(dilation(I)) \tag{2.43}$$

or, using the operator $\bullet$,

$$I \bullet M = (I \oplus M) \ominus M \tag{2.44}$$

It should be noted that multiple openings or closings have no effect after the operator has been applied once.

2.5 Image Segmentation

Image segmentation is one of the most important steps in the entire image processing technique, as subsequent extracted data are highly dependent on the accuracy of this operation. Its main aim is to divide an image into regions that have a strong correlation with objects or areas of interest (i.e., ROIs). Segmentation can be achieved by three different techniques: *thresholding*, *edge-based* segmentation, and *region-based* segmentation (Sonka et al. 1999; Sun 2000). These algorithms, their modifications, and combinations with other approaches are frequently used in food quality applications.

Thresholding or global thresholding is a simple and fast technique for characterizing image regions based on constant reflectivity or light absorption of their surfaces. Edge-based segmentation relies on edge detection by edge operators. Edge operators detect the sharp discontinuities in gray level, color, texture, etc., in the image. Region-based segmentation involves the grouping together of similar pixels to form regions representing single objects within the image (such as in the seeded region growing method). The segmented image may then be represented as a boundary or a region. Boundary representation is suitable for analysis of size and shape features, while region representation is used in the detection and evaluation of texture, defects or simply ROIs (Brosnan and Sun 2004).

Two examples are presented to illustrate the segmentation process. Figure 2.12 shows the image segmentation process for green and yellow bananas using a simple global thresholding. In general, successful results are found for the identification of the true area of both bananas since the intensity distribution of the banana peel and background pixels are sufficient distinct. The histogram is bimodal with the peaks for background being tall and narrow, and separated from the banana peaks by deep valleys (Fig. 2.12b).

Ideally a segmentation process should be fully automatic so that it can provide fully objective and consistent data. Several methods have been proposed for automatic threshold determination. An interesting alternative for that is the *Otsu method*, which selects the threshold based on the minimization of the within-group variance of the two groups of pixels separated by the thresholding operator. So, if the histogram is bimodal, the thresholding problem is to determine the best threshold t separating the two modes of the histogram from each other (as shown in Fig. 2.12). Each threshold t determines a variance for the group of values that are less than or equal to t ($\leq t$) and a variance for the group of values greater than t ($>t$). The definition for the best threshold suggested by Otsu is that threshold for which the weighted sum of within-group variances is minimized. The weights are the probabilities of the respective groups (Shapiro and Stockman 2001). The application of the Otsu method for segmenting the same images of bananas in ripening stages 1 and 5 (shown in Fig. 2.12) gave threshold values of 56 and 63, respectively.

Removal of the background should be fairly straight-forward but removing non-useful sub-regions of an object can be far more difficult. Difficulty in segmentation can vary with each task. Figure 2.13 compares the results of segmenting pores and cellular material of apple tissue from X-ray computed tomography (CT) images using a *global thresholding* and a combination with the *kriging based segmentation* method (Oh and Lindquist 1999). For automatic segmentation of apple pores from a stack of CT images, a common practice in using a simple global threshold is to choice a threshold value that would match a predetermined bulk measurement of porosity (e.g., around 28 % for 'Jonagold' apples). However, this procedure is very subjective and may lead to biases when one is trying to segment and extract reproducible quantitative information from multiple images (e.g., for binary 3D reconstructions from stacks of CT images). The distinction between the void and solid phases in radiographic images is frequently not sharp (i.e., do not show a bimodal distribution) due to the amount of peak overlapping in

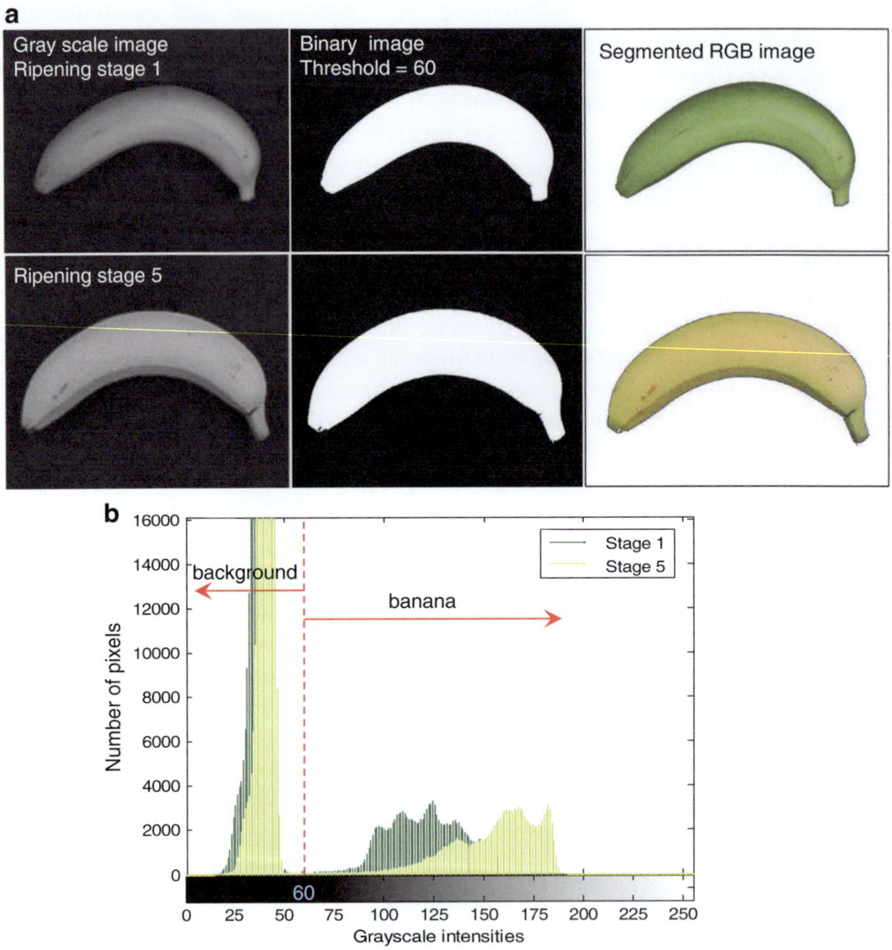

Fig. 2.12 Image segmentation process for bananas using a simple global thresholding: (**a**) pre-processing of a color image of bananas in ripening stages 1 and 5, including previous *grayscale* transformation and image smoothing using a *Laplacian-of-Gaussian filter* (LOG filter) [3 × 3] for easing the edge detection, binarization, and segmentation; (**b**) histogram of the grayscale images for both bananas showing the chosen threshold value of 60

the attenuation coefficient histogram and the nature of X-ray tomography (where processing and analysis are based on voxels instead of pixels). Moreover, the resulting binary image of apple tissue using global thresholds of 60 is noisy (Fig. 2.13b), and the average porosity is highly dependent on the selected threshold value. Figure 2.13e plots the distribution of the transmitted X-ray intensities (solid line, right axis) measured for the reconstructed tomography images of apple tissue, and it also shows the typical dependence of the porosity (open circles, left axis) of the resulting segmented image when a simple global threshold is chosen.

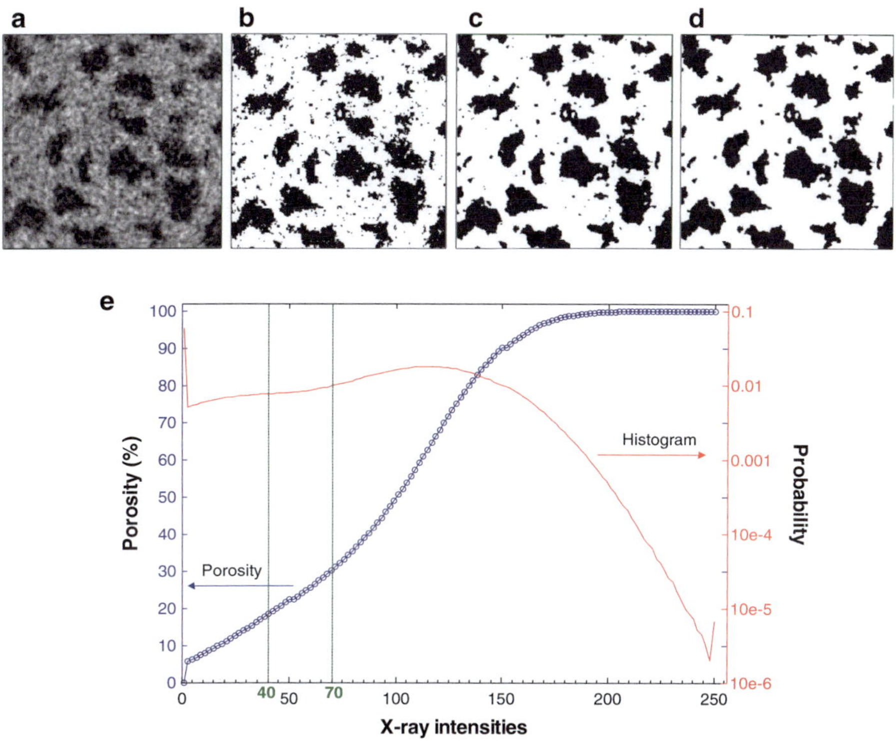

Fig. 2.13 Image segmentation process for apple tissue images using a simple global thresholding and krigking based segmentation methods: (**a**) original grayscale image; (**b**) segmented image using a simple global threshold of 60; (**c**) segmented image after using a thresholding window T0 (=40) and T1 (=70) and indicator krigking for edge detection; (**d**) segmented image after cleaning (**c**) by erosion and dilation operations; and (**e**) the distribution of the transmitted X-ray intensities (*solid line*, *right axis*) measured for the tomography images of apple tissue, and the typical dependence of the porosity (*open circles*, *left axis*) of the resulting segmented image when a simple global threshold is chosen (Reproduced from Mendoza et al. 2007b with permission from Springer-Verlag)

Alternatively, the thresholding algorithm developed by Oh and Lindquist (1999) is a non-parametric formulation able to analyze regions of uncertainty based on the estimation of the spatial covariance of the image in conjunction with indicator kriging to determine object edges (Mardia and Hainsworth 1988). Use of indicator kriging makes the thresholding local based on two threshold values, T_0 and T_1, and guarantees smoothness in the threshold surface. Implementation of the method requires a-priori population identification of some percentage of the image. Thus, for the thresholding step of apple tissue images, the gray threshold values were set at 40 and 70. According to this thresholding window, pores were identified as those voxels with gray values less than 40, and voxels of gray value greater than 70 were classified as cellular material (in general, non-edge). Finally, the remaining voxels of the population were identified by indicator kriging (Fig. 2.13c).

Small disconnected void or solid phase components which may be either physical or arise from segmentation error, can be cleaned up from each image using a morphological procedure called *opening* which involves a simple step of erosion followed by dilation (Fig. 2.13d).

2.6 Quantitative Analysis

A large part of computer vision is concerned with extracting features from images. Techniques by which numerical measurements are extracted from images vary considerably in technological complexity. The process of image measurement involves an enormous reduction in the amount of data, by selecting from the original image those objects and parameters that are important for further characterization and sorting of samples. This selection and reduction is at the heart of image analysis and measurement, and it is achieved by ignoring irrelevant information.

In food images, the most frequently occurring features are color, texture and morphology. These are the features that reveal information on palatability, quality and defects. They are easy to measure but can be difficult to characterize succinctly. After processing of the image and with the region of interest identified, the region must be described with measurable features based on standard chemical and physical measurements (e.g., color parameters by colorimeter or spectrophotometer, firmness by penetrometer, soluble solids content by refractometer, etc.), or empirical standards developed by experts (standard charts for color, gloss and texture of food surfaces, and sensorial evaluations). Many image processing techniques developed for 2D images can also be extended to analyzing multi-dimensional images, such as multispectral or hyperspectral images and tomographic images. In this section, we discuss in some detail the three main approaches and methods used in image analysis: color analysis, image texture analysis, and geometrical or morphological analysis, accompanied with specific application examples in food quality inspection.

2.6.1 Color Analysis

Color is specified by the geometry and spectral distributions of three elements: the light source, the reflectivity of the sample and the visual sensitivity of observer. Each of these was defined by the *Commission Internationale de I'Eclairage* (CIE) in 1931. CIE also defined, for the purpose of color measurement, the cone of spectral sensitivities of an average observer and introduced several methods to describe color objectively. The definition was aimed at stimulating the human color perception based on a 2° field of view, a set of primaries (red, green, blue), and color-matching functions (CIE 1986). Several standard illuminants which are specified by their color temperatures were also defined by CIE. The most common one is standard illuminant D_{65}, corresponding to the radiation of a black body at 6,500°K, which is intended to represent average daylight (Hunt 1991).

2.6.1.1 Color Evaluation

To accurately specify object colors and color differences, CIE recommended two color spaces that are less illumination-dependent, namely, CIELAB or $L^*a^*b^*$ and CIELUV $L^*u^*v^*$ (Robertson 1976). The CIELAB space has a function of correcting for chromatic adaptation to the white point, and is intended for object color displays. The CIELUV space is defined in a similar manner, and the coordinate (L^*, u^*, v^*) is calculated from the Y and (u', v') of the given light stimulus and the white point.

The color difference in the CIELAB space is calculated as the Euclidean distance between the points in this three-dimensional space, and is given by,

$$\Delta E^*_{ab} = \left[\left(\Delta L^*\right)^2 + \left(\Delta a^*\right)^2 + \left(\Delta b^*\right)^2 \right]^{1/2} \tag{2.45}$$

Equation 2.45 is called the CIE 1976 (L^*, a^*, b^*) color difference formula. Chroma C^*_{ab} and hue angle h^*_{ab} are also calculated from (L^*, a^*, b^*) by

$$C^*_{ab} = \left(a^{*2} + b^{*2}\right)^{1/2} \tag{2.46}$$

$$h^*_{ab} = tan^{-1}\left(b^*/a^*\right) \tag{2.47}$$

Procedures for calibrated color measurements and color reproduction using machine vision systems with color digital cameras were proposed and discussed in Mendoza et al. (2006) for the analysis of fruits and vegetables and in Valous et al. (2009a) for the analysis of pre-sliced hams.

2.6.1.2 Color Measurement on Curved Surfaces

Computer vision systems based on color digital cameras can quantify the color of any food easily and quickly (Segnini et al. 1999; Papadakis et al. 2000). Different color channels give different color information about the pixel intensity distribution of an image, and therefore, evaluations using different color components represent an advantage in further quality characterization and classification tasks. However, average color results from curved and/or uneven surfaces, as they are typical in many fresh foods, should be interpreted with caution. A sensitivity analysis of color measurements by a common machine vision system for samples with curved surfaces (bananas and red pepper) revealed that $L^*a^*b^*$ is more appropriate for

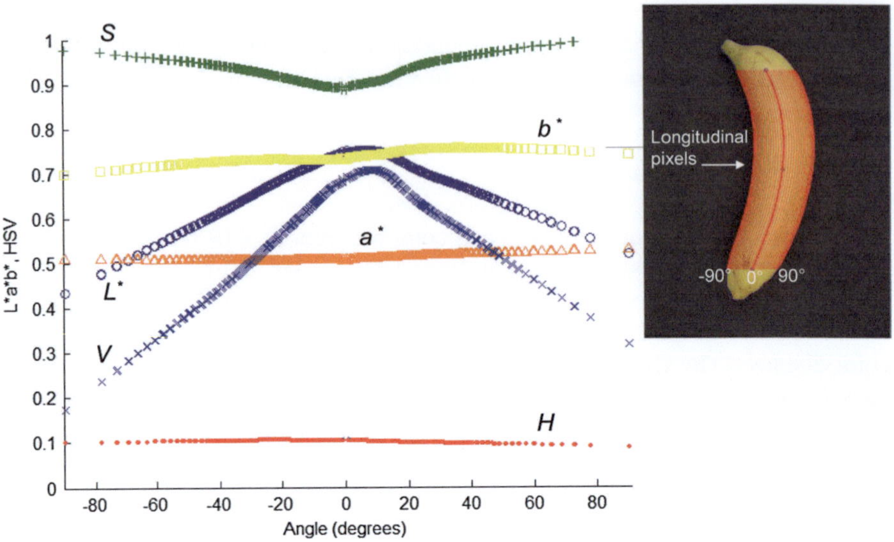

Fig. 2.14 Color profiles expressed in $L^*a^*b^*$ and *HSV* color scales for a yellow banana. The profiles were constructed using the average values of the pixels in the longitudinal direction of the segmented image and its angular positions varying from – 90° to 90° (Modified from Mendoza et al. 2006)

color representation of surfaces or materials illuminated by a light source. These color profiles are less affected by the degree of curvature, shadows and glossiness of the surfaces than the *RGB* and *HSV* color systems, and, therefore, more appropriate for color measurement of food surfaces (Mendoza et al. 2006). An example of the effect of the curvature on a yellow banana using the $L^*a^*b^*$ and *HSV* color scales is illustrated in Fig. 2.14. L^* and *V* color scales were highly sensitive to the curvature of the banana surface, while the *S* scale was also surface-curvature sensitive but to a lesser extent. For the a^*, b^* and *H* scales no or minimum color variations were observed.

2.6.2 Texture Analysis

The meaning of the term *texture* in image processing is completely different from the usual meaning of texture in foods. Image texture can be defined as the spatial organization of intensity variations in an image at various wavelengths, such as the visible and infrared portions of the electromagnetic spectrum (Haralick et al. 1973). Image texture is an important aspect of images and textural features play a major role in image analysis. In some images, it can be the defining characteristic of regions and critical in obtaining a correct analysis (Shapiro and Stockman 2001). These features provide summary information defined from the intensity maps of the scene, which may be related to visual characteristics (coarseness of the texture, regularity, presence of a privileged direction, etc.), and also to characteristics that cannot be visually differentiated.

Texture is a property of areas, so there is no texture for a point. Texture involves the spatial distribution of gray levels and there is a need for a significant number of intensity units (i.e., pixels) in a region to detect texture features. This is linked to the fact that texture involves gray levels in a spatial neighborhood. There are many texture analysis techniques that can be applied to images. Among the most popular methods used for the characterization and evaluations of food surfaces and biological structures are: *First-order Statistics*, *Gray Level Co-occurrence* and *Run Length Matrices*, and *Fractals*.

2.6.2.1 First-Order Statistics (FOS)

Image histogram gives primarily the global description of the image. The histogram of a gray-level image represents the relative frequency of occurrence of each gray-level in the image. The features of FOS are commonly derived from the normalized gray-level histogram of the digital image, which is built by counting the number of pixels (N) with the gray value of i (I) and can be written as:

$$H(i) = N\langle (x, y) | I(x, y) = i \rangle \tag{2.48}$$

The histogram, $H(i)$, is normalized using the function given below:

$$H'(i) = \frac{H(i)}{\sum_i H(i)} \tag{2.49}$$

The extracted statistical features for further partial least squares analysis included: *mean* of the pixel histogram (MV), *variance* (VA), *entropy* (ET), and *energy* (EN), which are defined as follows (Cernadas et al. 2005):

$$MV = \sum_i iH'(i) \tag{2.50}$$

$$VA = \sum_i (i - MV)^2 H'(i) \tag{2.51}$$

$$ET = -\sum_i H'(i)\log\left(H'(i)\right) \tag{2.52}$$

$$EN = \sum_i i^2 H'(i) \tag{2.53}$$

These FOS features, however, do not provide any insight about the possible textural differences in an image, because they do not extract any information about the relative position of pixels, and the correlation of their intensities.

2.6.2.2 Gray Level Co-occurrence Matrix (GLCM)

When a non-random spatial distribution (or more than one texture in an image) has to be characterized, second-order statistics are more suitable to describe these types of relationships. A GLCM is a matrix with a number of rows and a number of columns equal to the number of gray level intensities of the image, which contains textural features (spatial relationships) from an image obtained using second order statistics. Each element of the gray level co-occurrence matrix $P(i,j|\Delta x, \Delta y)$ is the relative frequency or probability of occurrence linked to the combination of intensities $I(i,j)$ in all pixel pairs of the image located at a distance $(\Delta x, \Delta y)$. This relative position is defined by a distance (d) and an angle ($\theta = 0°, 45°, 90°, 135°$). For a given directional orientation and distance of the patterns, 14 textural features can be extracted from a grayscale image using this matrix as proposed by Haralick et al. (1973), the most common being *energy*, *entropy*, *contrast*, *correlation*, *local homogeneity* and *variance*. They are computed by:

$$Energy = \sum_i \sum_j P_{d\theta}(i, j)^2 \tag{2.54}$$

$$Entropy = -\sum_i \sum_j P_{d\theta}(i, j) \cdot \log(P_{d\theta}(i, j)) \tag{2.55}$$

$$Contrast = \sum_i \sum_j (i - j)^2 \cdot P_{d\theta}(i, j) \tag{2.56}$$

$$Correlation = \frac{\sum_i \sum_j (i \cdot j) p(i, j) - \mu_i \cdot \mu_j}{\sigma_i \cdot \sigma_j} \tag{2.57}$$

$$Homogeneity = \sum_i \sum_j \frac{1}{1 + (i - j)^2} \cdot P_{d\theta}(i, j) \tag{2.58}$$

$$Variance = \sum_i \sum_j (i - u_{ij})^2 \cdot P_{d\theta}(i, j) \tag{2.59}$$

where μ_i and μ_j are the means, and σ_i and σ_j are the standard deviations. *Energy* measures the textural uniformity of the image, i.e., the repetition of pixel pairs.

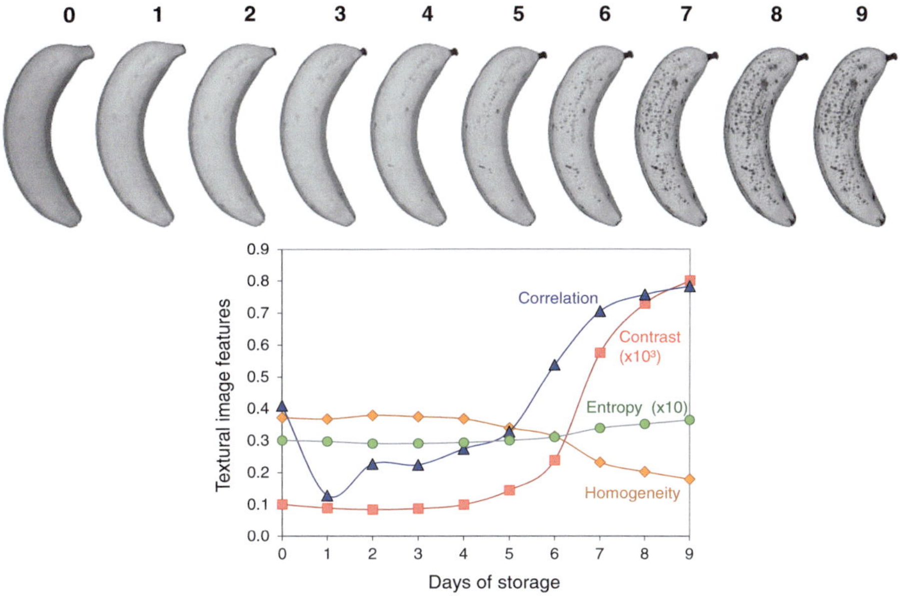

Fig. 2.15 Image texture analysis using the gray level co-occurrence matrix (GLCM) method for quantifying the ripening process of a banana (*Musa cavendish*)

Entropy measures the disorder or randomness of the image and it is an indication of complexity within the image, i.e., more complex images have higher entropy values. *Contrast* is a measure of the local variations present in the image; so higher contrast values indicate larger local variations. *Homogeneity* (also called an inverse difference moment) is inversely proportional to contrast at constant energy. Similarly at constant contrast, homogeneity is inversely proportional to energy. Finally, *correlation* is a measure of image linearity among pixels (Mendoza et al. 2007a). An application of GLCM features for the characterization of the surface appearance of a banana (*Musa cavendish*) during ripening is presented in Fig. 2.15.

2.6.2.3 Run Length Matrix (RLM)

First introduced by Galloway (1975), the gray-level RLM approach characterizes texture by the gray-level run, which is a set of consecutive pixels with the same gray-level. Run length is the number of pixels in a run. Therefore, the run length of coarse textures will be longer than that of fine textures. In RLM method, a matrix containing the information about the run length of images is constructed in terms of the brightness value and length of the runs (Fardet et al. 1998).

The run-length matrix $P(i, j)$ is defined by specifying direction (i.e., $0°$, $45°$, $90°$, $135°$) and then counting the occurrence of runs for each gray levels and length in this direction. i-dimension corresponds to the gray level (bin values) and has a length equal to the maximum gray-level (bin values), *j*-dimension corresponds to the run length and has a length equal to the maximum run length (bin values). Five features were proposed by Galloway (1975), namely, *short run emphasis* (SRE) which measures the distribution of short runs, *long run emphasis* (LRE) which measures the distribution of long runs, *gray-level non-uniformity* (GLNU) which measures the similarity of gray-level values throughout the image, *run length non-uniformity* (RLNU) which measures the similarity of the length of runs throughout the image, and *run length percentage* (RLP) which measures the homogeneity and the distribution of runs of an image in a given direction. These RLM features are calculated as follows:

$$SRE = \frac{1}{n_r}\sum_{i=1}^{M}\sum_{j=1}^{N}\frac{P(i,j)}{j^2} \tag{2.60}$$

$$LRE = \frac{1}{n_r}\sum_{i=1}^{M}\sum_{j=1}^{N}P(i,j)\cdot j^2 \tag{2.61}$$

$$GLNU = \frac{1}{n_r}\sum_{j=1}^{N}\left(\sum_{i=1}^{M}P(i,j)\right)^2 \tag{2.62}$$

$$RLNU = \frac{1}{n_r}\sum_{i=1}^{M}\left(\sum_{j=1}^{N}P(i,j)\right)^2 \tag{2.63}$$

$$RLP = \frac{n_r}{n_p} \tag{2.64}$$

in which n_r is the total number of runs and n_p is the number of pixels in the image, M is the number of gray levels (bins) and N is the number of run length (bins).

2.6.2.4 Fractal Methods

A fractal describes a rough or fragmented geometric shape that can be subdivided into parts, each of which is, at least approximately, a reduced-size copy of the whole. This means that they are generally self-similar and independent of scale (Mandelbrot 1983). Contrary to the classical Euclidean geometry, fractals are not regular and may have an integer or non-integer dimension. Thus, fractal dimensions offer a systematic approach in quantifying irregular patterns that contain an internal structure repeated over a range of scales. Self-similarity is not visually obvious, but there may be numerical or statistical measures that are preserved across scales.

Due to their statistical scaling invariance, natural objects may exhibit statistical fractality (Klonowski 2000).

The methods to calculate fractal dimensions can be divided into two types: spatial and spectral. The first type operates in the spatial domain, while the second type operates in the frequency domain, using the Fourier power spectrum. The two types are unified by the principle of fractional Brownian motion (Dougherty and Henebry 2001).

Box-Counting Method

Estimations of *fractal dimension* are all based on the *box-counting method*. The technique allows to obtain the scaling properties of two dimensional fractal objects (such as from binary images) by covering a measure with a range of boxes of size ε and counting the number of boxes containing at least one pixel representing the object under study. This means that the technique does not consider the amount of mass (density of pixels) inside each box. Hence, in a homogeneous system, the number N of features of a certain size ε varies as (Evertsz and Mandelbrot 1992):

$$N(\varepsilon) \propto \varepsilon^{-D_0} \tag{2.65}$$

where the fractal dimension D_0,

$$D_0 = \lim_{\varepsilon \to 0} \frac{log N(\varepsilon)}{log \frac{1}{\varepsilon}} \tag{2.66}$$

can be measured by counting number N of boxes needed to cover the object under investigation for increasing box sizes ε and estimating the slope of a *log-log* plot. Figure 2.16 illustrates the scaling properties of a binary image for apple tissue.

Variogram Method

The *directional fractal dimension* (DFD) of grayscale images using the *variogram* model is computed as follows (Kube and Pentland 1988; Mandelbrot 1983):

$$v(d) = c \cdot d^a \tag{2.67}$$

$$v(d) = \frac{1}{N_d} \sum_{N(d)} [y(s+d) - y(s)]^2 \tag{2.68}$$

$$DFD = 3 - 0.5\hat{a} \tag{2.69}$$

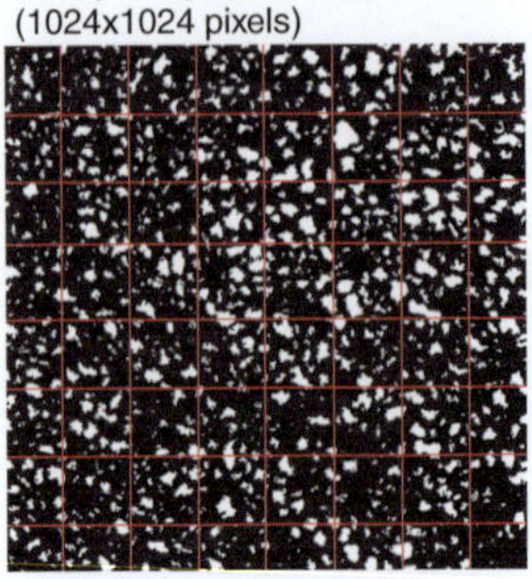

$\varepsilon = Lx2^{-k}$	$N(\varepsilon)=2^k$	Actual $N(\varepsilon)$
$1024x2^{-1}$	4	4
$1024x2^{-2}$	16	16
$1024x2^{-3}$	64	64
$1024x2^{-4}$	256	250
$1024x2^{-5}$	1024	995
$1024x2^{-6}$	4096	3084
$1024x2^{-7}$	16384	7236
$1024x2^{-8}$	65536	18282
$1024x2^{-9}$	262144	53050

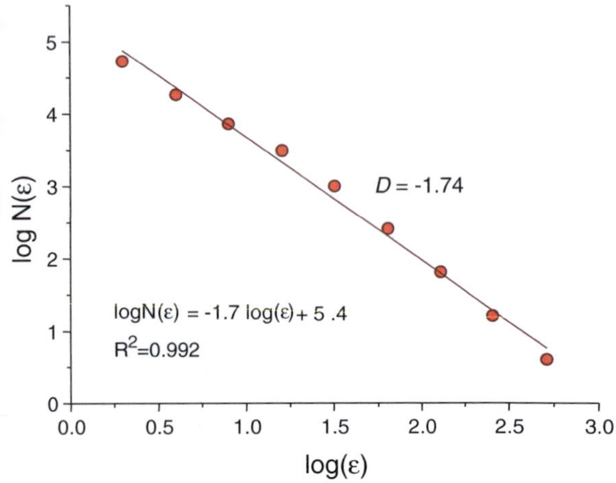

Fig. 2.16 Illustration of fractal *box-counting* theory applied to a binary image of fresh apple tissue (pores are represented by *white pixels*)

where $v(d)$ is the *variogram* of the image, a is termed the fractal index, c is a constant, d is the separation distance in pixels, $y(s)$ denotes the gray-level of the pixel at location s, and $N(d)$ denotes the cardinality of the set of pairs of observations. The image *variogram* $v(d)$ represents the variance or dispersion of the difference between two random variables. Thus, the relationship between a and $v(d)$ can also be represented using a linear regression model by applying log function to both sides of this equation to get an estimate of fractal index $\hat{a}$. Then, the directional fractal dimension (DFD) of the image can be directly computed.

The *variogram* of the image and hence the fractal dimension are estimated at a fixed image resolution level without specifying any spatial direction along which the set of pairs of observations is constructed. This means that the image is assumed to be isotropic. Since the fractal texture parameters of many images of foods and biological materials do not have isotropic patterns, four variograms should be computed along the directions $d = 0°$, $45°$, $90°$, and $135°$ (namely horizontal, first diagonal, vertical and second diagonal) respectively. These variograms should be analyzed and averaged for a given pixel location s of an image, and then be used in further analysis and image characterization.

This approach has been recently applied to the appearance characterization and classification of commercial pork, turkey and chicken ham slices (Mendoza et al. 2009). DFD features were extracted from the digitalized intensity images in grayscale, and $R, G, B, L^*, a^*, b^*, H, S$, and V color components were calculated for three image resolution levels (100, 50, and 25 %). Simulation results showed that in spite of the complexity and high variability in color and texture appearance, the modeling of ham slice images with DFD allowed the capture of differentiating textural features between the four commercial ham types. Independent DFD features entailed better discrimination than that using the average of four directions. However, the DFD features revealed a high sensitivity to color channel, orientation and image resolution for the fractal analysis. The classification accuracy using six DFD features was 93.9 % for the training data and 82.2 % for the testing data.

Fourier Fractal Texture (FFT)

In the calculation of FFT, the 2D Fourier transform of the grayscale image is first calculated and the 2D power spectrum is then derived. The 2D power spectrum is reduced to a 1D radial power spectrum (direction-independent mean spectrum, i.e., the average of all possible directional power spectra) by averaging values over increasingly larger annuli for each of the radial increments. The power spectrum, $P(f)$, varying with frequency f, is calculated as follows (Dougherty and Henebry 2001):

$$P(f) = k \cdot f^{(-1-2H)} \tag{2.70}$$

where k is a constant and H is the *Hausdorff-Besicovitch dimension.* When the log $[P(f)]$ is plotted against log[f], a straight line can be fitted. According to the Fourier slice theorem, the 1D Fourier transform of a parallel projection of an image along a line with the direction h is identical to the value along the same line in the 2D Fourier transform of the image. This means that the line through the spectrum gives the spectral information obtained from a projection with the same orientation in the spatial domain. FFT dimensions, D_f, are calculated as a function of orientation based on this theorem, with 24 being the frequently number of directions that the frequency space is uniformly divided. The data of magnitude vs. frequency are plotted in *log-log* scale and its slope is determined using linear least-squares regression. Thus, Hausdorff-Besicovitch dimension H is computed from the slope c of the straight line, $c=(-1-2H)$. The D_f dimension of the grayscale image is related to the slope c of the *log–log* plot by the equation below, with $H=D_f-3$, $2<D_f<3$ and $3<c<1$ (Geraets and Van der Stelt 2000):

$$D_f = \frac{7}{2} + \frac{c}{2} \tag{2.71}$$

For analysis the slope and intercept for all directions can be computed from each image and used for further image characterization and classification. This algorithm was proposed by Russ (2005) and was modified by Valous et al. (2009b) for hams processing.

Fractal Lacunarity (FL)

Various studies have shown that the fractal dimension alone is not a sufficient metric for the characterization of most textures, because fractal dimension only measures how much space is filled. A second-order fractal metric such as *Lacunarity* complements fractal dimension by measuring how the data fill the space, enabling the parsimonious analyses of textures (Tolle et al. 2008). In this method, a square structuring element or moving window of side length b is placed in the upper left-hand corner of the ham slice image of side length T (pixels), such that $b \leq T$. The algorithm records the number or mass m of pixels that are associated with the image underneath the moving window. The window is then translated by one pixel to the right and the underlying mass is again recorded. When the moving window reaches the right-hand side of the image, it is moved back to its starting point at the left-hand side of the image and is translated by one pixel downward. The computation proceeds until the moving window reaches the lower right-hand edge of the image, at which point it has explored every one of its $(T - b + 1)^2$ possible positions. Allain and Cloitre (1991) defined *lacunarity* Λ measured with a square moving window of side length b on an image of side length T (pixels), such that $b \leq T$, as:

$$\Lambda = \frac{\sigma^2}{\mu^2} + 1 \tag{2.72}$$

where the ratio of σ (standard deviation) to μ (mean) changes with window size which signifies that lacunarity depends on the scale (moving window size relative to image size). Lacunarity may assume any value between 1 and ∞. A lacunarity value of 1 indicates that the textural content is uniformly dispersed at a given scale, while values $\neq 1$ indicate non-uniformity.

The technique can be performed on binary and grayscale images. However, computing the lacunarity for characterizing binary images would result in the counting of every gliding box, thus the boxes would all be full and the output would be rather uninteresting. In contrast, when the algorithm measures the average intensity of pixels per gliding box in most cases the results of this textural analysis is more than adequate (Krasowska et al. 2004).

Valous et al. (2009b) applied the gliding box algorithm to characterize the texture appearance of ham slices based on segmented images of pores/defects and fat-connective tissue. Later, Valous et al. (2010) used the same approach to characterize ham slices based on grayscale information. Lacunarity was computed for each value of b between $b_{\min} = 2$ and $b_{\max} = 256$ with a step size of 1 for the 256×256 pixel intensity images. Once the computation was complete, the lacunarity as a function of moving window b was presented as a spatial response function with a 2D scatter plot, which illustrated the scale dependency of spatial non-stationary in the intensity images. As shown in Fig. 2.17, lacunarity plots revealed important textural content information that corresponds to degree of

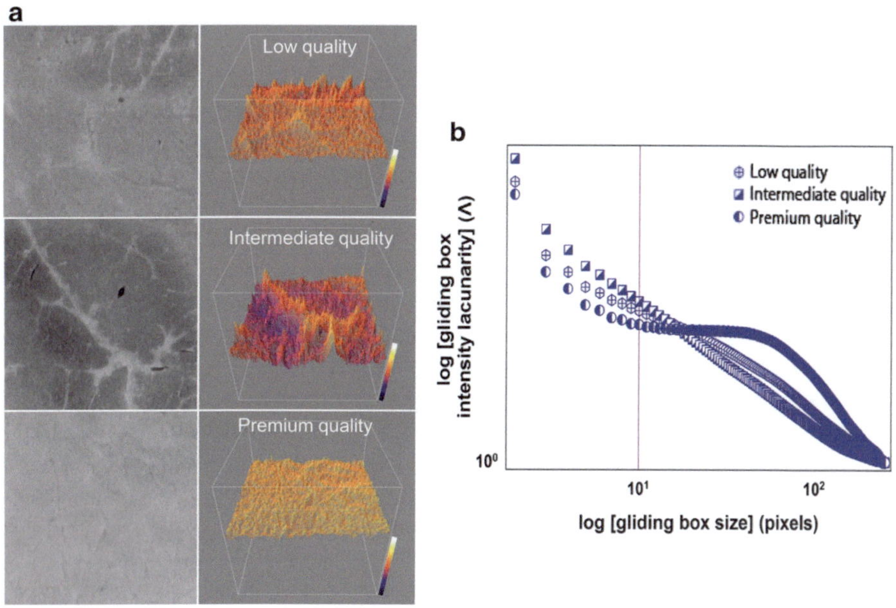

Fig. 2.17 Visualization of surface of three wet-cured cooked pork ham qualities (high yield or low quality, medium yield or intermediate quality, and low yield or premium quality ham) using different muscle sections, percentages of brine solutions and processing regimes. (**a**) intensity images of *B* color channel and their three-dimensional mesh plots. (**b**) two-dimensional *log–log* scatter plot of averaged values of intensity *lacunarity* as a function of moving window *b*. (Reproduced from Valous et al. 2010)

spatial heterogeneity of intensities and level of self-similar behavior. The results of intensity lacunarity suggested that window sizes up to 10 pixels may be adequate to cover textural features and produce meaningful results.

2.6.3 Morphology Analysis

Once a set of regions-of-interest has been identified by image segmentation, the individual regions (objects) or silhouettes can be described by their geometric and shape properties. Most image analysis systems offer a number of measures related to the size and shape and produce a numeric output suitable for further image characterization and classification. Size and shape are common object measurements for food quality evaluation; and compared with other features such as color and image texture, they are easier to measure using image processing techniques. Among these morphological properties are the counting of objects (such as particles, holes, granules, droplets, bubbles, among others) and defining their position, surface roughness and orientation in the image.

2.6.3.1 Particle Size

Many image analysis applications need to characterize or describe the distribution of particle sizes on binary images. The usual way of summarizing data for a particular system is to draw a frequency histogram of the number of particles in a certain size class. Furthermore, since many natural samples are broadly shaped like a *normal* or *Gaussian* distribution, this allows to apply statistical methods to evaluate confidence parameters and make recommendations on the minimum number of particles to analyze in order to achieve a reasonable level of statistical significance. Another widely used method of depicting data is to calculate a cumulative distribution, which shows what percentage of material lies above or below a particular size (Aguilera and Stanley 1999).

In food quality evaluations, there are three commonly used features for size measurement of an object: area, perimeter, and length and width. The most basic measurement for size is the area, which is represented by the number of pixels within the area and is straightforwardly determined by counting. The perimeter of an object is the length of its boundary, and it is particularly useful for discriminating between objects with simple and complex shapes. Area and perimeter are easily computed from a segmented image, however the quality of the measurements is highly dependent of the complexity of the measured objects and the efficiency of the segmentation method. The length and width of an object can also be used to measure the size of an object. It is necessary to locate the major axis of the object and measure its relative length and width (Du and Sun 2004).

Although relative measurements of the percentage of pixels of selected areas on an image could be done for various particle size evaluations, in some applications we need to know the x and y dimensions of an image in its real dimensional units. Hence, before making any measurements, the relationships between the size of a pixel to a size of an object of known length (in mm for example) that is visible in the image must be first specified. A scale calibration factor (f) for the x and y directions can be determined by

$$f = \frac{real\ distance\ (mm)}{image\ distance\ (pixels)} \tag{2.73}$$

2.6.3.2 Shape Descriptors

Measuring size alone is sometimes insufficient to detect important but subtle differences between samples. This is because particles or grains with similar composition and/or structure could measure the same area or perimeter, but have different shapes.

Frequently, the objects of one class can be distinguished from the others by their shapes, which are physical dimensional measurements that characterize the appearance of an object. Shape features can be measured independently and by combining

Table 2.2 Common shape descriptors

Statistical parameter	Calculation
Area ratio	$= \dfrac{Area}{Max\ Diameter \cdot Min\ Diameter}$
Aspect ratio	$= \dfrac{Max\ Diameter}{Min\ Diameter}$
Compactness	$= \dfrac{Perimeter^2}{Area}$
Circularity	$= \dfrac{4\pi \cdot Area}{Perimeter^2}$
Diameter range	$= Max\ Diameter - Min\ Diameter$
Eccentricity	$= \sqrt{1 - \dfrac{Semi\ Minor^2}{Semi\ Major^2}}$
Elongation	$= 1 - Aspect\ Ratio$
Roundness	$= \dfrac{4\pi \cdot Area}{\pi \cdot Max\ Diameter^2}$
*Shape factor*1	$= \dfrac{4\pi \cdot Area}{Perimeter^2}$
*Shape factor*2	$= \dfrac{Max\ Diameter}{Area}$

size measurements. Table 2.2 summarizes some of the most widely used shape features with combinations of size measurements for food products.

Of particular interest in particle analysis is *circularity*, which is a good measure of deviation from a perfect circle. However it is important to remember that it is unlikely that one single shape descriptor will perfectly discriminate and characterize all applications and different combinations of shapes. *Circularity* has values in the range of $0 - 1$. A perfect circle has a circularity of 1, while a very *spiky* or irregular object has a circularity value closer to 0. Circularity is sensitive to both overall form and surface roughness.

2.7 Concluding Remarks

The field of image processing and analysis has experienced dramatic growth and its application has become increasingly widespread in recent years. The development of new and more effective algorithms for image processing and analysis, along with advances in image acquisition, computer, data storage, and the Internet, has made it possible to handle an increasingly large volume of image data. Digital image processing has become economical in many fields of research and in industrial applications. While each application is different or unique from the others, they are all concerned about speed, affordability and performance or accuracy. More and more research activities and applications in machine vision are being focused on real-time and interactive operations, where image acquisition, processing, and analysis and decision-making are almost carried out simultaneously or in parallel.

For the past decade, we have also witnessed an exponential increase in research and development activities in multispectral and hyperspectral imaging technology for food and agriculture applications. While many of the basic image processing and analysis methods described in this chapter are still useful for processing and analyzing two or three-dimensional multispectral and hyperspectral images, it also entails new challenges in handling these types of images. Methods and techniques specific to processing and analyzing multispectral and hyperspectral images are presented in Chaps. 6 and 7.

References

Aguilera JM, Stanley DW (1999) Microstructural principles of food processing and engineering, 2nd edn. Aspen, Gaithersburg

Allain C, Cloitre M (1991) Characterizing the lacunarity of random and deterministic fractal sets. Phys Rev A 44(6):3552–3558

Brosnan T, Sun D-W (2004) Improving quality inspection of food products by computer vision—a review. J Food Eng 61:3–16

Castleman KR (1979) Digital image processing. Prentice-Hall, Englewood Cliffs

Cernadas E, Carrión P, Rodriguez PG, Muriel E, Antequera T (2005) Analyzing magnetic resonance images of Iberian pork loin to predict its sensorial characteristics. Comput Vis Image Und 98:344–360

CIE (1986) Colorimetry, Official recommendations of the International Commission on Illumination, CIE Publication No. 15.2. CIE Central Bureau, Vienna

Dougherty G, Henebry GM (2001) Fractal signature and lacunarity in the measurement of the texture of trabecular bone in clinical CT images. Med Eng Phys 23:369–380

Du C-J, Sun D-W (2004) Recent developments in the applications of image processing techniques for food quality evaluation. Trends Food Sci Technol 15:230–249

Du C-J, Sun D-W (2005) Comparison of three methods for classification of pizza topping using different color space transformations. J Food Eng 68:277–287

Du C-J, Sun D-W (2007) Quality measurement of cooked meats. In: Sun D-W (ed) Computer vision technology for food quality evaluation. Elsevier/Academic, London, pp 139–156

Evertsz CJG, Mandelbrot BB (1992) Multifractal measures. In: Peitgen H-O, Jurgens H, Saupe D (eds) Chaos and fractals. New frontiers of science. Springer, New York, pp 921–953

Fardet A, Baldwin PM, Bertrand D, Bouchet B, Gallant DJ, Barry J-L (1998) Textural images analysis of pasta protein networks to determine influence of technological processes. Cereal Chem 75:699–704

Galloway MM (1975) Texture analysis using grey level run lengths. Comput Graph Image Process 4:172–179

Geraets WGM, Van der Stelt PF (2000) Fractal properties of bone. Dentomaxillofac Radiol 29:144–153

Gonzales RC, Woods RE (2008) Digital image processing. Prentice-Hall, Englewood Cliffs

Gunasekaran S, Ding K (1994) Using computer vision for food quality evaluation. Food Technol 6:151–154

Haralick RM, Shanmugan K, Dinstein I (1973) Textural features for image classification. IEEE Trans Syst Man Cybern 3:610–621

Hunt RWG (1991) Measuring of color, 2nd edn. Ellis Horwood, New York

Klinger T (2003) Image processing with LabVIEW and IMAQ vision. Prentice Hall Professional Technical Reference, Upper Saddle River

Klonowski W (2000) Signal and image analysis using chaos theory and fractal geometry. Mach Graphics Vis 9(1/2):403–431
Krasowska M, Borys P, Grzywna ZJ (2004) Lacunarity as a measure of texture. Acta Phys Pol B 35:1519–1534
Kube P, Pentland A (1988) On the imaging of fractal surfaces. IEEE Trans Pattern Anal Mach Intell 10:704–707
Mallat S (1989) A theory for multiresolution signal decomposition: the wavelet representation. IEEE Trans Pattern Anal Mach Intell 11:674–693
Mandelbrot BB (1983) The fractal geometry of nature. W.H. Freeman, New York
Marchant JA (2006) Machine vision in the agricultural context. In: Munack A (ed CIGR) Precision agriculture, CIGR handbook of agricultural engineering, vol VI. Information technology. The International Commission of Agricultural Engineering, St. Joseph, MI (Chapter 5)
Mardia KV, Hainsworth TJ (1988) A spatial thresholding method for image segmentation. IEEE Trans Pattern Anal Mach Intell 6:919–927
Macaire L, Postaire JG (1989) Real-time adaptive thresholding for on-line evaluation with line-scan cameras. In: Proceedings of computer vision for industry, Society of Photooptical Instrumentation Engineers, Boston, MA, pp 14–25
Mendoza F, Dejmek P, Aguilera JM (2006) Calibrated color measurements of agricultural foods using image analysis. Postharvest Biol Tech 41:285–295
Mendoza F, Dejmek P, Aguilera JM (2007a) Color and texture image analysis in classification of commercial potato chips. Food Res Int 40:1146–1154
Mendoza F, Verboven P, Mebatsion HK, Kerckhofs G, Wevers M, Nicolaï B (2007b) Three-dimensional pore space quantification of apple tissue using X-ray computed microtomography. Planta 226:559–570
Mendoza F, Valous NA, Allen P, Kenny TA, Ward P, Sun D-W (2009) Analysis and classification of commercial ham slice images using directional fractal dimension features. Meat Sci 81:313–320
Oh W, Lindquist W (1999) Image thresholding by indicator kriging. IEEE Trans Pattern Anal Mach Intell 21:590–602
Piñuela JA, Andina D, McInnes KJ, Tarquis AM (2007) Wavelet analysis in a structured clay soil using 2-D images. Nonlin Process Geophys 14:425–434
Puglia S (2000) Technical primer. In: Sitts MK (ed) Handbook for digital projects: a management tool for preservation and access, 1st edn. Northeast Document Conservation Center, Andover
Papadakis S, Abdul-Malek S, Kamdem RE, Jam KL (2000) A versatile and inexpensive technique for measuring color of foods. Food Technol 54:48–51
Prats-Montalbán JM, de Juan A, Ferrer A (2011) Multivariate image analysis: a review with applications. Chemometr Intell Lab 107:1–23
Rec. ITU-R BT.709-5 (2002) Parameter values for the HDTV standards for production and international programme exchange (1990, revised 2002). International Telecommunication Union, 1211 Geneva 20, Switzerland
Robertson AL (1976) The CIE 1976 color difference formulae. Color Res Appl 2:7–11
Russ JC (2005) Image analysis of food microstructure. CRC, New York
Segnini S, Dejmek P, Öste R (1999) A low cost video technique for color measurement of potato chips. Lebensm-Wiss U-Technol 32:216–222
Shapiro LG, Stockman GC (2001) Computer vision. Prentice-Hall, Upper Saddle River
Soille P (1999) Morphological image analysis. Springer, Berlin
Sonka M, Hlavac V, Boyle R (1999) Image processing, analysis, and machine vision. PWS, Pacific Grove
Sun D-W (2000) Inspecting pizza topping percentage and distribution by a computer vision method. J Food Eng 44:245–249
Tolle CR, McJunkin TR, Gorsich DJ (2008) An efficient implementation of the gliding box lacunarity algorithm. Phys D 237:306–315

Valous NA, Mendoza F, Sun D-W, Allen P (2009a) Colour calibration of a laboratory computer vision system for quality evaluation of pre-sliced hams. Meat Sci 42:353–362

Valous NA, Mendoza F, Sun D-W, Allen P (2009b) Texture appearance characterization of pre–sliced pork ham images using fractal metrics: Fourier analysis dimension and lacunarity. Food Res Int 42:353–362

Valous NA, Sun D-W, Mendoza F, Allen P (2010) The use of lacunarity for visual texture characterization of pre-sliced cooked pork ham surface intensities. Food Res Int 43(3):87–395

Wikipedia (2012) Binary image. http://en.wikipedia.org/wiki/Binary_image. Accessed 12 Dec 2012

Chapter 3
Basics of Spectroscopic Analysis

Stephen R. Delwiche

3.1 Vibrational Spectroscopy Defined

The region of the electromagnetic spectrum that draws our interest is between the ultraviolet (lower end of 400 nm wavelength) and the far-infrared (upper end of 50,000 nm). This region encompasses the visible (400–780 nm), near-infrared (780–2,500 nm), and mid-IR (2,500–25,000 nm) regions. Flanking this large swath of wavelengths are gamma rays (~0.001 nm) and X-rays (~0.01 nm) on the short end, and microwaves (~10^7 nm) radio waves (~10^{10} nm) on the long end (Fig. 3.1). The fact that information on molecular structure is contained in this region, particularly that in the mid-IR, can be deduced by the wave-particle principles of quantum theory, starting with the expression for the energy of a photon,

$$E = hv \tag{3.1}$$

Where E is the photon's energy, v is the frequency of the wave, and h is Planck's constant. We see that the energy of a photon is directly proportional to its frequency.

We also recall that the wavelength (λ) and frequency (v) are inversely related to each other, with their product being the speed of light (c) in the medium that the light is passing through,

$$c = \lambda v. \tag{3.2}$$

S.R. Delwiche (✉)
U.S. Department of Agriculture, Agricultural Research Service, Food Quality Laboratory, 10300 Baltimore Ave., Bldg. 303 BARC-East, Beltsville, MD 20705-2350, USA
e-mail: stephen.delwiche@ars.usda.gov

B. Park, R. Lu (eds.), *Hyperspectral Imaging Technology in Food and Agriculture*, Food Engineering Series, DOI 10.1007/978-1-4939-2836-1_3

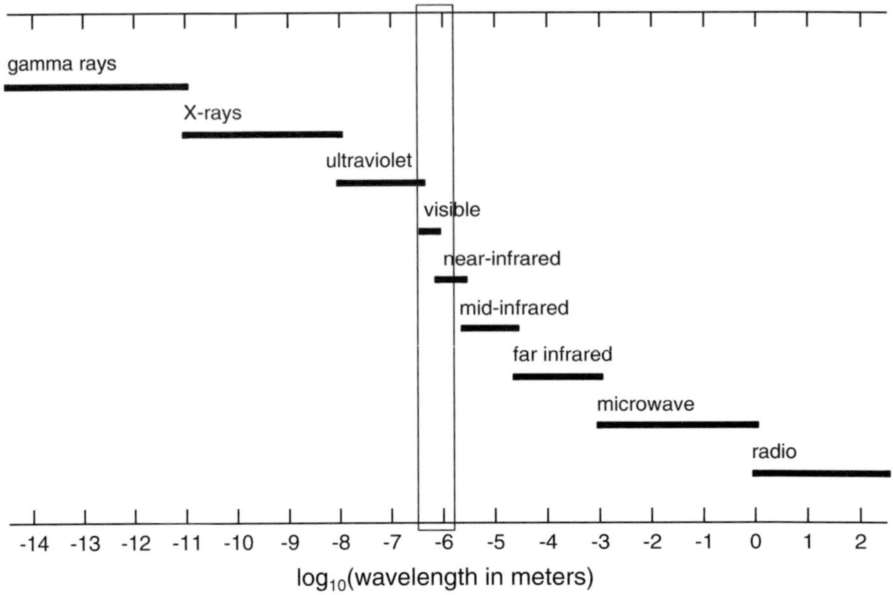

Fig. 3.1 Electromagnetic spectrum, highlighting the region used in hyperspectral imaging (400–1,700 nm)

Because of the underlying quantum theory of band vibrations, spectroscopists typically identify band locations in terms of a modified form of frequency defined as the number of wave cycles within a fixed distance. By convention, the distance is a centimeter, so that the term, wavenumber having units of cm^{-1}, can be thought of as the number of complete wave cycles in a 1 cm thickness. Physicists and engineers, on the other hand, typically speak in terms of wavelength, and the unit of choice for the visible and near-infrared region is the nanometer, which is one billionth (10^{-9}) of a meter. Because of their reciprocal relationship, conversion between wavelength and wavenumber or vice versa is a matter of multiplying the reciprocal by 1×10^7. Because the popularization of near-infrared measurement and analysis arose from the physicist/engineering community, whereas qualitative analysis using the mid-IR region arose from the spectroscopist, we continue today with this dichotomy in absorption band assignment. Although the conversion between the two scales is routine, it is important to remember that if the increment between neighboring readings from an instrument is uniform in one scale, it will not be in the other. With the gaining popularity of Fourier transform (FT) near-infrared spectrometers, whose scale is based on wavenumbers, it is especially important to keep this in mind when comparing FT spectra with conventional monochromator-based dispersive spectrometers whose basis is uniform spacing in the wavelength domain.

Quantum theory dictates that the absorption of light by a molecule comes about by discrete changes in energy levels (quantum levels) that, for the mid-IR region, happen when an inter-atomic bond within the molecule absorbs energy that equals the difference between two adjacent quantum levels. Taking a diatomic (two-atom) molecule such as carbon monoxide as an example, the vibrational frequency at which the bond expands and contracts is set by the selection rules of quantum theory. These rules also apply to more complex, polyatomic molecules.

3.2 Inter Atomic Bond

3.2.1 Theory

The starting point for modeling atomic bond vibrations is usually the harmonic oscillator described by classical mechanics. In this model, two atoms are bonded by a restoring force that is linearly related to their bond distance. In its simplest form, a bond between two atoms is modeled as a spring connecting two spherical masses, m_1 and m_2. The potential energy of this two ball assembly, V, depends on the displacement of the masses with respect to their rest positions, caused by either compression or elongation of the spring,

$$V = \frac{1}{2}k(x - x_{rest})^2 \tag{3.3}$$

where $(x - x_{rest})$ is the distance between the centers of masses and k is force constant of the spring. In this simple model potential energy varies in a quadratic relation with distance to form a parabolic shape, as demonstrated in Fig. 3.2. Two problems become readily apparent when using this model to approximate molecular behavior. First, limits must be placed on the distance of compression, as atoms are of physical mass and dimension, such that it is not possible for the atoms to have a zero compression distance. Second, a bond between atoms may only elongate so far before the atoms disassociate.

A third problem, which was not adequately addressed until the introduction of quantum mechanical theory in the 1920s, is explained by first considering the total energy of the system, which is the sum of the potential energy (V) and kinetic energy. With the latter written in terms of momentum (p), the total energy (E) of the system is

$$E = \frac{p^2}{2m} + V \tag{3.4}$$

where m is the total mass of the system. Classical mechanics allows the energy to take on a continuum of values, but this turns out to be impermissible in nature. This is explained by the Heisenberg uncertainty principle, part of which states that for a given direction it is not possible to know position and momentum simultaneously. Related to this is the restriction that energy is quantized, which means that at a

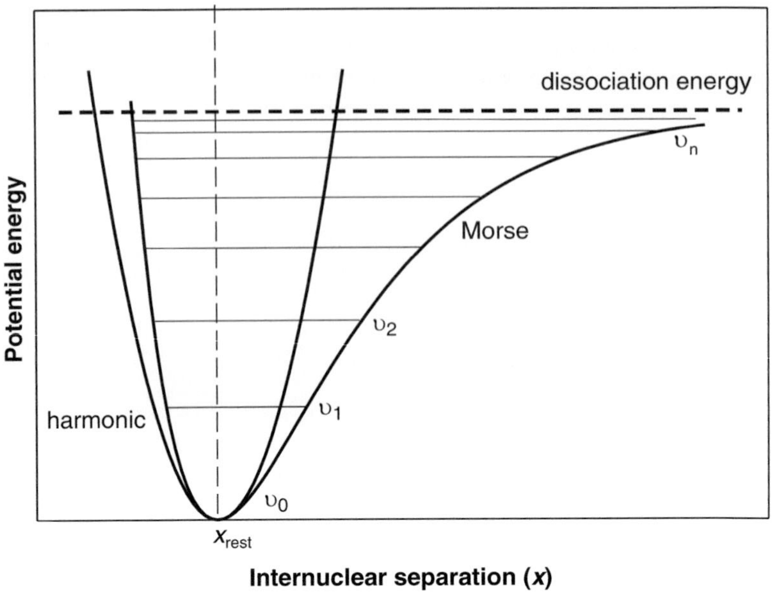

Fig. 3.2 Potential function for two bonded atoms

specific frequency the energy of the oscillator is limited to discrete, *i.e.*, *quantum*, levels, υ. Solution of the wavefunction form of the harmonic oscillator becomes,

$$E_{\upsilon} = h\nu\left(\upsilon + \frac{1}{2}\right) \tag{3.5}$$

in which E_{υ} is the energy of the υth quantum level ($\upsilon = 0, 1, 2, \ldots$) and ν is the fundamental frequency of the vibration, which is related to the force constant of the bond (k) and the *reduced* mass (μ) by

$$\nu = \frac{1}{2\pi}\sqrt{\frac{k}{\mu}} \tag{3.6}$$

recalling that the reduced mass of a diatomic molecule defined as $1/\mu = 1/m_1 + 1/m_2$, where m_1 and m_2 are the masses of the atoms. Classical mechanics theory produces the result that like atomic bonds within a molecule vibrate in phase at these fundamental or normal frequencies, with the number of unique vibrational frequencies related to the size (*i.e.*, number of atoms $= N$) of the molecule by the relation ($3N-6$). Taking a triatomic molecule such as water for example as shown in Fig. 3.3, three modes of vibration are possible: symmetric stretching (both hydrogen atoms moving toward and away from the central oxygen atom in tandem), asymmetrical stretching (one hydrogen moving away at the same time as the other moving closer to the oxygen), and bending (hydrogen atoms moving toward and

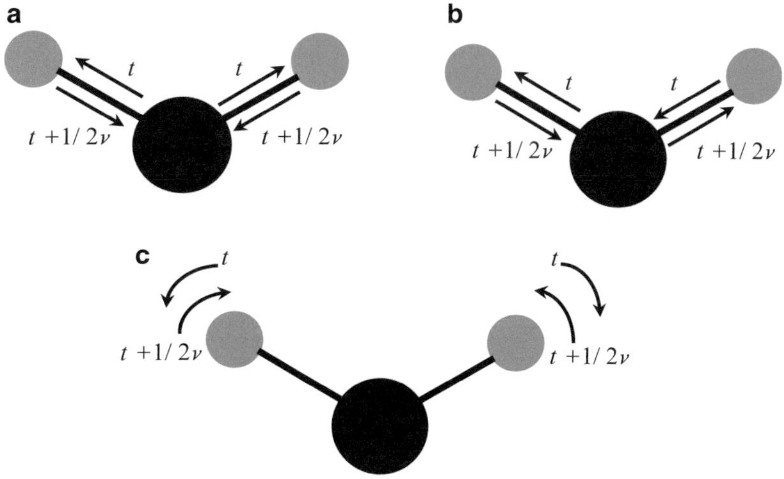

Fig. 3.3 Modes of vibration for a single water molecule

away from each other). Actual vibrational behavior of water is far more complicated, as we shall see below.

Vibrations between bonded atoms occur when the energy of the photon matches that of the difference between energy levels of two sequential quantum levels of the bond. For the electrical field to impart its energy into the molecule a polar distribution of charge, or dipole, must exist or be induced to exist across the bond. The jump between the ground state ($\upsilon = 0$) and the first level of excitation ($\upsilon = 1$) characterizes the fundamental vibrations across the mid-infrared region, this being from 4,000 cm^{-1} to 400 cm^{-1} (2,500–25,000 nm).

It turns out that the energy relationship of Eq. 3.5 can be used to describe bond behavior for small values of the vibrational quantum number, corresponding to the bottom region of the energy curve (Fig. 3.2) where there is near symmetry between left and right sides. For larger quantum levels, the energy relation is more complicated, such that the nonsymmetrical Morse-type function $\left[\left(1 - e^{-c(x-x_{rest})}\right)^2\right]$, also shown in Fig. 3.2, is used to incorporate the features of mechanical and electrical anharmonicity. Mechanical anharmonicity arises from the fact that because of the atoms' dimensions and mass there are physical limits to the separation distance between bonded atoms that preclude them to being too close (overlapping) or too distant (disassociating). Electrical anharmonicity arises from a nonuniform change in dipole moment with change in distance between bonded atoms. Unlike the parabolic nature of the mechanical model, the Morse function allows for dissociation of the two atoms as the energy level is increased. The solution to the wavefunction becomes

$$E_{\upsilon} = h\nu\left(\upsilon + \frac{1}{2}\right) - xh\nu\left(\upsilon + \frac{1}{2}\right)^2 \tag{3.7}$$

The variable x is the anharmonicity constant. The presence of anharmonicity allows for (1) overtone transitions, which arise from a change between nonadjacent vibrational quantum levels (e.g., $|\Delta\upsilon| > 1$); (2) combination bands, which occur when the energy from one photon produces simultaneous changes in quantum levels of two or more different vibrational modes; and (3) unequal differences between energy levels of the quantum states as described by Miller (2001). These allowances would otherwise be forbidden under a set of conditions known as *selection rules* that arise from *group theory* in quantum mechanics (Wilson et al. 1985). The significance of these allowances becomes apparent when we shift away from the fundamental vibrations of the mid-IR region to the overtone and combination vibrations of the near-IR. To a first approximation, the frequencies of the overtone bands are integer multiples of the corresponding fundamental frequency, with each higher overtone (first, second, ...) being weaker than the preceding. Thus, absorptions from overtone vibrations of the same bond become progressively weaker as wavelength is decreased. Combination bands involving CH, NH, and OH tend to be longer in wavelength than overtones, though with overlap between the two band types. Taking water in the liquid state as an example, its two most prominent bands in the near-IR region are a combination band ($\upsilon_2 + \upsilon_3$ = asymmetric stretch + bending) occurring at ~1,910 nm and a first overtone of OH ($\upsilon_1 + \upsilon_3$) occurring at 1,460 nm, where it is noted that the location of these bands as well as the other combination bands and higher overtones are strongly influenced by temperature through changes in hydrogen bonding. Further complication arises with water absorbed in biological matrices whereby hydrogen bonding can occur between water, polysaccharide, lipid, and protein molecules. Considering wheat starch and microcrystalline cellulose separately equilibrated to 53 % RH as examples (Delwiche et al. 1992), the wavelength location of the prominent 1,900 water combination band peak decreases by ~17 and 11 nm, respectively, as temperature increases from −80 to 60 °C, which causes a reduction in the hydrogen bond strength between water and the matrix (Fig. 3.4).

3.2.2 Practical Ramifications for the Near-IR Region

Because of the departures from the ideal case of the harmonic oscillator we are left with the inherent complexities, which are fortunate opportunities of near-IR spectroscopy. Three general statements are made to underscore the limits and power of the near-IR:

1. Compared to the fundamental absorption bands of the mid IR region, absorption bands of the near-IR region are weak. What appears to be a detriment for the NIR analyst on first glance is actually a boon because materials that are examined in

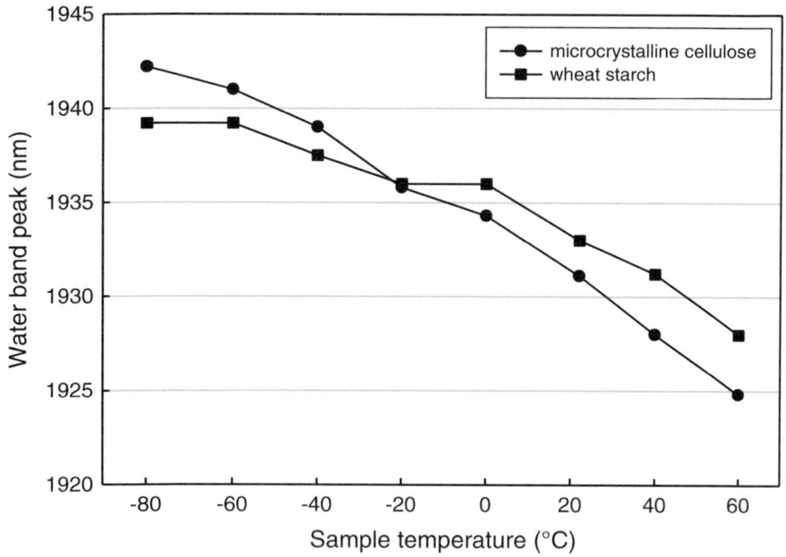

Fig. 3.4 Temperature effect on the 1,940 nm water combination band in moisture containing starch and cellulose

either transmission or reflection modes do not have to be diluted beforehand, as is the typical procedure for mid IR analysis. At most, especially for reflectance measurement, the material is ground into fine particles thereby reducing the heterogeneity of the sample caused by spatial differences in either chemical or physical structure.

2. The near-IR region is primarily composed of the overtone and combination bands arising from bonds involving the lightest atom, hydrogen. Typically, these include the bonds C–H, O–H, and N–H, all of which are prevalent in organic molecules. Thus, near-IR analysis is especially well suited for the studies involving agricultural, biomedical, pharmaceutical, and petrochemical materials.
3. Because hydrogen is much lighter than any other atom, its bonds with carbon, oxygen, and nitrogen produce vibrational movements that cause the largest motion for the hydrogen atoms, thus making vibrational movements localized to the functional group. Intrachain bond vibrations such as C–C are not active in the near-IR region.
4. Due to the multitude of overlapping bands in the near-IR region, the exact assignment of a bond's vibration to a wavelength or frequency is seldom possible, thus making near-IR spectroscopy a poor candidate for qualitative analysis. By the same token, however, quantitative analysis such as the concentration of chemical group is possible through the power of advanced regression algorithms.

The exact location and magnitude of overtone and combination bond vibrations for functional groups are very difficult to determine because of the effects from anharmonicity and dipole moment changes. As a rule of thumb, bond strength and reduced mass determine the band frequency location, while the dipole moment and anharmonicity affect the band's magnitude. Other factors such as hydrogen bonding and neighboring groups will have secondary effects on location and magnitude.

3.3 Light Absorption in a Scattering Medium

The realization that electromagnetic radiation behaves in both corpuscle and waveform conditions has led to several theoretical models to describe the infrared spectral response. We will consider some of these in brief format, with the reader directed to particular seminal texts for more detail.

3.3.1 Light Without Scattering

This is the most common model used in calculating concentration of a solute in gases and clear liquids, in other words media with negligent scattering. The theory was originally developed by Pierre Bouguer (1729) and Johann Lambert (1760) independently, and later expanded by August Beer (1852) to include substances of varying concentrations within the media. In the translated words (from Latin) of Lambert's Theorem 68, "The logarithm of the remaining light, when it is weakened in a less transparent medium, is related by a ratio to the maximum of all the intermediary obstructing material of that medium, which it encounters along its path, and to whatever manner in which the obstructing material may be disseminated in the medium and whatever the curvature of the path is." (p. 391). With inclusion of Beer's contribution, the law states that the intensity I of light decreases exponentially with penetration distance d and the concentration of the compound of interest $[J]$,

$$\frac{I(d)}{I(d=0)} = e^{-k[J]d} \tag{3.8}$$

where k is an absorption coefficient (formerly called the extinction coefficient when dealing with molar concentrations). Equation 3.8 is universally referred to as the Beer-Lambert Law, secondarily as Beer's law, and lastly, but seldom, the Beer-Lambert-Bouguer Law. A more familiar format appears when the base 10 logarithm of each side is taken,

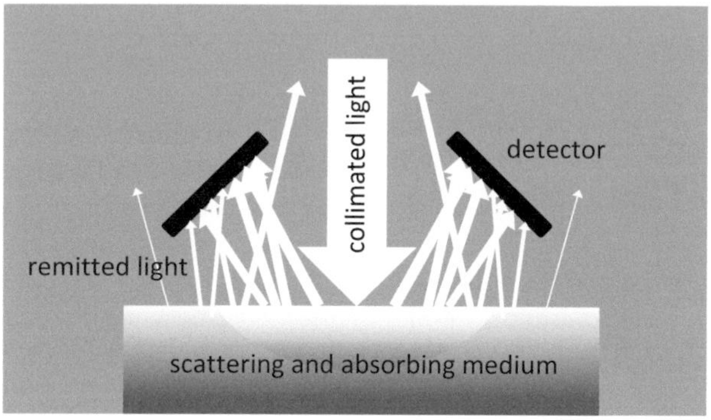

Fig. 3.5 Schematic of light remittance

$$A = \log\left(\frac{1}{T}\right) = k^{'}[J]d \tag{3.9}$$

in which we write transmittance (T) as the ratio of the intensity of the light at depth d to that at the surface. In practice, the concentration c may be written in terms of molarity ($n_{\text{solute}}/\text{volume}_{\text{solution}}$), mole fraction ($n_{\text{solute}}/n_{\text{solvent}}$), or mass fraction ($\text{mass}_{\text{solute}}/\text{mass}_{\text{solution}}$) with the units for $k^{'}$ selected accordingly so that the right hand side product is dimensionless. Direct application of Eq. 3.9 occurs with the use of simple spectrophotometers in which cuvettes of precise dimension are used to measure the intensity of transmitted light through a clear solution at a single wavelength, typically in the UV region. Upon the development of the calibration curve, absorbance is directly related to the concentration of a conjugate from a biochemical assay. The complexities of natural materials, such as plant and animals, leads to a stretching of the rules for the Beer-Lambert equation, yet conveniently and frequently with success.

In diffuse reflectance analysis, liberties are taken with the Beer-Lambert law to allow the substitution of *reflected* or, in Dahm and Dahm's (2007) terminology, *remitted* energy for transmitted energy. A simplified representation of diffuse reflectance is shown in Fig. 3.5. In this case, white or monochromatic light is collimated and then directed onto a the surface of a sample, whereupon the light may (1) be directly reflected from the surface of the first particle that it encounters; (2) penetrate the first surface, followed by additional internal reflections and transmissions with other particles; (3) be remitted light from the surface that it entered; or (4) be absorbed by an atomic bond. Mathematical modeling of these phenomena is the subject of ongoing investigation (Dahm and Dahm 2007). In practice, terms on the right side of Eq. 3.9 are commonly lumped together and collectively referred to as the concentration of the compound of interest. Hence,

$$\log\left(\frac{1}{R}\right) \approx [analyte] \tag{3.10}$$

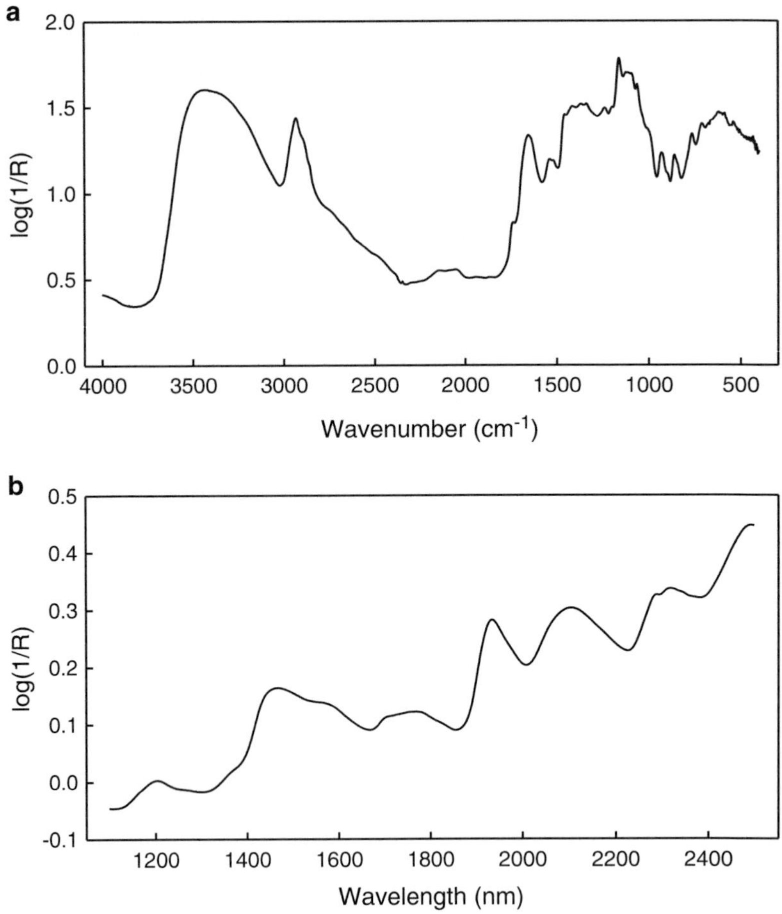

Fig. 3.6 Mid IR and near-IR spectra of ground wheat

with square brackets indicating concentration. The implementation of this relationship implies that pathlength is constant across samples and other compounds or analytes are not interfering. Using ground wheat as an example, typical uncorrected log(1/*R*) spectra are shown for in Fig. 3.6a, b for the mid-IR and NIR regions, respectively. What is obvious from either plot is the lack of a clear baseline response despite the existence of wavenumber or wavelength regions of low spectral absorption. The non-horizontal behavior of these spectra, especially noticeable in the NIR region is caused by scatter. Because pathlength in a scattering medium is extremely difficult, if not impossible, to determine, the assumption of constant pathlength is favored when samples are of the same distribution in particle size makeup. A workaround to the particle size problem is to apply a mathematical correction to the log(1/*R*) spectrum, typically a multiplicative scatter (signal) correction (Martens and Næs 1989), a standard normal variate transformation

(Barnes et al. 1989), or a first or higher order derivative, as explained in Sect. 3.4. The problem of interfering absorbers is addressed by considering responses at more than one wavelength, whereupon by using linear modeling methods (multiple linear regression, principal component, partial least squares) or nonlinear methods (artificial neural networks, support vector machines) quantitative models of sufficient accuracy are possible. Known collectively as chemometrics, extensive details on such algorithms for quantitative and qualitative analysis of spectra are the subject of several texts (Mark and Workman 2007; Naes et al. 2002; Varmuza and Filzmoser 2009; Jolliffe 2002; Cristianini and Shawe-Taylor 2000). One further simplification that is often employed in remote sensing hyperspectral analysis is the use of reflectance R directly, in which it is assumed that the degree of nonlinearity between R and its log reciprocal transform is negligible [for example, over the reflectance range 0.2–0.8, the coefficient of determination with $\log(1/R)$ is 0.97].

3.3.2 Kubelka-Munk

Unlike the Beer-Lambert-Bouguer theory that was based on transmission and adopted for diffuse reflectance, the theory known as Kubelka-Munk (K-M) is fundamentally based on reflectance from a scattering medium, with its application primarily relegated to the paper and paints industries. As with Beer-Lambert-Bouguer, K-M theory is most appropriate for media and analytes of low absorption (Olinger et al. 2001). Originally developed by Kubelka and Munk (1931), the theory assumed light to be traveling through a continuum, a medium with no internal boundaries such as particle surfaces. Further, the light is modeled as having a forward flux and a backward flux. Diffuse illumination at the surface is also assumed and the medium scatters the radiation isotropically. The behavior of radiation is written as a combination of two constants, K and S, which have analogies to absorption and scatter. With solution of coupled differential equations for forward and backward radiations, the well known Kubelka-Munk function was derived (Kortüm 1969)

$$\mathrm{F}(R_\infty) = \frac{K}{S} = \frac{(1 - R_\infty)^2}{2R_\infty} \tag{3.11}$$

In this equation, R_∞ is the remitted radiation from a medium of infinite thickness, which may be experimentally determined by observing when the addition of depth to a sample produces no change in R. For the mid-IR and near-IR regions the infinite thickness is a reasonable assumption for thicknesses greater than several millimeters. As explained by Dahm and Dahm (2007), the problem with Kubelka-Munk equation in practice arises with the attempts at disentangling K and S. Ideally, one would like to treat K as a pure absorption coefficient in the same manner as Beer-Lambert so that concentrations of absorbing compounds can be accurately modeled. In reality, the equation falls short for reasons of (1) a two flux model overly simplifying light direction; (2) specimens are not diffusely illuminated, but instead illuminated with collimated light; (3) the medium is not a continuum but instead consists of discrete particles that individually reflect and refract light; and (4) given that the instrument measures remitted energy, the expression does not inherently provide a means to separate absorption, hence concentration of a component, from scatter. With respect to log(1/*R*), F(*R*) is more greatly affected by baseline errors (Griffiths 1995). Also, from experimental measurements of reflectance from a three component mixture of varying proportions (NaCl as the nonabsorbing matrix, graphite as a general absorbing compound, and carbazole as a typical organic analyte possessing both C–H and N–H bonds), Olinger and Griffiths (1988) reasoned that the linearity of F(R_∞) with concentration of an absorbing compound (carbazole in this case) is highest when the matrix is nonabsorbing because the photons have more opportunity to undergo interactions with many particles before leaving the sample surface. Linearity drops off as the matrix becomes more absorbing, as they observed when 5 % graphite by weight was added to the NaCl matrix. Conversely, the theoretical lack of a dependency of the linearity of log(1/*R*) with many particle interactions for a photon is the reasoning behind the general better performance of log(1/*R*) in diffuse reflectance spectroscopy of powdered materials. Because of these limitations, the Kubelka-Munk theory is not commonly applied to near-IR diffuse reflectance spectroscopy of biological and agricultural materials.

3.3.3 Diffusion Theory

This is also a continuum approach that has gained use in modeling the decay of light in biological tissue and uses a mathematical model to derive an expression for remitted light as a function of a coefficient due to absorption (μ_a) and another coefficient due to scattering, called the transport scattering coefficient (μ_s'), in which scattering is assumed to be isotropic (Farrell et al. 1992). An assumption of a highly scattering matrix allows for the development of a diffusion equation for photon propagation, as derived from the Boltzmann radiative transport equation. For the special condition of considering light as a point source directed onto a semi-infinite medium at a direction normal to the surface, the remitted radiation $R(r)$, where r is the radial distance from the point of entry, can be derived (Farrell et al. 1992),

$$R(r) = \frac{a'}{4\pi}\left[\frac{1}{\mu_1'}\left(\mu_{\mathrm{eff}} + \frac{1}{r_1}\right)\frac{e^{-\mu_{\mathrm{eff}} r_1}}{r_1^2} + \left(\frac{1}{\mu_1'} + \frac{4A}{3\mu_t'}\right)\left(\mu_{\mathrm{eff}} + \frac{1}{r_2}\right)\frac{e^{-\mu_{\mathrm{eff}} r_2}}{r_2^2}\right] \qquad (3.12)$$

where the transport albedo, $a' = \mu_s'/(\mu_a + \mu_s')$, the effective attenuation coefficient, $\mu_{\mathrm{eff}} = \left[3\mu_a(\mu_a + \mu_s')\right]^{1/2}$, $r_1 = \left[(1/\mu_t')^2 + r^2\right]^{1/2}$, and $r_2 = \left[(1/\mu_t' + 4A/3\mu_t')^2 + r^2\right]^{1/2}$.

Further, the total interaction coefficient, $\mu_t' = \mu_a + \mu_s'$, and A is a parameter that is related to the internal reflection and is derived from the Fresnel reflection coefficients. In practice, A may be determined empirically as a function of the relative refractive index (Groenhuis et al. 1983), and with additional simplification it may be treated as a constant. With this assumption the right side of Eq. 3.12 becomes an expression of only two terms, the absorption coefficient μ_a and the transport scattering coefficient μ_s'. Among other features, the diffusion theory approach differs from K-M in that absorption and scattering are mathematically decoupled.

From experimental measurements of $R(r)$ at various radial positions and inverse application of Eq. 3.12, separate values for μ_a and μ_s' are determined over the wavelength range of interest, thus producing separate absorption spectra and scattering spectra. Lu and coworkers (Qin and Lu 2008; Lu et al. 2010) developed this approach using line scan hyperspectral imaging ($\lambda = 500$–1,000 nm) to nondestructively examine ripeness in tomatoes (Qin and Lu 2008), bruising in apples (Lu et al. 2010), and mechanical damage in pickling cucumbers (Lu et al. 2011).

3.4 Practical Outcomes for Near Infrared Reflectance

Because log(1/*R*) is most commonly used in NIR spectroscopy the following discussion will assume this format, though many of the transformations may be applied to the other formats of reflectance data mentioned in the previous section. Broadly termed as spectra preprocessing, these transformations are performed for improving signal-to-noise and minimizing the effect of scatter often with the expectation that band intensities become more linearly related to the concentration of the absorbing compound. Improvement in signal-to-noise is typically performed by a smoothing operation, such as a running mean,

$$\overline{A}_j = \frac{A_{j-l} + \ \ldots + A_{j-1} + A_j + A_{j+1} + \ldots + A_{j+l}}{2j+1} \tag{3.13}$$

where A_j is the original spectral value at wavelength j and $\bar{A}_j$ is the mean value as determined from the original value as well as the neighboring l points on the left and the same number of points on the right. Selection of the size for l should be based on the inherent bandpass of the spectrometer (typically 10 nm for dispersive scanning monochromators) and the size of the absorption band of interest. In practice, the value is selected by trial and error, with too small a value yielding insufficient noise reduction and too large a value attenuating the higher frequency absorption bands. Using simulated spectral data, Brown and Wentzell (1999) warned of the deleterious effect of smoothing on multivariate calibrations such as principal component (PC) regression. Also, they noted that smoothing has the greatest chance of being beneficial under conditions of high measurement noise and wavelength-to-wavelength correlation, in which case the improvement in characterization of the spectral subspace through PC reduction offsets the losses caused by spectral distortion.

Spectral derivatives, or more accurately stated as spectral differences, usually of the first and second order are applied for removal of vertical offset and slope effects. In simplest form, these are two-point (first) and three-point (second) central finite difference expressions. Although the points need not be consecutive (in which case the difference becomes a poor approximation to the true derivative but nevertheless may produce a better calibration), the intervals between the end points and the central point should be equal. With these $\Delta y/\Delta x$ and $\Delta^2 y/\Delta x^2$ difference expressions, it is common to omit the denominator term when chemometric modeling is the goal. The omission is not a problem unless one is trying to accurately show the values for the derivative spectrum for its own sake or one is attempting to compare derivative spectra possessing different values for Δx.

A more common form of spectral differentiation is the Savitzky-Golay polynomial approximation procedure, as first popularized by the authors (Savitzky and Golay 1964). Using a sliding window along the wavelength axis, and assuming a constant wavelength spacing, the window of points (typically an odd number between 5 and 25) is fitted using least squares regression to a polynomial of second through sixth order, whereupon the analytical derivative of the polynomial function is evaluated at each point. The procedure can be computationally simplified to a convolution operation using the same number of points in the window and using published coefficient values, as reported in their original paper and later corrected by Steinier et al. (1972). Spectral differentiation is shown by example in Fig. 3.7. A 'spectrum' has been created by adding two Gaussian functions, with one being twice the magnitude and twice the width of the other, to a sloped line and then adding random noise. In this simple example, one sees a replacement of the vertical offset in the absorbance curve with a small offset (=1/1,500 on average, *i.e.*, the slope of the upwardly trending line in the absorbance curve) in the first derivative curve, which disappears altogether in the second derivative. The two local absorbance maxima become zero crossing points in the first derivative curve but reappear as local minima in the second derivative curve. This is typical behavior and explains why the second derivative is generally easier to interpret than the first derivative. However, interpretation of the second derivative is tricky even without considering the complexities of vibrational physics. For instance, in the simple example of Fig. 3.7, the Gaussian band at the lower wavelength, despite being half the magnitude of the other Gaussian band, appears as having a larger absolute value in second derivative. Because the lower wavelength band has half the width of the upper band its curvature and hence second derivative magnitude is greater. Secondly, derivatives have the tendency to amplify noise, as seen in the progression of a smooth appearance in the absorbance curve (Fig. 3.7a) to a first derivative curve with noise of pen width in magnitude (Fig. 3.7b), and on to a second derivative curve with very noticeable noise (Fig. 3.7c). In practice, the noise amplification effect is not as pronounced

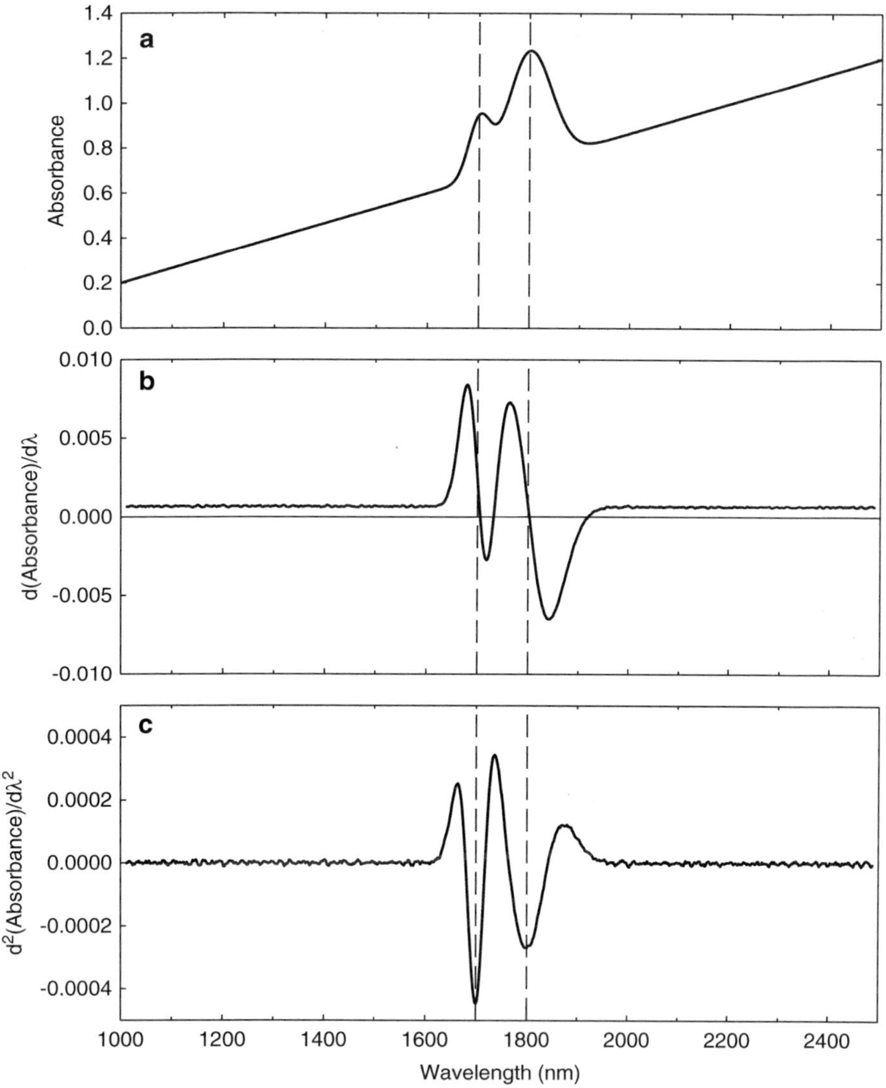

Fig. 3.7 Demonstration of spectral differentiation. (**a**) Artificial spectrum created by combining two Gaussian bands (full widths at half maximum of 50 and 100 nm, with peak values of 0.25 and 0.50, respectively, and centered at 1,700 and 1,800 nm, respectively) to a *sloped line* [y(1,000 nm) = 0.2, y(2,500 nm) = 1.2], then adding random noise (−0.0004 to 0.0004 peak to peak, uniformly distributed). (**b**) Savitzky-Golay first derivative (cubic polynomial, 11 point convolution window). (**c**) Savitzky-Golay second derivative (cubic polynomial, 11 point convolution window)

because the 'noise' in a spectrum is not entirely random as in the artificial spectrum of Fig. 3.7, but instead may largely consist of baseline drift, which has low-frequency dominance in the noise power spectrum. The low-frequency nature means that noise levels of neighboring wavelengths are not fully independent. Brown et al. (2000) investigated drift noise through simulation and found that derivatives may reduce drift noise, but at the same time spectra can be distorted with respect to the underlying chemical constituents, thus making it difficult to predict the benefit of this preprocessing technique in multivariate calibrations. For example, if the left Gaussian band in Fig. 3.7 was centered at 1,750 nm rather than 1,700 nm, the resulting absorbance spectrum (Fig. 3.8a) would appear as one broad but asymmetrical band superimposed on the upwardly trending baseline curve. Upon first differentiation using the same SG convolution function the two zero crossing points are replaced by one located at approximately 1,770 nm (Fig. 3.8b), which is between the absorbance peak positions of 1,750 and 1,800 nm. The second derivative (again with the same function as originally used) has local minima at 1,750 (the same as original) and 1,814 nm (Fig. 3.8c), which is 14 nm longer than the absorbance peak position. This helps to explain why NIR calibrations are often trial and error operations that are enhanced by *a priori* knowledge of the locations and magnitudes of bands associated with the analyte of interest.

Because of the complexities of the scatter-absorption effect, separation of these components through theoretical means as described earlier is often replaced by working corrections as part of spectral preprocessing. The two most common full spectrum approaches are the multiplicative signal (scatter) correction (MSC) and the standard normal variate transformation (SNV). With MSC, as popularized by Martens (see Geladi et al. 1985), a sample's reflectance spectrum is corrected to have roughly the same degree of scatter as the other samples within the calibration set. The usual procedure is to calculate the mean spectrum of the calibration set and then for every sample within the set, a least squares correction (most often a first order polynomial but this can be of higher order) is developed by regressing the spectrum's points onto those of the mean spectrum. The regression coefficients are then used to 'correct' the spectrum to the mean spectrum. This has the noticeable effect of collapsing spectra together so that under ideal conditions all sample-to-sample spectral differences are attributed to chemical absorption. This transformation requires the retention of the reference (mean) spectrum in order to correct future spectra before the calibration equation is applied. An example of this transformation is shown in Fig. 3.9, which consists of spectra of 198 samples of ground wheat, first with no transform (Fig. 3.9a) and next with MSC (Fig. 3.9b). Alternatively, one may conduct a scatter correction that is based on each spectrum independently. Known as the standard normal variate (SNV) transformation, this correction has a format similar to that of standard error in statistics, this being that within each spectrum the mean value (over the wavelength region) is subtracted from each spectral value and this difference is then divided by the standard deviation of the spectrum's values (Barnes et al. 1989), as shown by example in

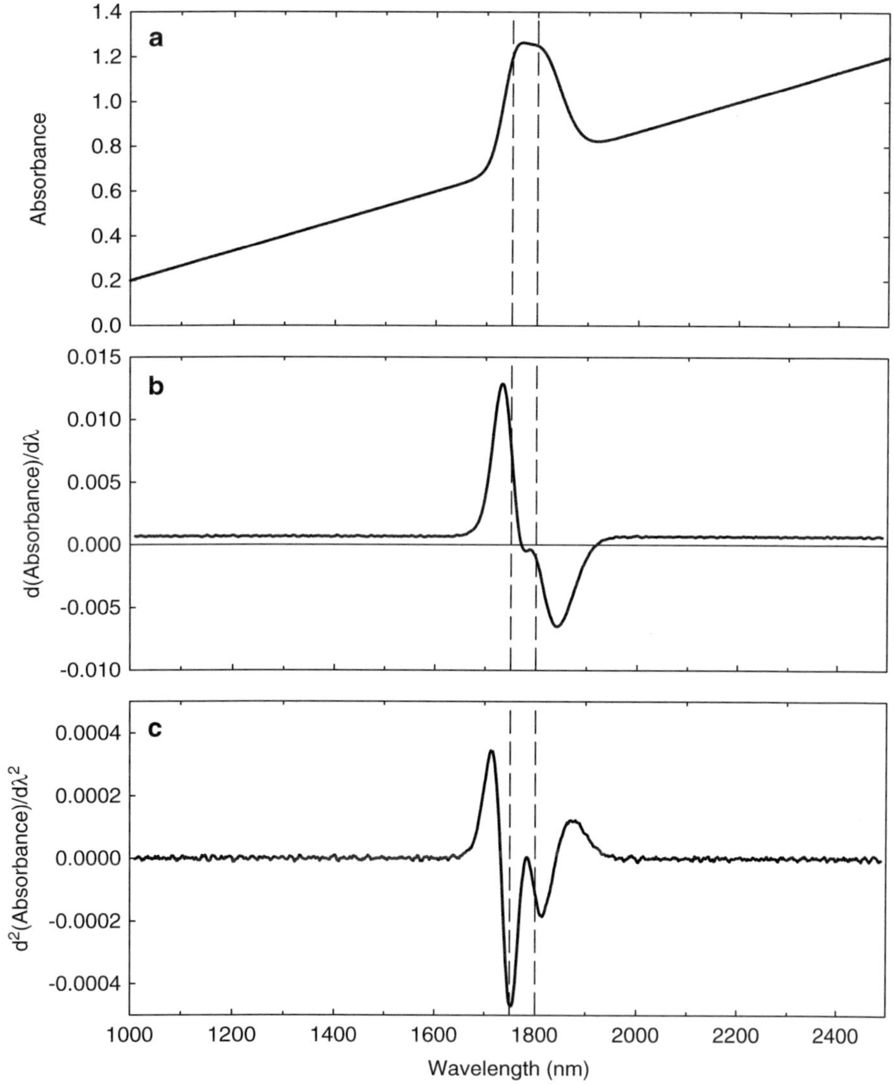

Fig. 3.8 Demonstration of spectral distortion by differentiation. The conditions that produced graphs (**a**–**c**) are identical to those of Fig. 3.7 with exception that the low wavelength absorption peak is located at 1,750 nm and the dashed lines are moved to his location

Fig. 3.9c. This results in each transformed spectrum having a mean of zero and a standard deviation of unity. As with MSC, the intention is that the benefits of the SNV correction through reduction of variation from scatter outweigh whatever losses in chemical information that result from spectral distortion.

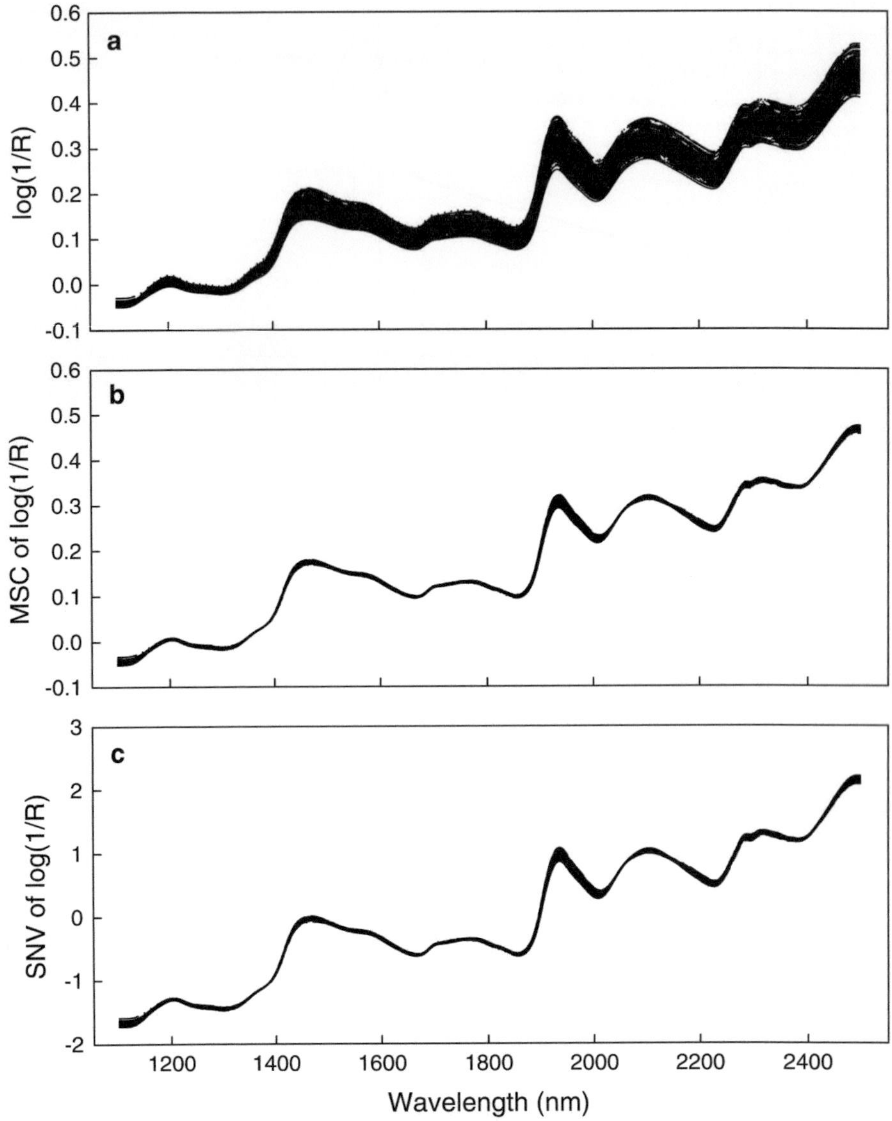

Fig. 3.9 Example of spectral scatter removal techniques using a set of 198 ground wheat samples. (**a**) Raw log(1/R) spectra. (**b**) With multiplicative scatter correction (MSC). (**c**) With standard normal variate (SNV)

3.5 Application to Imaging

The principles of NIR spectroscopy carry over to NIR hyperspectral imaging. With the latter, the measurement of energy remitted from a broad surface, as read by one or a set of detectors, is replaced by image measurement from a camera sensor array in which each element or pixel captures energy from a small region of the sample surface. The spectral dimension arises from one of two general formats, a liquid crystal tunable filter employed to capture two-dimensional images at a series of tuned wavelengths, or a dispersion device called a spectrograph that is placed between the lens and body of the camera. Between the spectrograph and the lens is a slit that reduces the focused image to that of a narrow line. The line of light is passed onto the spectrograph, which then disperses the light from each 'point' along the line to a series of wavelengths. By methodically advancing the location of the line on the object, either by moving the camera or moving the object, additional lines are imaged until the entire object has been scanned and reproduced as a mosaic of lines.

In its simplest and most common form, camera array readings are referenced to a highly reflecting Lambertian material such as 'Spectralon' (Labsphere, North Sutton, NH) while also being corrected for dark current of the sensor. In such case reflectance (R) becomes

$$R = \frac{E_{sample} - E_{dark}}{E_{reference} - E_{dark}} \tag{3.14}$$

where E_x is the energy from each x component. The reference material is treated as being 100 % reflective and Eq. 3.14 assumes a linear response for sample reflectance. Alternatively, sample reflectance may be determined using a higher order polynomial to describe the response (Burger and Geladi 2005). In such cases, a set of reflectance standards with traceable reflectance values (typically 3–8 samples of Spectralon doped with carbon black) whose reflectances span the anticipated range of the samples is used to develop a calibration equation. For example, assuming a quadratic response, reflectance is written (Burger and Geladi 2005),

$$R = b_0 + b_1 E + b_2 E^2. \tag{3.15}$$

During calibration, the left side values of Eq. 3.15 are known for the reflectance standards, E is measured for each standard, and the coefficients b_0, b_1, and b_2 are determined by least squares regression. The regression procedure is performed at each wavelength, which may be done on a pixel-by-pixel basis or globally using the median spectrum, as determined from pixels within a region of interest.

3.5.1 Collection of the Hypercube

Hyperspectral systems fall into two broad categories depending on the method of light dispersion. As illustrated in Fig. 3.10a, the tunable filter system collects two dimensional spatial images at each 'tuned' wavelength. This results in a stack of spatial images with each page in the stack representing a separate wavelength. The push broom system (Fig. 3.10b) builds up one dimensional spatial spectral pages.

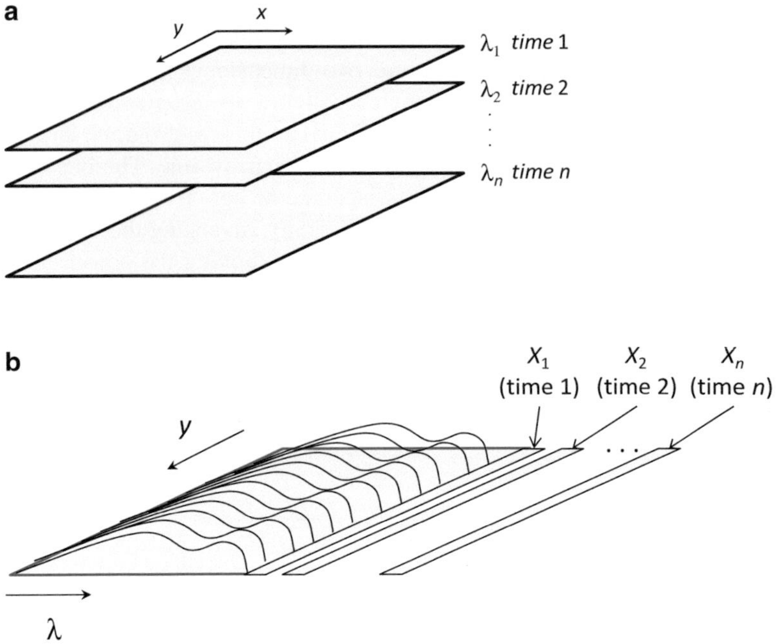

Fig. 3.10 Schematic of the two modes for hyperspectral image collection. (**a**) Stacked wavelength—at a given time instant the camera records two spatial dimensions (*x* and *y*) at one passband (λ) of a liquid crystal tunable filter. Recording continues with the next passband. (**b**) Pushbroom—at a given time instant the camera records one spatial (*y*) and one spectral (λ) dimension, where the spectral component is created by radiation dispersion through a spectrograph located between a slit and the camera body. Recording continues with relative movement of the object in a direction (*x*) perpendicular to the other spatial direction

Regardless of mode of instrument operation, the stored data, known as a hypercube, consists of one spectral and two spatial dimensions.

By way of example, a set of 81 spectra collected from an approximately square (9 pixel × 9 pixel) region of a wheat kernel using a push broom hyperspectral imaging system is shown in Fig. 3.11. (Details of the system and settings are found in Delwiche et al. 2012). A digital photograph of a wheat kernel is added as an inset in Fig. 3.11 for the purpose of showing the approximate location and size of the square region. The 9-element width is just a small portion of the line, which consists of 320 elements. Individual pixel spectra have a much higher level of noise than those from a conventional spectrometer, as seen by comparing Figs. 3.11 and 3.9a. Averaging of all pixel spectra within the square region results in reduction in noise (Fig. 3.11, solid black curve), albeit at the expense of fine feature detail of the individual pixels. It should be noted that whether it be on an individual pixel level or on a regional level, the principals of spectroscopy as well as the mathematical transformations leading up to and including qualitative and quantitative modeling hold true for hyperspectral imaging.

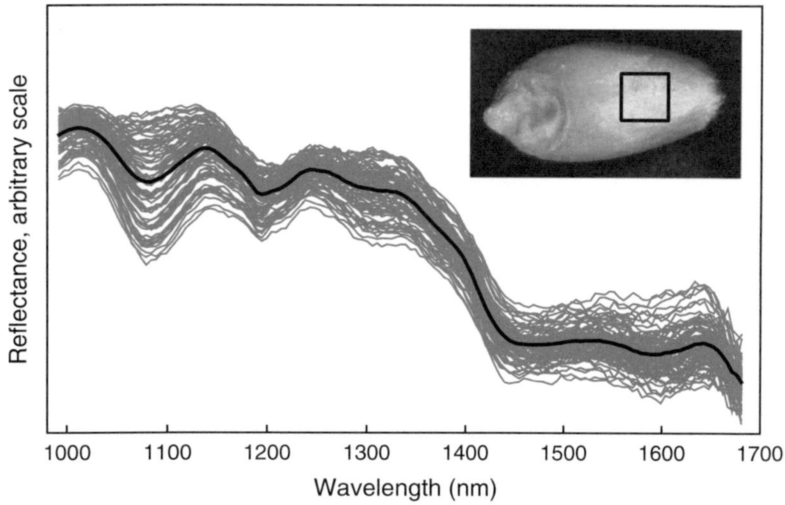

Fig. 3.11 Pixel reflectance spectra and average spectrum from a 9 × 9 pixel square region of a wheat kernel

3.6 Raman

Raman spectroscopy is based on the property of the photons from light of a very narrow frequency (e.g., a laser) striking a sample whereupon while most of the photons' energy momentarily raises the energy state of the molecules but then is released as the molecule returns to its ground state. A small fraction of photons, however, release a portion of their energy to the molecule whereby the bond is not returned to its ground state, and further the photon emerges at a lower energy and hence lower frequency. This phenomenon is known as Stokes scattering. The Stokes shift is a measurement of the difference in frequency of the incident and emergent photon. Oppositely, photons may pick up energy from molecules already at a higher than ground state in the matrix as they return to a lower state. In this case when the photon is released from the medium after scattering, the frequency is greater than the incident frequency. This is known as anti-Stokes scattering and occurs at an even lower prevalence than Stokes scattering because there are relatively few molecules already in an excited state. As with infrared spectroscopy selection rules exist for a Raman transition, these based on the requirement that the polarizability of light changes as the molecule vibrates. Traditionally, Raman spectroscopy has been used in determining force constants, dissociation energies, and bond lengths. The change in polarizability determines the intensity of the bands in the Raman spectrum. Further, the intensity is proportional to the fourth power of monochromatic light of excitation upon shifting,

$$I_{Raman} \propto (v_0 \pm v_j)^4 \left(\frac{d\alpha}{dQ}\right)^2 \tag{3.16}$$

where v_0 and v_j are the original and scattered light frequencies, respectively, and the squared term is the change in polarizability that occurs during vibration. Knowledge of this relationship is useful for two reasons. First, it shows that the frequency of monochromatic light is not fixed by the Raman effect, but rather Raman spectra may be obtained at any number of frequencies. In practice, the monochromatic sources are supplied by lasers, for which the two most popular are the infrared diode laser at 785 nm (12,740 cm^{-1}) and Nd:YAG at 1,064 nm (9,400 cm^{-1}). Second, Raman intensity diminishes with increase in the wavelength of the laser source by a fourth order relationship. Hence, without complicating factors, a shorter wavelength source would be preferable. In reality, fluorescence, which is often prevalent at low wavelengths, becomes the complicating factor. Because Raman signals are inherently weak, fluorescence emission can at times overpower the Stokes lines. This is particularly problematic with botanical samples. Conversely, fluorescence can be avoided by exciting at longer wavelengths, such as with a Nd:YAG laser, but at the expense of reduced Raman intensity.

Raman spectroscopy and infrared spectroscopy, though both based on molecular vibration, are complementary to one another. Bonds that exhibit strong absorption in the infrared, such as water, will typically be weak in the Raman and vice versa. Thus for biological samples, which typically have more than 50 % water by mass, Raman spectroscopy offers a means to examine molecular structure *in situ*.

References

Barnes RJ, Dhanoa MS, Lister SJ (1989) Standard normal variate transformation and detrending of near infrared diffuse reflectance. Appl Spectrosc 43:772–777

Beer A (1852) Bestimmung der absorption des rothen Lichts in farbigen Flüssigkeiten (Determination of the absorption of red light in colored liquids). Annalen der Physik und Chemie 86:78–88

Bouguer P (1729) Essai d'Optique Sur la Gradation de la Lumiere (Test of optics on the gradation of light) Claude Jombert, Paris, 164pp

Brown CD, Wentzell PD (1999) Hazards of digital smoothing filters as a preprocessing tool in multivariate calibration. J Chemometrics 13:133–152

Brown CD, Vega-Montoto L, Wentzell PD (2000) Derivative preprocessing and optimal corrections for baseline drift in multivariate calibration. Appl Spectrosc 54:1055–1068

Burger J, Geladi P (2005) Hyperspectral NIR image regression part I: calibration and correction. J Chemometrics 19:355–363

Cristianini N, Shawe-Taylor J (2000) An introduction to support vector machines and other kernel-based learning methods. Cambridge University Press, Cambridge, 189pp

Dahm DJ, Dahm KD (2007) Interpreting diffuse reflectance and transmittance: a theoretical introduction to absorption spectroscopy of scattering materials. NIR, Chichester, 286pp

Delwiche SR, Norris KH, Pitt RE (1992) Temperature sensitivity of near-infrared scattering transmittance spectra of water-adsorbed starch and cellulose. Appl Spectrosc 46:782–789

Delwiche SR, Souza EJ, Kim MS (2012) Near-infrared hyperspectral imaging for milling quality of soft wheat. Trans ASABE, submitted

Farrell TJ, Patterson MS, Wilson B (1992) A diffusion theory model of spatially resolved, steady-state diffuse reflectance for the noninvasive determination of tissue optical properties in vivo. Med Phys 19:879–888

Geladi P, McDougel D, Martens H (1985) Linearization and scatter-correction for near-infrared reflectance spectra of meat. Appl Spectrosc 39:491–500

Griffiths PR (1995) Practical consequences of math pre-treatment of near infrared reflectance data: log(1/*R*) vs F(*R*). J Near Infrared Spectrosc 3:60–62

Groenhuis RAJ, Ferwerda HA, Bosch JJT (1983) Scattering and absorption of turbid materials determined from reflection measurements. 1: theory. App Opt 22:2456–2462

Jolliffe IT (2002) Principal component analysis, 2nd edn. Springer, New York, 487pp

Kortüm G (1969) Reflectance spectroscopy: principles, methods, applications. Springer, Berlin, 366pp

Kubelka P, Munk F (1931) Ein beitrag zur optik der farbanstriche. Zeitschrift fur Technische Physik 12:593–601

Lambert JH (1760) Photometria sive de Mensura et gradibus Luminis, Colorum et Umbrae (Photometria or of the measure and degrees of light, colors, and shade) Augustae Vindelicorum Eberhardt Klett, Germany

Lu R, Cen H, Huang M, Ariana DP (2010) Spectral absorption and scattering properties of normal and bruised apple tissue. Trans ASABE 53:263–269

Lu R, Ariana DP, Cen H (2011) Optical absorption and scattering properties of normal and defective pickling cucumbers for 700–1000 nm. Sens Instrum Food Qual 5:51–56

Mark H, Workman J Jr (2007) Chemometrics in spectroscopy. Academic, Amsterdam, 526pp+24 color plates

Martens H, Næs T (1989) Multivariate calibration. Wiley, Chichester, 419pp

Miller CE (2001) Chemical principles of near-infrared technology. In: Williams PC, Norris KH (eds) Near-infrared technology in the agricultural and food industries, 2nd edn. American Association of Cereal Chemists, St. Paul, pp 19–37

Naes T, Isaksson T, Fearn T, Davies T (2002) A user-friendly guide to multivariate calibration and classification. NIR, Chichester, 344pp

Olinger JM, Griffiths PR (1988) Quantitative effects of an absorbing matrix on near-infrared diffuse reflectance spectra. Anal Chem 60:2427–2435

Olinger JM, Griffiths PR, Burger T (2001) Theory of diffuse reflection in the NIR region. In: Burns DA, Ciurczak EW (eds) Handbook of near-infrared analysis, 2nd edn. Marcel Dekker, New York, pp 19–51

Qin J, Lu R (2008) Measurement of the optical properties of fruits and vegetables using spatially resolved hyperspectral diffuse reflectance imaging technique. Postharvest Bio Technol 49:355–365

Savitzky A, Golay MJE (1964) Smoothing and differentiation of data by simplified least squares procedures. Anal Chem 36:1627–1639

Steinier J, Termonia Y, Deltour J (1972) Comments on smoothing and differentiation of data by simplified least squares procedure. Analytical Chem 44:1906–1909

Varmuza K, Filzmoser P (2009) Introduction to multivariate statistical analysis in chemometrics. CRC, Boca Raton, 321pp

Wilson EB Jr, Decius JC, Cross PC (1985) Molecular vibrations: the theory of infrared and Raman vibrational spectra (388pp.), originally published in 1955 by McGraw Hill and republished by Dover, New York

Chapter 4
Hyperspectral Image Processing Methods

Seung-Chul Yoon and Bosoon Park

4.1 Introduction

Hyperspectral image processing refers to the use of computer algorithms to extract, store and manipulate information from visible near-infrared (VNIR) or near-infrared (NIR) hyperspectral images for various information processing and data mining tasks, such as analysis, classification, regression, target detection, and pattern recognition (Chang 2013; Eismann 2012; Sun 2010; Thenkabail et al. 2011; Plaza et al. 2009; Gowen et al. 2007; Landgrebe 2003; Shaw and Manolakis 2002). Many disciplines have contributed to the advances in hyperspectral image processing technology. The fundamental theories and techniques for hyperspectral image processing are deeply rooted in the optics, the digital signal processing dealing with one-dimensional time- and frequency-domain signals and the digital image processing dealing with multidimensional space- and space-time-domain signals such as images and videos. Because hyperspectral images are inherently multidimensional signals, many techniques developed for digital image processing were directly applicable to hyperspectral image processing without modifications. However, hyperspectral images contain the spectral domain signals, i.e., spectral information at each image pixel. Thus, special tools and techniques have been developed to process both spatial and spectral information of hyperspectral images. Most tools and techniques for hyperspectral image processing have been initially developed by the remote sensing community for target detection (Manolakis et al. 2003; Manolakis and Shaw 2002), change

S.-C. Yoon (✉)
U.S. Department of Agriculture, Agricultural Research Service, US National Poultry Research Center, 950 College Station Road, Athens, GA 30605, USA
e-mail: seungchul.yoon@ars.usda.gov

B. Park
U.S. Department of Agriculture, Agricultural Research Service, Athens, GA, USA
e-mail: bosoon.park@ars.usda.gov

B. Park, R. Lu (eds.), *Hyperspectral Imaging Technology in Food and Agriculture*, Food Engineering Series, DOI 10.1007/978-1-4939-2836-1_4

detection (Rogers 2012; Eismann et al. 2008; Radke et al. 2005), anomaly detection (Stein et al. 2002), spectral unmixing (Keshava and Mustard 2002), and classification (Harsanyi and Chang 1994; Melgani and Bruzzone 2004; Chang 2003). On the contrary, the demand for new hyperspectral image processing tools and techniques more appropriate for near sensing in laboratories or fields of various science and engineering communities has been increasing in more recent years (Sun 2010; Gowen et al. 2007; Lorente et al. 2012; Ruiz-Altisent et al. 2010; Davies 2009; Dale et al. 2013). From the perspective of multivariate statistics, hyperspectral images are multivariate data as well. Thus, in recent years, chemometrics and multivariate analysis techniques started being applied to process hyperspectral images (Geladi and Grahn 1997; Grahn and Geladi 2007; Prats-Montalbán et al. 2011; Gendrin et al. 2008). Naturally, the spectroscopy and Chemometrics communities contributed many analysis tools and methods such as principal component analysis and partial least squares regression analysis that could be applied to the process of spectral information within hyperspectral images. The machine learning and data mining communities also contributed to the advances. For example, as the computing resources to process hyperspectral images are becoming more readily accessible to the science and engineering communities, new machine learning techniques and algorithms have been used to tackle more complex problems such as real-time detection of complex materials or targets that were impossible in the past (Stevenson et al. 2005; Stellman et al. 2000; Yoon et al. 2011; Tarabalka et al. 2009; Heras 2011).

Hyperspectral image processing workflows are fundamentally different from the conventional color image processing workflows although both data types are multidimensional and multivariate. The typical hyperspectral image processing workflow includes calibration and atmospheric correction (only for remote sensing), creation of a reflectance data cube, dimensionality reduction, spectral library and data processing. It is unique to require both spectral and spatial preprocessing in the workflow. However, many researchers in food and agricultural research communities have difficulty in grasping the entire workflow of hyperspectral image processing and modifying the workflow appropriate for their applications, partially due to the multidisciplinary aspect and the lack of appropriate tools and resources. In this chapter, the recent advances in hyperspectral image processing algorithms and workflows for hyperspectral image processing are discussed. The contents of the chapter consist of the basics of hyperspectral image processing techniques such as calibration, spectral and spatial preprocessing, dimensionality reduction including feature extraction and selection. The typical hyperspectral image processing workflow is described in the below figure.

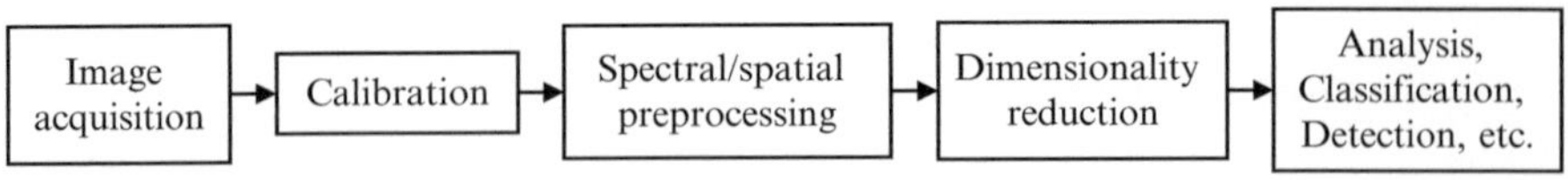

4.2 Hyperspectral Image Acquisition

Hyperspectral imaging systems typically acquire several hundreds of discrete wavelength data points at each image pixel, and thus produce three-dimensional (3D) data cubes with xyz coordinates where x and y are the spatial coordinates and z is the wavelength (i.e., spectral) coordinate. The detectors commonly used in the hyperspectral imagers are two-dimensional (2D) devices such as CCD (charge-coupled device) and CMOS (complementary metal-oxide semiconductor) image sensors (CCD vs. CMOS). Therefore, in order to acquire a complete hyperspectral image data cube, a hyperspectral imaging instrument usually requires to acquire the 2D array data either in xz (or yz) or xy coordinates several hundred times by a 2D image sensor. One type of hyperspectral imaging instruments uses dispersive optical components such as diffraction grating, prism and mirror, installed behind the front lens but in front of the detector. The most common dispersive hyperspectral imaging sensor is the pushbroom line scanner, where one line scan produces all wavelength data points along the same spatial coordinate as the line direction (Lawrence et al. 2003; Robles-Kelly and Huynh 2013). Another type of hyperspectral imaging uses electro-optical filters such as liquid crystal tunable filter (LCTF) and acousto-optical tunable filter (AOTF) for simultaneous acquisition of a spatial image with xy coordinates at each z coordinate (Robles-Kelly and Huynh 2013).

Regardless of the type of hyperspectral imaging, the resulting 3D data cube typically is constructed either in BSQ (band sequential), BIL (band interleaved by line), or BIP (band interleaved by pixel) format. These formats are known as ENVI file formats, where ENVI is a commercial software product for hyperspectral image analysis and processing (ENVI). In the BSQ format, each image of a wavelength band (or known as spectral band) that represents a narrow range (several nanometers) of the electromagnetic spectrum is followed by the next band image. In one spectral band, each line is stored from top to bottom and from left to right on the image. The BSQ format is good for spatial processing at a spectral band. Therefore, the BSQ format is recommended for creation of regions of interest, spatial feature extraction, and image processing. In the BIL format, starting from the first line (row) of the first band image, the lines located at the same top row in the subsequent bands are stored. Then, the second rows of all bands are stored in the same order as the first rows, and so on. This format is recommended for multivariate classification and chemometrics that use both spatial and spectral processing. In the BIP format, the first pixels of all spectral bands are stored in the wavelength order, and followed by the second pixels, the third pixels, and so on. This format is the best for spectral processing such access to z-profiles (spectra). In a summary, when saving or before processing the hyperspectral data cubes, the selection of the optimal format is desirable for each task. However, the BIL format usually provides a good compromise on most hyperspectral image processing tasks.

4.3 Calibration

Calibration of measured hyperspectral image data is important to ensure the accuracy and repeatability of results produced by a hyperspectral image system. This section provides a brief overview of the hyperspectral image calibration in the spectral, spatial and radiometric domains.

Spectral (or wavelength) calibration is a process that links band numbers with wavelengths. Pencil-style calibration lamps or monochromatic laser sources that produce narrow and intense peaks at a few known wavelengths are widely used for the wavelength calibration. The popular gas types of commercial calibration lamps are argon, krypton, neon, xenon, mercury-neon, and mercury-argon. A linear or non-linear regression analysis is used to predict the wavelengths at the unknown spectral bands. The wavelength calibration is typically done by a product vendor before the hyperspectral image camera is shipped to a customer. Thus, an end user needs to perform the wavelength calibration only on an occasional basis or only if necessary.

Spatial (or geometric) calibration (or correction) is a process that correlates each image pixel to known units such as meters or to known features such as grid patterns. Spatial calibration provides the information about the spatial dimensions of each sensor pixel on the surface of the material or the absolute location and size of the material. Spatial calibration also corrects optical aberrations (smile and keystone effects). Smile and keystone effects refer to the curvatures in two-dimensional image detectors. With an ideal imaging spectrometer, the spectral and spatial line data would be straightly projected on the area detector. However, imaging spectrometers suffer from smile and keystone effect. The "smile" refers to the curved spectral information along the spatial direction whereas the "keystone" refers to the curved spatial lines along the spectral direction. The optical aberrations cause blurred images and thus reduce the optical resolutions. In recent years, the manufactures of imaging spectrometers have improved their designs to minimize smile and keystone effects well below the tolerances of the spatial resolutions necessary for most applications.

Radiometric calibration, in remote sensing, refers to the conversion of the digital numbers (raw data) to physical units of radiance and the subsequent calculation of reflectance at the surface on the ground by atmospheric correction. In food and agricultural applications that do not use remote sensors, radiometric calibration often means reflectance (or transmittance) calibration or flat-field correction without the conversion to radiance and atmospheric correction. Reflectance (or transmittance) correction is a process that converts the measured digital numbers to percent reflectance (or transmittance) values with standard (known) materials that are spectrally flat and spatially homogenous. Because a flat-surface material is used, a term, flat-field calibration (or correction) is also used instead of reflectance or transmittance calibration. The reflectance calibration calculates relative (percent) reflectance values at each pixel with a known diffuse reflectance material (white or gray) in the field of view, such as Spectralon (Labsphere) and

Teflon. The transmittance calibration can be done similarly with transparent or semi-transparent known materials such as Teflon and neutral-density filters. However, transmittance calibration is more difficult to estimate accurate transmittance values in relative or percentage term, especially when measuring turbid media such as biological materials, because of the strong scattering phenomenon. The well known percent-reflectance calibration equation is the following.

$$R(x, y, \lambda) = \frac{I_{white}(x, y, \lambda) - I_m(x, y, \lambda)}{I_{white}(x, y, \lambda) - I_{dark}(x, y, \lambda)} * \mathbf{C}(\lambda)$$

where I_{white}, I_{dark}, and I_m are a white (or gray) reference, a dark current, and a measured image. The x and y are the spatial coordinates, and λ is the wavelength. $\mathbf{C}(\cdot)$ is a vector of multiplication scale factors (e.g., 100 and 40 %) defined at each wavelength, which is typically provided by the manufacture. If $\mathbf{C}(\cdot)$ is either not available or almost constant across all wavelengths, a nominal reflectance value or an average of $\mathbf{C}(\cdot)$ can be used as a constant. In practice, the constant 99 (or 100) is widely used for calibrating hyperspectral images with a 99 % Spectralon reflectance panel (Lawrence et al. 2003).

Flat-field correction is another way to calculate apparent (relative) reflectance. Flat-field correction is useful when the image includes a uniform area that has a relatively flat spectral curve and a topographically flat surface. The scene is converted to "relative" reflectance (or transmittance) by dividing each image spectrum by the flat field mean spectrum. Average relative reflectance conversion also normalizes image spectra by dividing a mean spectrum that is computed from the entire image.

4.4 Spatial Preprocessing

Spatial preprocessing refers to a process to enhance or manipulate the information in the spatial image domain. Any conventional image processing techniques for filtering and enhancement can be applied here. Spatial preprocessing for denoising and sharpening is not usually applied to raw or calibrated hyperspectral images unless a specific task requires it because spatial preprocessing may affect the spectral signatures. Spatial postprocessing is more popular than preprocessing because classification or prediction images are ordinary images that need spatial interpretation, manipulation, and pattern recognition. The details about these spatial processing methods were already covered in Chap. 4. Thus, we omit the details here. Instead, we introduce spatial preprocessing methods that are important for spectral preprocessing and later processing tasks such as classification and regression in the image domain.

4.5 Spatial Sampling and ROIs

One of the early steps for hyperspectral image processing is to determine spatial locations to be examined. This process typically starts from an image binarization (or segmentation) with a thresholding operation on a spectral image band. The thresholding operation can be replaced with any type of an image segmentation method, if necessary. The image binarization produces a binary image where the background is masked out. So, this binarization process is also known as background masking (Yoon et al. 2009). A heuristic but practically useful method for background masking is to browse the spectral image bands and pick the heuristically best wavelength band, and then all spectral bands are binarized with the same segmentation process. Another useful method is to find an imaging condition that provides a simpler solution for background masking. For example, when imaging white objects, a black background is better. A more systematic method to find the band for background masking is to use factor analysis such as principal component analysis (PCA) to find the band with the largest variation in reflectance (or absorbance) values. The spurious noise, unwanted holes, or excessive boundary edges on the binary mask are often removed by spatial image processing such as median filtering and morphological filtering.

After background masking, the remaining foreground pixels are examined for specular reflection. Specular pixels, typically with reflectance values approaching or exceeding 100 %, produce highly invalid spectral responses. Hence, it is not recommended to include the specular pixels in the data set for any spectral processing. The specular reflection is often caused by wet or glossy surface features on the scene and is more pronounced when the angle between the incident light and the camera is small or too much of the light is incident upon the surface. Therefore, regardless of the source of the specular reflection, any pixels with reflectance values close to the saturation should be dealt with a caution. If necessary, these pixels should be masked out by either a thresholding operation or a classification method. The reflectance values associated with specular reflection are significantly higher than the normal scene features and can thus be differentiated.

The creation of ground-truth regions-of-interest (ROIs) in hyperspectral image processing is similar to sampling or sample design in statistics, which is concerned with the selection of subset of individual samples to extrapolate the population from the selected samples (Yoon et al. 2013). The key idea of ROI creation is to represent the population of each material with the samples within each ROI. Typically, if there are multiple materials to examine, ROIs for all materials are prepared in such a way that the pixels showing pure spectra are included in the ROIs. Glints and shadows are often excluded in the ROIs. Pixels along the boundaries of each material are also not included in the ROIs. A rule of thumb is to exclude any pixels with mixed spectral responses unless it is necessary. Either a binary mask (1 for each ROI pixel and 0 for the others) or a grayscale image with class labels associated with each ROI type is also created. The ROIs are used for building spectral libraries, designing classification models and evaluating the performance of the classification models.

4.6 Image Mosaicing

A mosaic-based image representation can be used to facilitate data analysis and classification algorithm development, where a mosaic of multiple hyperspectral images are merged into a single hyperspectral image (Yoon et al. 2009, 2013). The calibrated hyperspectral images are tessellated without overlap into an image mosaic according to a pre-defined rule such as replicates, duplicates, material types, etc. For example, all images obtained from each replicate experiment can be shuffled into two neighboring columns if the measurement was duplicated twice. Then, the next replicates can be arranged chronologically from left to right. The image mosaic can expedite the image analysis and classification development and evaluation using one single hyperspectral data cube.

4.7 Spectral Preprocessing

Most spectral preprocessing of hyperspectral images can be roughly grouped into two categories according to their associated tasks. The first category is for endmember extraction, where an endmember refers to a pure spectral signature. Endmember extraction is an important task in remote sensing, greatly influenced by mineralogy. Endmembers can be obtained on the ground or in laboratories by spectrometers in order to build a spectral library of pure signatures. However, if endmembers should be extracted from a given image, they are typically obtained by spectral preprocessing methods such as pixel purity index (PPI) and N-finder algorithm (N-FINDR) (Chang 2013). Extracted endmembers are widely used in spectral unmixing, target detection and classification (Eismann 2012). The second category is for chemometrics, where chemometrics refers to chemometric analysis of spectral data. Chemometric spectral preprocessing is an important task in spectroscopy. Spectral data obtained by chemometric spectral preprocessing are typically used in multivariate analysis such as PCA and partial least squares (PLS) (Ozaki et al. 2006; Williams and Norris 2001; Rinnan et al. 2009; Vidal and Amigo 2012). From a slightly different perspective, spectral unmixing using endmembers, called abundance estimation in remote sensing, is similar to concentration estimation using PLS regression model in spectroscopy because both tasks, when applied to a hyperspectral image, attempt to predict how much of pure materials or chemical components is measured at each image-pixel sample. This perspective implies the selection of appropriate preprocessing methods is highly data-driven and task-specific. Nonetheless it is a good practice to map the preprocessed spectra back to the (spatial) image domain so that spatially-bound processing algorithms can be applied. In this chapter, we focus on the second category: chemometric spectral preprocessing methods because they can be applied to any spectral data as needed.

4.8 Transformation to Absorbance

For spectroscopic data analysis, reflectance or transmittance is transformed to absorbance that is a logarithmic ratio of the radiation incident on a material to the radiation reflected or transmitted through the material. The formula for the transformation is given by

$$A = \log_{10}(1/R)$$

where R is the reflectance value (a fractional point value between 0 and 1). Absorbance can be also transformed to reflectance by $R = 10^{-A}$. Note that if the percent reflectance is used, R should be properly scaled by **C**/R where **C** is the percentage scale factor, such as 100. Similarly, the transmittance, T can be converted to the absorbance by replacing R with T. The transformation to absorbance has been known to decrease the non-linearity in reflectance or transmittance measurements, to some degree (Burns and Ciurczak 2007).

4.9 Noise Reduction

Hyperspectral data typically suffer from noise. In spectroscopy, it is common to measure multiple scans to reduce the noise. In hyperspectral imaging, it is not common to measure multiple scans because of the restricted scan time in many applications or the limitation of the acquisition software. If possible, it is a good practice to measure multiple scans and take an average or median to reduce the noise. Even after the multiple scans to reduce the noise, the data may still contain the noise. In that case, a denoising algorithm can reduce the noise in the data. Denoising in the spectral domain of the hyperspectral image data is an essential preprocessing method for producing high quality data. Most denoising methods using smoothing filtering such as moving average and Savitzky-Golay filters are based on an assumption that signals are locally smooth so that neighbors are spectrally similar (Savitzky and Golay 1964; Press et al. 2007). The selection of the proper filter type and size is important because otherwise the original spectral signals may be distorted due to the filtering.

A low pass filter that passes low-frequency signals is the simplest of all smoothing filters with windows of fixed size such as 3×3, 5×5, and 1×3. The net effect of low pass filtering is obviously smoothing the data. The low pass filter is fast and simple to implement by a convolution. Let's assume that x is a measured spectrum, h is a filter, and y is the smoothed output spectrum. Spectral smoothing with a filter can be generalized by a 1D convolution,

$$y[i] = x[i]*h[i] = \sum_{m=-k}^{k} x[m] \cdot h[i+m]$$

where i is the band number and h is a finite impulse response (FIR) filter, usually called impulse function, kernel, filter, or window (Proakis and Manolakis 2006). The implementation of the convolution is done by a sliding window method in the spectral domain or a fast Fourier transform (FFT) in the frequency domain. Various filters such as moving average with equal weights, weighted moving average, and Savitzky-Golay filter were developed and applied to spectral denoising. The moving average filter with equal weights has the same coefficients with a box shape. The weighted moving average gives different weights at different positions in the windows and thus produces various filter shapes such as triangle and exponential. The Savitzky-Golay filer is a FIR filter that performs a local least-squares polynomial regression (or approximation) to obtain the filter coefficients that are constant for all outputs. Thus, the Savitzky-Golay filer is a weighted moving average, where their coefficients are already calculated for the different sizes of the windows and the different orders of polynomials. The advantage of the Savitzky-Golay filter for spectral smoothing is that it tends to preserve the original signals by removing noise only to some degree, which is better than moving average with equal weights. However, as in any smoothing filtering, aggressive smoothing with the increased windows size lowers the peak values and broadens the shapes. Other techniques for spectral denoising are wavelet transform, moving median filtering, and minimum noise fraction transform. De-noising methods using both spatial and spectral domain signals were also developed.

The signal-to-noise ratio (SNR) is a metric to measure the performance of denoising filters, defined as the ratio of mean to standard deviation of an image.

$$SNR(\lambda) = \frac{\mu(\lambda)}{\sigma(\lambda)}$$

where $\mu(\lambda)$ is the mean of the image pixel values and $\sigma(\lambda)$ is the standard deviation at the wavelength λ.

4.10 Baseline Correction

A baseline shift (or offset) refers to a non-zero background signal that is a slowly varying low-frequency signal, observed without measurements of actual samples (Ozaki et al. 2006; Williams and Norris 2001). A non-zero baseline shift introduced into the measured spectra is caused by many different sources such as the instrument, light scattering, and different spectral path lengths through the sample. The goal of the baseline correction is to put the baseline back into the zero-absorbance (or reflectance) baseline by subtracting the shifted baseline from the measured spectral data. In general, it is difficult to model the baseline shifts with a unified theoretical framework. A simple manual approach for baseline correction is for the user to pick a few representative points from the baseline and interpolate the other points along the baseline using linear, polynomial, or spline functions. A regression fit can be also used. There are several automated baseline correction methods such as derivatives (discrete differentiation) or high-pass filtering, detrending, multivariate or polynomial baseline modeling, iterative weighting, etc. The baseline correction can be integrated as a part of the data normalization such as multiplicative scatter correction and standard normal variate with detrending, which will be discussed in a separate section later on. Detrend is a baseline correction technique that removes the overall trend or slope in the data. Detrend fits the entire data with a polynomial function such a line and a quadratic, etc. and subtracts this polynomial line or curve from the data.

4.11 Derivatives

Derivative computes simply the slope of a signal at a certain point. Differentiation is a process of applying the derivative to the signal. Differentiation can remove the baseline shift and also extract spectral features such as spectral shape and peak width (Ozaki et al. 2006; Williams and Norris 2001; Rinnan et al. 2009; Vidal and Amigo 2012). Differentiation reduces the effect of baseline shift. In practice, second derivatives and higher-order derivatives tend to remove the baseline shift. A spectral shape such as Gaussian and sigmoid can be predicted by examining the derivatives. For example, a Gaussian-type spectral signal has a peak. A positive derivative means that the signal slope goes up. A negative derivative means that the signal slope goes down. A zero derivative means a zero slope at the peak location. Thus, the location of the maximum (peak) can be computed by the location of the zero-crossing in its first derivative. The peak (crest) of the signal becomes a trough (valley) in its second derivative. The second example is a sigmoid-type spectral signal with an "S" shape. The inflection point where the slope becomes the maximum corresponds to the maximum in its first derivative and the zero-crossing in its second derivative. Thus, the exact location of the inflection point can be computed by the zero-crossing in its second derivative. Another important feature to predict by differentiation is that a peak width in a Gaussian-type or any peak-type signal is inversely proportional to the amplitude of derivatives. The narrower peak produces the larger derivative amplitude and vice versa. Therefore, when a signal has a wide peak, the amplitude of its derivative at the peak location is small. Note that differentiation reduces the signal-to-noise ratio if smoothing is not properly applied before the differentiation. Any de-noising algorithm can be used but the Savitzky-Golay differentiation filter is one of the popular methods because it combines differentiation and smoothing into one convolution algorithm.

4.12 Normalization and Scatter Correction

Data normalization is a process to correct multiplicative scale effects (i.e., scale differences) in spectral data. Light scattering is one of the main causes for multiplicative scale differences. This is why some normalization methods such as multiplicative scatter correction (MSC) and standard normal variate (SNV) that reduce scatter effects are also called scatter correction methods (Ozaki et al. 2006; Williams and Norris 2001; Rinnan et al. 2009; Vidal and Amigo 2012). The simplest form of normalization is to normalize each spectrum with a constant weight that is the sum of all sample values (reflectance, transmittance, or absorbance). Normalization itself is a simple division by a constant.

MSC was developed to remove both scaling effects (a multiplicative factor, a) and baseline shift effects (an additive factor, b) with the below linear regression equation.

$$\mathbf{x}_i = a\overline{\mathbf{x}} + b$$

where $\mathbf{x}_i$ is the spectrum of a sample and $\overline{\mathbf{x}}$ is a reference spectrum (typically a mean spectrum in the calibration data set). For each sample, a and b are estimated by least-squares regression of a measured spectrum $\mathbf{x}_i$ and the mean spectrum $\overline{\mathbf{x}}$. Each $\mathbf{x}_i$ is corrected by the following formula

$$\text{MSC} : \widetilde{\mathbf{x}}_i = \frac{\mathbf{x}_i - b}{a}.$$

SNV is also widely used for scatter correction. The SNV transformation does not need a reference spectrum because it is applied to individual samples whereas MSC needs a calibration set to compute a reference (mean) spectrum. For the SNV transformation, the spectrum of each sample is mean-centered and scaled by the values derived from the sample spectrum itself.

$$\text{SNV} : \widetilde{\mathbf{x}}_i = \frac{\mathbf{x}_i - m_i}{\sigma_i}$$

where m_i and σ_i are the mean and standard deviation values of the sample spectrum $\mathbf{x}_i$, respectively. The SNV corrected spectrum has always zero-mean and unit-variance. In practice, the SNV transformation is typically followed by detrending. De-trending also corrects individual spectra without a reference spectrum. Detrending subtracts a quadratic (i.e., 2nd degree) polynomial baseline, calculated by linear least squares regression, from the signals.

4.13 Dimensionality Reduction

It has been known that the high dimensionality of hyperspectral images does not always provide the abilities and effectiveness necessary for hyperspectral image processing tasks because of the curse of dimensionality. The curse of dimensionality is also known as the Hughes phenomenon that when the same size is fixed, the increased dimensionality does not increase the classification accuracy (Landgrebe 2003). A solution to this problem is to increase the number of sample size. However, in most applications, it is simply not easy to determine the statistically sufficient sample size beforehand and collect these samples. Thus, one widely used method to overcome the curse of dimensionality is to reduce the data dimension and to extract spatial and/or spectral features in a lower dimensional space. We introduce a few popular dimensionality reduction methods for hyperspectral image processing: data binning, feature extraction and feature selection.

4.13.1 Binning

Binning reduces the spatial and spectral resolution but increases the signal to noise ratio (SNR). The CCD-based hyperspectral image sensors perform on-chip binning at the hardware level inside the camera in order to reduce the amount of readout data, which increases the frame rate and at the same time improves the SNR. The on-chip binning performed on CCD devices is a summation operation that combines the photon charges of neighboring pixels and thus no noise is added to the binned signal. The relationship between the improved SNR and binning on CCD sensors is linear in that N-pixel binning provides N-fold SNR improvement. However, no true charge binning is possible in conventional CMOS detectors because the voltage signals with random noise are added into the binned signal. In general, binning of N adjacent pixels in a CMOS sensor may improve the SNR to the square root of the number of binned pixels, i.e., only $\sqrt{N}$-fold SNR improvement. Therefore, hardware binning in CMOS image sensors is similar to software binning done in computer hardware, except the binning, either summation or average, is performed inside a CMOS camera. In hyperspectral imaging, if spatial binning is applied to one direction (x or y direction) on an image sensor, spectral binning is defined on the other direction. The binning size at each direction is typically the power of 2, e.g., 1, 2, 4, etc.

4.14 Feature Extraction

Hyperspectral images carry a lot of redundant information in the spatial and spectral domains. Feature extraction refers to a linear or non-linear transformation procedure that lowers the data dimensionality and thus reduces the data redundancy in the spatial and/or spectral domain. A feature extraction process should be carefully designed to extract only the information relevant to desired applications such as classification and regression. A rule of thumb for finding the best features is to extract features specific to the desired application, which involves the domain knowledge about the data. If the domain knowledge is not available, general feature extraction methods can be used. The general feature extraction methods include PCA (Pearson 1901), PLS (Wold et al. 2001), independent component analysis (ICA) (Comon 1994), kernel PCA (Schölkopf et al. 1998), and spatial image processing to detect spatial features edges, corners, blobs and shapes, etc. These feature extraction methods can be also used to extract the features in an ad hoc matter with the domain knowledge. If an application requires real-time image acquisition and processing, multispectral imaging with several band ratios and/or vegetation indices can be a practically viable solution. In determining the feature extraction methods, there are several important factors to consider.

Regardless of how the features are extracted, the feature extraction methods should preserve or reveal the information necessary for the chosen application such as classification and detection. For classification tasks, the ideal features must minimize the classification error. However, it is usually difficult to estimate or predict the classification error directly from the raw data. Instead, class separability that can be directly measured from the data is widely used to predict the classification or discriminative power of the feature. Some common separability measures are Euclidean distance (within class and between-class), Mahalanobis distance (Gaussian density model), Bhattacharyya distance (similarity of distributions), Kullback–Leibler (KL) divergence (expectation of likelihood ratio), and entropy. The class separability problem can be formulated with

$$\max_{\vec{w}} J\left(\vec{w}\right)$$

where $J\left(\vec{w}\right)$ is an objective (or called a criterion) function of the separability measured between classes and $\vec{w}$ is a transformation (e.g., a linear projection) that projects the raw feature space to another feature space (usually, a reduced dimensional space). Thus, the feature extraction is to find a vector $\vec{w}$, or called a linear predictor, that maximizes the class separability. The Fisher's linear discriminant analysis (LDA) (Duda et al. 2000; Hastie et al. 2009) considers the following objective function, called the Fisher criterion

$$J\left(\vec{w}\right) = \frac{\vec{w}^T S_B \vec{w}}{\vec{w}^T S_W \vec{w}}$$

where S_B is the between-class scatter matrix and S_W is the within-class scatter matrix. After mathematical manipulations, the optimal solution maximizing the Fisher criterion $J\left(\vec{w}\right)$, obtained by solving the generalized eigenvalue problem is a projection vector, $\vec{w}^* = S_w^{-1}(\mu_1 - \mu_2)$ where μ_1 and μ_2 are mean vectors of each class. The projection vector $\vec{w}$ is perpendicular to the hyperplane that partitions the vector space into two sets, one for each class. A limitation of LDA is that LDA fails if discriminatory information is embedded in the variance of the data. In this case, PCA is a better choice. Also, LDA is a parametric method that assumes unmoral Gaussian distribution of the data. Therefore, if the data distribution is non-Gaussian, LDA is not effective in separating the overlapped distributions. Support vector machines are worthy to be mentioned here and compared with the standard feature extraction methods such as LDA and PCA. PCA is a linear transformation that projects the original data space into an orthogonal vector space with uncorrelated features, i.e., principal components. In a nutshell, the features in the PCA domain are optimal for data representation but the features obtained by LDA are optimal for data classification.

4.14.1 Support Vector Machine and Feature Extraction

A linear support vector machine (SVM) (Hastie et al. 2009) is built upon the well-defined statistical learning theory that separates the linearly separable feature space into two classes with the maximum margin. If the feature space is not linearly separable, a non-linear SVM can be used or the linear SVM is modified to a soft margin classifier in order to allow the misclassification error by adding error variables, called slack variables, to its objective function. If the non-linear SVM is used, the basic concept is to represent the data in a higher (even infinite) dimensional feature space such that the non-linearly mapped data are linearly separable in the high dimensional feature space and thus the linear SVM can be applied to the newly mapped data. However, the non-linear mapping via a dot product in the high dimensional space is computationally very expensive. Alternatively, the non-linear SVM using a kernel trick provides a computationally feasible solution that produces the same results as what the non-linear mapping functions would. The kernel trick with a kernel function eliminated the need to know explicitly about the required higher dimensionality and the non-linear mapping function. Popular kernel functions are polynomial, radial basis function of Gaussian, and sigmoidal. The non-linear SVM finds nonlinear decision boundaries in the original feature space. Although a limitation of SVM is in that the best kernel function for a given problem is typically found by trial and error, the kernel trick alone or the non-linear SVM itself has been combined with many popular feature extraction methods to make an effective framework for classification and regression analysis, such as kernel-PCA (Schölkopf et al. 1998; Zhang et al. 2012; Zhu et al. 2007), kernel discriminant analysis (Mika et al. 1999), and PCA-SVM (Zhang et al. 2012) and LDA-SVM.

4.15 Feature Selection

4.15.1 Search Algorithms and Selection Criteria

In machine learning, feature extraction refers to a process to transform the input features to a set of new features in a lower dimensional space, whereas feature selection refers to a process to select a subset of the input features without a transformation. That's why feature selection is also called feature subset selection. Like feature extraction, feature selection requires the optimization of a objective function, J. For feature selection, the optimization finds a subset over the set of all possible subsets. Thus, the best subset X is found by optimizing the following objective function,

$$\max_{X \in X_d} J(X)$$

where X_d is the set of all possible subset combinations obtained from the input features. To solve this optimization problem, feature selection typically requires a search strategy to select candidate feature subsets and an objective function to evaluate the feature subset candidates. Some examples of search algorithms (Webb 2002) are branch and bound, greedy hill climbing, exhaustive search (usually computationally most expensive), sequential forward selection, sequential backward selection, bidirectional search, projection pursuit, etc. When a search algorithm suffers from local minima, a randomized search strategy such as simulated annealing and genetic algorithms can be used by incorporating randomness in the search procedure to escape from the local minima (Liu and Motoda 2007). The objective functions for feature selection fall into three groups: filter, wrapper, and embedded methods (Guyon and Elisseeff 2003; Molina et al. 2002). Filter methods evaluate the candidate subsets by the information content such as distance, separability, correlation and mutual information. Wrapper methods use a classifier to evaluate the candidate subsets by the classification performance (accuracy or error rate) on test data generated by statistical resampling or cross-validation. The wrapper methods usually provide the explicit best feature subset optimized for a given learning model, whereas the filter methods typically provide a feature ranking. The wrapper methods may suffer from the risk of overfitting and are computationally expensive. The filter methods, however, are much faster to compute than the wrappers. Embedded methods incorporate the feature selection process into a learning model. The computational complexity of embedded methods is between filters and wrappers.

4.15.2 Band Selection

Feature selection has been proven to be effective in many multivariate data analysis and classification tasks such as DNA microarray analysis and hyperspectral image classification, especially when there are many more features (i.e., variables) than samples. Although the "feature selection" terminology has been widely used and recognized in many fields, this feature selection problem is also commonly known as a band selection problem in the hyperspectral imaging literature (Bajwa et al. 2004; Nakariyakul and Casasent 2004; Martínez-Usó et al. 2006; Su et al. 2008). In fact, a band selection problem is a particular case of feature selection, specifically to reduce the dimensionality of hyperspectral images and find the important and useful features, i.e., wavelengths, for analysis, classification and regression. Therefore, any feature selection methods mentioned previously can be used to solve the band selection problem. In this section, we introduce a band selection method with PCA that provides a practically feasible solution.

Band selection refers to a process to identify a few wavelengths that provide discriminative information. Although PCA is widely used as a feature extraction method, it can be used for feature selection (Yoon et al. 2009; Koonsanit et al. 2012; Song et al. 2010; Malhi and Gao 2004; Cataltepe et al. 2007; Cohena et al. 2013). The basic idea of using PCA for feature selection is to select variables (i.e., wavelengths) according to the magnitude of the coefficients of loadings because the loadings can be understood as the weights (or the amount of contribution) of each input variable to the principal component. To determine the contribution of each of all n bands to the new features (PC score image bands in this case), the squares of the loading vector coefficients at each band are computed and normalized to the sum of 1 by the following equation:

$$W_k(i) = \frac{P_k(i)^2}{\sum_{i=1}^{n} P_k(i)^2}, \quad i \in \{1, \ldots, n\}$$

where $W_k(i)$ is a weighting factor of the i-th band of the k-th principal component, $P_k(i)$, i.e., a loading vector. We call all of $W_k(i)$ the PCA-band weightings. The PCA-band weightings of the first several principal components can be examined to select the wavelengths.

4.16 Fusion of Spatial and Spectral Information

In this section, we briefly introduce a new trend in hyperspectral image processing: fusion of spatial and spectral information for hyperspectral image processing under a unified mathematical framework or an integrated processing algorithm. The widely accepted practice in processing hyperspectral images is the separate application of spatial and spectral processing methods (Plaza et al. 2009). No matter how the spectral data was processed, separate spatial processing methods were typically used to extract and enhance spatial features and suppress spatial noise, all independent of spectral processing. However, there is a growing need for processing spectral data in a spatial context or vice versa. The review paper of Plaza et al. (2009) provides an overview of recent developments in using both spatial and spectral information for hyperspectral image processing. These developments roughly fall into two categories. The first category is composite kernel methods utilizing cross-information between kernels for spatial (textual) and spectral feature extraction. A composite kernel machine developed by Camps-Valls et al. (2006) was applied to hyperspectral image classificaiton. The second category is the integration of spatial and spectral information for classification, segmentation and unmixing. Mathematical morphology (Fauvel et al. 2013) and Markov random fields (Jackson and Landgrebe 2002) were used for classification that integrated spectral features with spatial features such as size, orientation, contrast, and local homogeneity. For hyperspectral image segmentation, an iterative algorithm, called a butterfly approach, was developed by extracting both spatial topology and spectral latent variable (Gorretta et al. 2009). Mendoza et al. (2011) used 294 spatial and spectral features for predicting apple quality attributes with a PLS regression model. Martín and Plaza (2012) combined spatial homogeneity information and spectrally pure pixels for endmember extraction and spectral unmixing.

Although the resolution and quality of hyperspectral images generated for food and agricultural applications have been improved in recent years, it is still a difficult task to define pixels or regions that are both spatially and spectrally homogenous and thus to extract useful or relevant spectral features in a spatial context. Nonetheless, the incorporation of the spatial information and spectral signatures into a unified hyperspectral image processing algorithm is highly desirable. Hence, we foresee that the number of publications in this area will be increasing in near future.

4.17 Summary and Discussion

In this chapter, we described a variety of different methods and techniques for hyperspectral image processing from acquisition to dimensionality reduction. There are many more topics not covered in this chapter. For example, pattern recognition techniques such as supervised classification and unsupervised clustering algorithms were not discussed in this chapter. Instead, we focused on preprocessing and feature extraction and selection methods that can be used as inputs to a pattern recognition algorithm. It is not an easy task to determine the best pattern recognition algorithm for a given application because there are many pre-steps affecting the performance of the final algorithm. These pre-steps mentioned in this chapter are mainly about how to acquire, preprocess hyperspectral images and which features are used for the given application. If these pre-steps are carefully designed and executed, the selection of the best pattern recognition algorithm often becomes trivial. Although, in some applications, it is not possible to separate the feature selection and/or extraction from the classifier design, it is advisable to design a pattern recognition algorithm with the data in the reduced dimensional space.

References

Bajwa SG, Bajcsy P, Groves P, Tian LF (2004) Hyperspectral image data mining for band selection in agricultural applications. Trans ASAE 47(3):895–907

Burns DA, Ciurczak DW (2007) Handbook of near-infrared analysis. CRC, Boca Raton

Camps-Valls G, Gomez-Chova L, Munoz-Mari J, Vila-Frances J, Calpe-Maravilla J (2006) Composite kernels for hyperspectral image classification. IEEE Geosci Remote Sens Lett 3(1):93–97

Cataltepe Z, Genc HM, Pearson T (2007) A PCA/ICA based feature selection method and its application for corn fungi detection. In: 15th European signal processing conference, pp 970–974

CCD vs. CMOS http://www.teledynedalsa.com/imaging/knowledge-center/appnotes/ccd-vs-cmos/

Chang C-I (2003) Hyperspectral imaging: techniques for spectral detection and classification. Springer, New York

Chang C-I (2013) Hyperspectral data processing: algorithm design and analysis. Wiley-Interscience, Hoboken

Cohena S, Cohen Y, Alchanatis V, Levia O (2013) Combining spectral and spatial information from aerial hyperspectral images for delineating homogenous management zones. Biosyst Eng 114(4):435–443

Comon P (1994) Independent component analysis: a new concept? Signal Process 36(3):287–314

Dale LM, Thewis A, Boudry C, Rotar I, Dardenne P, Baeten V, Piern JAF (2013) Hyperspectral imaging applications in agriculture and agro-food product quality and safety control: a review. Appl Spectrosc Rev 48(2):142–159

Davies ER (2009) The application of machine vision to food and agriculture: a review. Imaging Sci J 57(4):197–217

Duda RO, Hart PE, Stork DH (2000) Pattern classification, 2nd edn. Wiley–Interscience, Hoboken

Eismann MT, Meola J, Hardie RC (2008) Hyperspectral change detection in the presence of diurnal and seasonal variations. IEEE Trans Geosci Remote Sens 46(1):237–249

Eismann MT (2012) Hyperspectral remote sensing, vol PM210. SPIE, Bellingham

ENVI file formats. http://geol.hu/data/online_help/ENVI_File_Formats.html

Fauvel M, Tarabalka Y, Benediktsson JA, Chanussot J, Tilton JC (2013) Advances in spectral-spatial classification of hyperspectral images. Proc IEEE 101(3):652–675

Geladi P, Grahn H (1997) Multivariate image analysis. Wiley, New York

Gendrin C, Roggo Y, Collet C (2008) Pharmaceutical applications of vibrational chemical imaging and chemometrics: a review. J Pharmaceut Biomed Anal 48(3):533–553

Gorretta N, Roger JM, Rabatel G, Bellon-Maurel V, Fiorio C, Lelong C (2009) Hyperspectral image segmentation: the butterfly approach. IEEE first workshop on hyperspectral image and signal processing: evolution in remote sensing, WHISPERS '09, pp 1–4

Gowen AA, O'Donnell CP, Cullen PJ, Downey G, Frias JM (2007) Hyperspectral imaging—an emerging process analytical tool for food quality and safety control. Trends Food Sci Technol 18:590–598

Grahn H, Geladi P (2007) Techniques and applications of hyperspectral image analysis. Wiley, New York

Guyon I, Elisseeff A (2003) An introduction to variable and feature selection. J Mach Learning Res 3:1157–1182

Harsanyi JC, Chang C-I (1994) Hyperspectral image classification and dimensionality reduction: an orthogonal subspace projection approach. IEEE Trans Geosci Remote Sens 32(4):779–785

Hastie T, Tibshirani R, Friedman J (2009) The elements of statistical learning: data mining, inference, and prediction, 2nd edn. Springer, New York

Heras DB (2011) Towards real-time hyperspectral image processing, a GP-GPU implementation of target identification. In: IEEE 6th international conference on intelligent data acquisition and advanced computing systems

Jackson Q, Landgrebe DA (2002) Adaptive Bayesian contextual classification based on Markov random fields. IEEE Trans Geosci Remote Sens 40(11):2454–2463

Keshava N, Mustard JF (2002) Spectral unmixing. IEEE Signal Process Mag 19(1):44–57

Koonsanit K, Jaruskulchai C, Eiumnoh A (2012) Band selection for dimension reduction in hyper spectral image using integrated information gain and principal components analysis technique. Int J Mach Learn Comput 2(3):248–251

Landgrebe DA (2003) Signal theory methods in multispectral remote sensing. Wiley, Hoboken

Lawrence KC, Park B, Windham WR, Mao C (2003) Calibration of a pushbroom hyperspectral imaging system for agricultural inspection. Trans ASAE 46(2):513–521

Liu H, Motoda H (2007) Computational methods of feature selection. Chapman & Hall, Boca Raton

Lorente D, Aleixos N, Gómez-Sanchis J, Cubero S, García-Navarrete OL, Blasco J (2012) Recent advances and applications of hyperspectral imaging for fruit and vegetable quality assessment. Food Bioprocess Technol 5(4):1121–1142

Malhi A, Gao RX (2004) PCA-based feature selection scheme for machine defect classification. IEEE Trans Instrum Meas 53(6):1517–1525

Manolakis D, Shaw GA (2002) Detection algorithms for hyperspectral imaging application. IEEE Signal Process Mag 19(1):29–43

Manolakis D, Marden D, Shaw GA (2003) Hyperspectral image processing for automatic target detection applications. Lincoln Lab J 14(1):79–116

Martín G, Plaza A (2012) Spatial-spectral preprocessing prior to endmember identification and unmixing of remotely sensed hyperspectral data. IEEE J Sel Topics Appl Earth Observ Remote Sens 5(2):380–395

Martínez-Usó A, Pla F, García-Sevilla P, Sotoca JM (2006) Automatic band selection in multispectral images using mutual information-based clustering. In: Progress in pattern recognition, image analysis and applications. Springer, Berlin, pp 644–654

Melgani F, Bruzzone L (2004) Classification of hyperspectral remote sensing images with support vector machines. IEEE Trans Geosci Remote Sens 42(8):1778–1790

Mendoza F, Lu R, Ariana D, Cen H, Bailey B (2011) Integrated spectral and image analysis of hyperspectral scattering data for prediction of apple fruit firmness and soluble solids content. Postharvest Biol Technol 62(2):149–160

Mika S, Ratsch G, Weston J, Scholkoph B, Mullers KR (1999) Fisher discriminant analysis with kernels. In: Proceedings of the 1999 I.E. signal processing society workshop, neural networks for signal processing IX, pp 41–48

Molina LC, Belanche L, Nebot A (2002) Feature selection algorithms: a survey and experimental evaluation. In: Proceedings of IEEE international conference on data mining, pp 306–313

Nakariyakul S, Casasent D (2004) Hyperspectral ratio feature selection: agricultural product inspection example. SPIE Proc Nondestructive Sens Food Safety Qual Natural Res 5587:133–143

Ozaki Y, McClure WF, Christy AA (2006) Near-infrared spectroscopy in food science and technology. Wiley, Hoboken

Pearson K (1901) On lines and planes of closest fit to systems of points in space. Philos Mag 2(11):559–572

Plaza A, Benediktsson JA, Boardman JW, Brazile J, Bruzzone L, Camps-Valls G, Chanussotg J, Fauvelg M, Gambah P, Gualtierii A, Marconcinie M, Tiltoni JC, Triannih G (2009) Recent advances in techniques for hyperspectral image processing. Remote Sens Environ 113:S110–S122

Prats-Montalbán JM, de Juan A, Ferrer A (2011) Multivariate image analysis: a review with applications. Chemometr Intell Lab Syst 107(1):1–23

Press WH, Teukolsky SA, Vetterling WT, Flannery BP (2007) Numerical recipes: the art of scientific computing, 3rd edn. Cambridge University Press, Cambridge
Proakis JG, Manolakis DK (2006) Digital signal processing, 4th edn. Prentice Hall, New York
Radke RJ, Andra S, Al-Kofah O, Roysam B (2005) Image change detection algorithms: a systematic survey. IEEE Trans Image Process 14(3):294–307
Rinnan A, van den Berg F, Engelsen SB (2009) Review of the most common pre-processing techniques for near-infrared spectra. Trend Anal Chem 28(10):1201–1222
Robles-Kelly A, Huynh CP (2013) Imaging spectroscopy for scene analysis. Springer, London
Rogers J (2012) Change detection using linear prediction in hyperspectral imagery. ProQuest, UMI Dissertation
Ruiz-Altisent M, Ruiz-Garcia L, Moreda GP, Lu R, Hernandez-Sanchez N, Correa EC, Diezma B, Nicolaï B, García-Ramos J (2010) Sensors for product characterization and quality of specialty crops—a review. Comput Electron Agric 74(2):176–194
Savitzky A, Golay MJE (1964) Smoothing and differentiation of data by simplified least squares procedures. Anal Chem 36(8):1627–1639
Schölkopf B, Smola A, Müller K-R (1998) Nonlinear component analysis as a kernel eigenvalue problem. Neural Comput 10(5):1299–1319
Shaw GA, Manolakis D (2002) Signal processing for hyperspectral image exploitation. IEEE Signal Process Mag 19(1):12–16
Song F, Guo Z, Mei D (2010) Feature selection using principal component analysis. In: IEEE international conference on system science, engineering design and manufacturing informatization (ICSEM), vol 1, pp 27–30
Stein DWJ, Beaven SG, Hoff LE, Winter EM, Schaum AP, Stocker AD (2002) Anomaly detection from hyperspectral imagery. IEEE Signal Process Mag 19(1):58–69
Stellman CM, Hazel GG, Bucholtz F, Michalowicz JV, Stocker A, Schaaf W (2000) Real-time hyperspectral detection and cuing. Opt Eng 39(7):1928–1935
Stevenson B, O'Connor R, Kendall W, Stocker A, Schaff W, Alexa D, Salvador J, Eismann M, Barnard K, Kershenstein J (2005) Design and performance of the civil air patrol ARCHER hyperspectral processing system. Proc SPIE 5806:731–742
Su H, Sheng Y, Du PJ (2008) A new band selection algorithm for hyperspectral data based on fractal dimension. Proc ISPRS, pp 3–11
Sun D-W (2010) Hyperspectral imaging for food quality analysis and control. Academic, San Diego
Tarabalka Y, Haavardsholm TV, Kasen I, Skauli T (2009) Real-time anomaly detection in hyperspectral images using multivariate normal mixture models and GPU processing. J Real-Time Image Process 4(3):287–300
Thenkabail A, Lyon PS, Huete JG (2011) Hyperspectral remote sensing of vegetation. CRC, Boca Raton
Vidal M, Amigo JM (2012) Pre-processing of hyperspectral images. Essential steps before image analysis. Chemometr Intell Lab Syst 117:138–148
Webb A (2002) Statistical pattern recognition, 2nd edn. Wiley, New York
Williams P, Norris K (2001) Near-infrared technology: in the agricultural and food industries, 2nd edn. American Association of Cereal Chemists, St. Paul
Wold S, Sjöström M, Eriksson L (2001) PLS-regression: a basic tool of chemometrics. Chemometr Intell Lab Syst 58(2):109–130
Yoon SC, Lawrence KC, Siragusa GR, Line JE, Park B, Feldner PW (2009) Hyperspectral reflectance imaging for detecting a foodborne pathogen: campylobacter. Trans ASABE 52 (2):651–662
Yoon SC, Park B, Lawrence KC, Windham WR, Heitschmidt GW (2011) Line-scan hyperspectral imaging system for real-time inspection of poultry carcasses with fecal material and ingesta. Comput Electron Agric 79:159–168
Yoon SC, Windham WR, Ladely S, Heitschmidt GW, Lawrence KC, Park B, Narang N, Cray W (2013) Hyperspectral imaging for differentiating colonies of non-O157 Shiga-toxin producing

Escherichia coli (STEC) serogroups on spread plates of pure cultures. J Near Infrared Spectrosc 21:81–95

Zhang X, Liu F, He Y, Li X (2012) Application of hyperspectral imaging and chemometric calibrations for variety discrimination of maize seeds. Sensors 12:17234–17246

Zhu B, Jiang L, Luo L, Tao Y (2007) Gabor feature-based apple quality inspection using kernel principal component analysis. J Food Eng 81(4):741–749

Chapter 5
Classification and Prediction Methods

James E. Burger and Aoife A. Gowen

Dedication
In the memory of Dr. James Burger
Jim Burger was a great friend, teacher and mentor. He brought life into hyperspectral imaging through his unyielding enthusiasm and positivity. Indeed, this paper would never have come to life without his encouragement and persistence. Jim was unique in that he looked at science through the eyes of a child—always full of wonder and joy. This was manifested in his hyperspectral research workshops, most notably "Hyperfest 2010 and 2012" and in the IASIM (International Association for Spectral Imaging) meetings he championed. His passing in September 2014 has left a huge gap, both in our hearts and in the wider hyperspectral imaging community.

5.1 Introduction

The primary goal of hyperspectral imaging (HSI) is to obtain quantitative or qualitative information from a scene or object in a non-destructive manner, based on its optical properties. Visual inspection of such data is limited, due to the large number (>100) of spectral channels and spatial locations (>10,000 pixels) in each hyperspectral image. Moreover, in the process control environment, where thousands of hyperspectral images are generated in minutes, visual inspection is

J.E. Burger (deceased)

A.A. Gowen
School of Biosystems Engineering, University College Dublin,
Belfield, Dublin 4, Ireland
e-mail: aoife.gowen@ucd.ie

B. Park, R. Lu (eds.), *Hyperspectral Imaging Technology in Food and Agriculture*,
Food Engineering Series, DOI 10.1007/978-1-4939-2836-1_5

impossible. Indeed, the reason for implementing HSI in such settings is to facilitate *automated* quality control. Thus the need for automatic techniques to extract meaningful information from HSI arises. This often means the derivation of very few summary values from the massive amount of data contained within individual hyperspectral images.

Luckily, due mainly to advances in digital signal/image processing techniques in the past 50 years, there now exists a huge variety of methods with which to mine data. These have been assimilated into, among other fields, the estimation of chemical properties from multivariate data, *chemometrics*, which forms the basis of multivariate image analysis (MIA). Hyperspectral images can be viewed as multivariate in two senses; each pixel is represented by multiple wavelength variables, equally each wavelength is represented by multiple intensity values (pixels). Therefore the tools of chemometrics and MIA are highly suitable for analysis of HSI data.

A selection of the most commonly used data analysis techniques applied to classification and prediction of HSI data is presented in Table 5.1. It is evident from this table that a multitude of techniques exist with which to tackle HSI data, whose variety seems to be ever expanding. This stems from the fact that there is no "best" method to deal with such data. Generally, problems nonlinear in nature benefit from the use of nonlinear techniques, such as support vector machines (SVMs), while linear techniques such as principal components analysis (PCA) are better suited to linear data. However, care must be taken to include sufficient independent datasets to avoid model overfitting. The fundamental theories behind many of the methods listed in Table 5.1, while outside of the scope of this chapter, can be found in various excellent books (Brown et al. 2009; Otto 1999).

Chemical quantification or classification is the basis of HCI. In this chapter, steps involved in the development of classification and prediction models from HCI data will be explained by examining results of their application to an example image. This sample image represents a familiar yet challenging real world example containing common objects—a raisin, a piece of wood, string, rice, a walnut and a paperclip (Fig. 5.1)—that we know and can easily recognize. This image is presented to illustrate and explore some of the basic concepts of MIA and the visual interpretation of subsequent results. These familiar objects illustrate some of the common challenges faced in HCI. Walnut and raisin have interesting textures with deep shadows, while wood and string are materials with similar compositions but with differing physical properties. The grains of rice exhibit heterogeneity—the embryo being clearly distinguished from the endosperm. To facilitate the isolation of the individual objects in this example image, all objects were placed on a background of black silicon carbide sandpaper, since this material is a very low absorber of NIR light. In addition to their differing chemical and physical makeup, all objects are non-flat and exhibit different morphologies; this results in shadow effects, non-uniform light distribution, and variable sample to detector distances. These effects all add variability to the spectra of each sample.

Table 5.1 A selection of signal processing techniques employed in hyperspectral image analysis

Classification/prediction method	Frequency[a]
Support vector machines (SVMs)	20
Principal components analysis (PCA)	8
Fuzzy clustering	5
Partial least squares	4
Artificial Neural Networks	4
Multiple Linear Regression	4
Band arithmetic	3
Spectral Angle Mapper	3
Random forest	2
Kernel PCA	2
Linear Discriminant Analysis	2
Spatial/spectral approach	2
Maximum Likelihood Classification	2
Combination of magnitude and shape parameters	2
Minimum Noise Fraction	2
Classification trees	1
Genetic Algorithms	1
Local manifold learning	1
Markov Random Fields	1
Tensor modeling	1
Independent Components Analysis	1
Random Fields	2
Supervised local tangent space alignment	1
Multiscale approach	1
k-Nearest Neighbors	1
Fast Fourier Transform	1
Canonical Transformation	1
Nearest component analysis	1
Swarm Intelligence/Wavelet analysis	1
Soft Independent Modelling of Class Analogy	1
Morphological profiles	1
Adaptive Cosine Estimator	1
Matched filtering	1
Regularized Maximum Likelihood Clustering	1
Bayesian learning with Gaussian processes	1

[a]Frequency estimated as number of times employed within the first 100 publications base on WOS search with title = (hyperspectral and (classification or prediction)), timespan = all years. Databases = SCI-EXPANDED, SSCI, A&HCI. Carried out on 03/02/2011

The example HCI hypercube used in this discussion has spatial dimensions of 600 × 318 pixels (each 100 × 100 μm in size) and 207 wavelength channels (962–1,643 nm), acquired with a BurgerMetrics HyperPro NIR imaging system (Martens and Dardenne 1998).

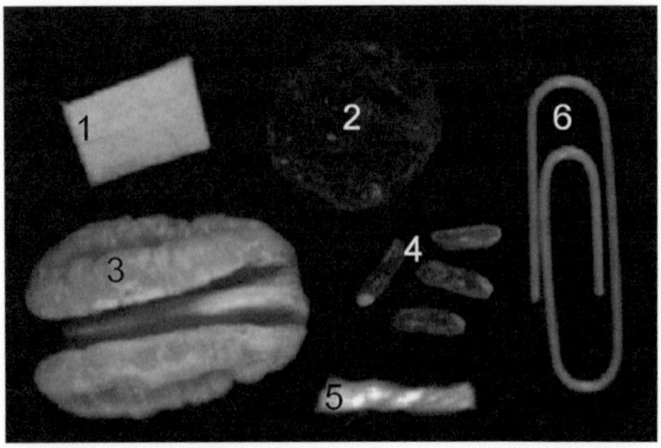

Fig. 5.1 Mean wavelength image of an example hypercube (obtained by averaging each pixel over the wavelength dimension), where *1* = wood, *2* = raisin, *3* = walnut, *4* = rice, *5* = string, *6* = paperclip. The image background can also be included as an image feature or class

5.2 Multivariate Image Analysis Applied to HSI

HSI hypercubes are three dimensional data structures (spatial × spatial × wavelength); however, most chemometric techniques are based on operations of two dimensional structures (matrices). Consequently, the three dimensional hypercube must first be rearranged prior to analysis. This is typically achieved by *unfolding* the three-dimensional hypercube into a two-dimensional matrix of spectra by stacking each pixel spectrum, one on top of the other, as shown in Fig. 5.2. This may be followed by a chemometric technique that favors two-dimensional structures, for example, PCA. The processing results, in this case score vectors, can then be *refolded* back to obtain spatial image representations, in this case score images. Such *unfolding* > *processing* > *refolding* operations are common throughout many HCI—chemometric operations.

5.2.1 *Principle Components Analysis*

PCA is a cornerstone of chemometrics. This unsupervised multivariate technique is the basis for many exploratory and prediction techniques. In NIR spectroscopy, some wavelength regions contain important information, while other regions are redundant and contribute primarily noise. One of the main features of PCA is that of data compression: high dimensional spectral data space is projected into a lower dimensional latent variable space. Effectively, signal is separated from noise. PCA bilinear decomposition provides a set of *principle components* (PCs) consisting of *loading* and *score* vectors. These PCs are ordered in such a way as to explain a sequentially decreasing amount of signal variance. What does all this mean? When the sample data set contains only tens or perhaps hundreds of spectra, this PCA compression may be difficult to understand. However this data compression is easily understood in a visual examination of score images obtained by applying PCA to HCI.

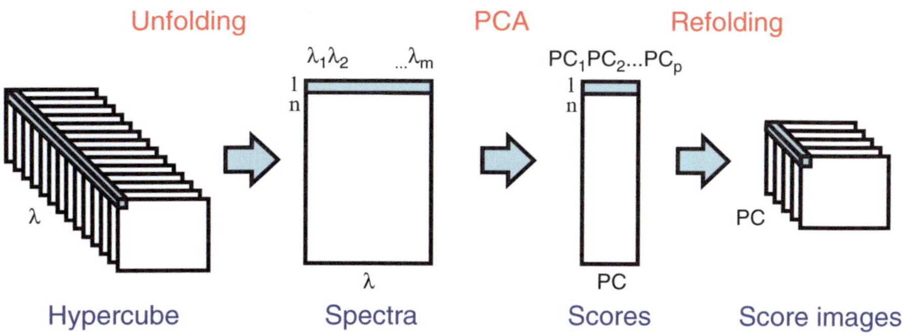

Fig. 5.2 Fundamental unfolding, processing, and refolding of HCI data

Our original HCI image contains 207 wavelength bands. Examination of the data at any individual wavelength band will undoubtedly yield a spatial image with a combination of identifiable features as well as background noise—every single wavelength band contains both signal and noise components. Moreover, contiguous wavelength band images in the hypercube are highly correlated. Our 207 wavelength bands equates to a 207 dimensional spectral space. Is there a way to separate the signal from noise? Are 207 dimensions required to explain the relevant information within the full hypercube? PCA enables a lower dimensional representation of the data by forming linear combinations of the original wavebands in directions of maximal variance, but how many dimensions are really required to represent our spectral data? Visual analysis of score images helps us to find the answers to these questions.

Figure 5.3 shows PCA score images of our example image corresponding to latent variables 1–12. The PC-1 projection axis represents the maximum variation in data signal, and consequently score image 1 shows the maximum contrast between pixel intensities in the image. We can see that physical variation due to the non-flat surfaces of the samples contributes greatly to the variance described by PC-1. For instance, the central region of the walnut is a similar gray level intensity to the raisin and rice, while the edges of the walnut have a similar grey level to the wood. Rather than contrasting between the different substances, this score image gives us contrast based on the physical texture of the samples. This demonstrates that PCA finds the directions of maximal variance in the data, regardless of their source. This may imply a need for spectral pretreatments such as standard normal variate or derivative transforms to reduce the influence of light scattering effects brought on by the physical properties of these kinds of samples. Score image 2 represents the HCI data projected to the PC-2 axis orthogonal to PC-1. In this image, the paperclip and the cellulose based objects (string and wood) are distinguishable from the food products (raisin, walnut and rice). Score image 3 distinguishes the walnut and paperclip from the other objects, while score image 4 distinguishes the paperclip from the other objects. One of the predominant questions of PCA is how many PCs to retain in order to fully explain the data space. Paging through sequential score images can contribute to a further

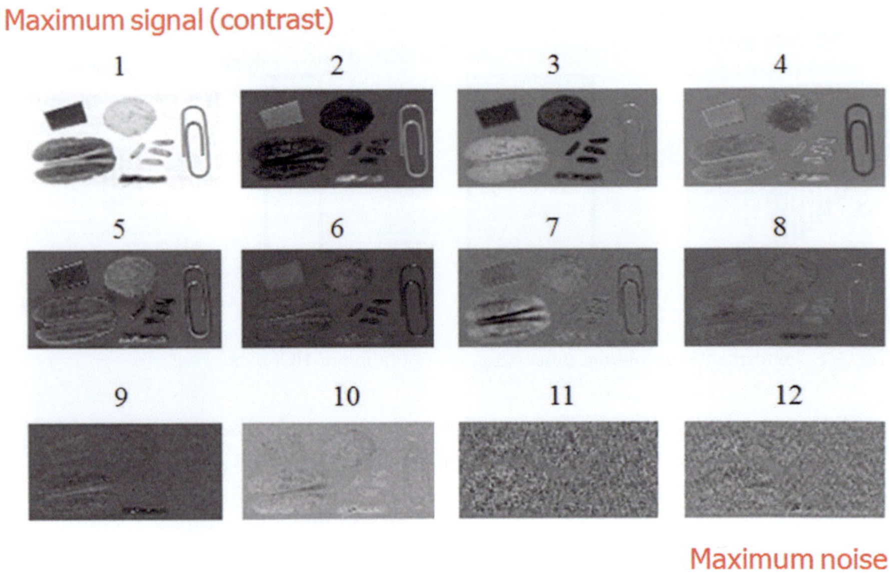

Fig. 5.3 Principal component (PC) score images for test hypercube. Beyond 11 PCs the contribution is largely noise

understanding of data space dimensionality. This is a distinct example of the *image advantage* of HCI. As the number of the PC increases, noise becomes more prevalent in the images. For example, the PC-11 score image appears extremely noisy, containing very little signal from the object features. PCs and score images beyond this essentially contain only noise, and can be excluded from further analysis. In such an image sequence we can clearly visualize the effect of PCA data compression.

It should also be noted that, in this example, the gray levels of the displayed score images have been autoscaled, i.e., the minimum gray level in the image corresponds to the minimum score value and likewise for the maximum. In some applications, PCs may not be monotonic in terms of information content. Significant noise present in the data can often contribute to a large portion of the variance manifesting in major PCs, while smaller contributions to variance from objects in an image may not appear until later PCs. In this case, autoscaling the score images is especially beneficial and may help to expose these minor object components to visual inspection. It is equally important, however, that autoscaling not be driven by extreme outlier values. Any extreme outliers in the score space should be removed before autoscaling, otherwise their inclusion would expand the scale of the image and thereby suppress finer details. It is also important to consider the eigenvalue corresponding to each score image. Image details apparent in low eigenvalue images may actually correspond to primarily only noise. If the score images are scaled by their respective eigenvalues, the sequential decrease in signal content is even more visually apparent.

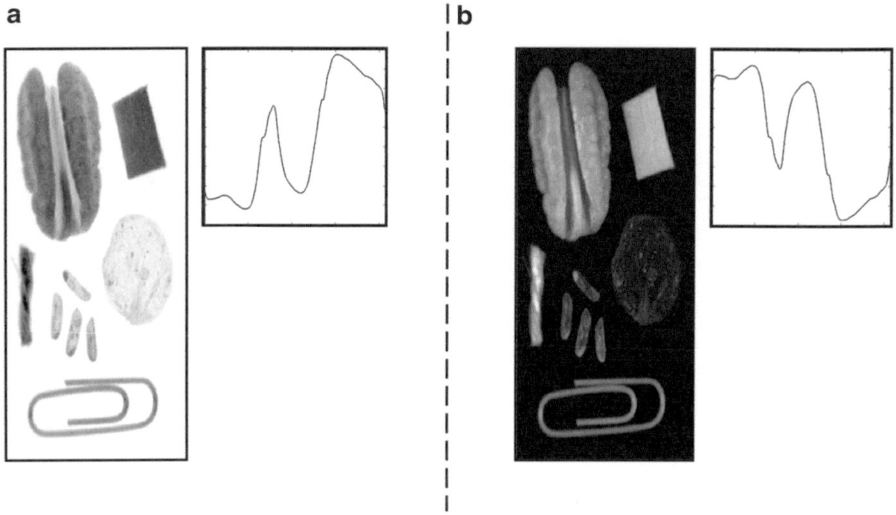

Fig. 5.4 Sign ambiguity of loading vectors and consequently of score images. (**a**) Original score image and loading vector and (**b**) inverse of image and loading vector shown in (**a**)

Another important point to note is that the sign of PCA loading vectors is ambiguous. This means that the score images shown for example in Fig. 5.4a, b are equally relevant. Even though they are simply the inverse of each other, it may be possible to make distinctly different visual interpretations of the two inverted score images. For example, the raisin appears to have greater detail in Fig. 5.4b than a, while the opposite is true for the piece of string. It is therefore important to always consider this inverse of a score image as a source of additional information.

Although scrolling though individual score image is instructive, there are other, more intuitive ways in which to visualize PCA results. Some examples are given in Fig. 5.5: false color red-green-blue ('RGB') mapping; scatter plots of all pixels in 2-dimensions and 3-dimensional scatter plot showing scores of spectral subsets of each object. The false color 'RGB' mapping is obtained by concatenating the autoscaled PC score 1, 2 and 3 images, so that each image pixel has an R, G and B channel intensity corresponding to scores 1, 2 and 3. This image shows clear 'color' distinction between some objects, but also some similarity problems between objects of different composition (e.g., wood and string, raisin and rice). The 2D scatter plot of PC1 and PC2 scores from each pixel of the entire image is colorized to represent cloud density (i.e., the number of pixels with overlapping score 1 and score 2 values). With this approach, clusters corresponding to individual objects can be seen: the main clusters are identified as wood, walnut, raisin, paperclip and background, as shown in Fig. 5.5. In addition, a random selection of 500 spectra (mean centered) from the six featured classes is plotted in 3D score space. The colors of classes within the various data-clouds show both separation and overlap of classes, as well as differences in within class variance. In particular, the string class (shown in magenta) has huge variance, most likely due to regions of specular

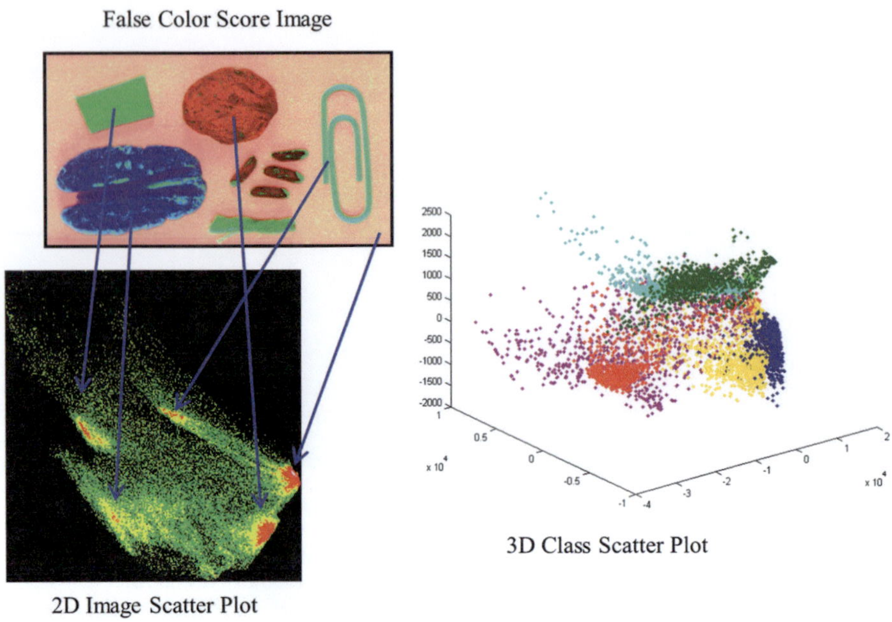

Fig. 5.5 Alternate strategies for visualizing PC scores: false color 'RGB' mapping is obtained by concatenating the auto-scaled score 1, 2 and 3 images; 2D scatter plot of PC1 and PC2 scores from each pixel of the entire image and 3D scatter plot of a random selection of 500 spectra (mean centered) from the six featured classes (*dark green* = walnut, *blue* = raisin, *magenta* = string, *cyan* = paperclip, *red* = wood, *yellow* = rice)

reflection which can be clearly seen in Fig. 5.1. Although hypercubes contain a massive quantity of data, this assortment of viewing tools coupled with PCA should always be considered to fully explore the basic variance structure of sample data.

5.2.2 Partial Least Squares: Discriminant Analysis

Classification of hyperspectral images can be achieved by identifying regions or objects of similar characteristics using the spectral and spatial information contained in the hypercube. In contrast to unsupervised methods such as PCA, supervised classification methods require the selection and labeling of representative calibration and training sets for classifier optimization. One of the major advantages of HSI in this respect is the sheer volume of data available in each hypercube with which to create robust calibration and training sets. Since PLS is one of the most widely used chemometric methods in HSI quantification and classification, we will present the example of PLS discriminant analysis (PLS-DA) here to demonstrate the use of HCI in interpretation of a supervised chemometric method (and vice versa). Although alternate chemometric methods may differ in the 'nuts and bolts' of how

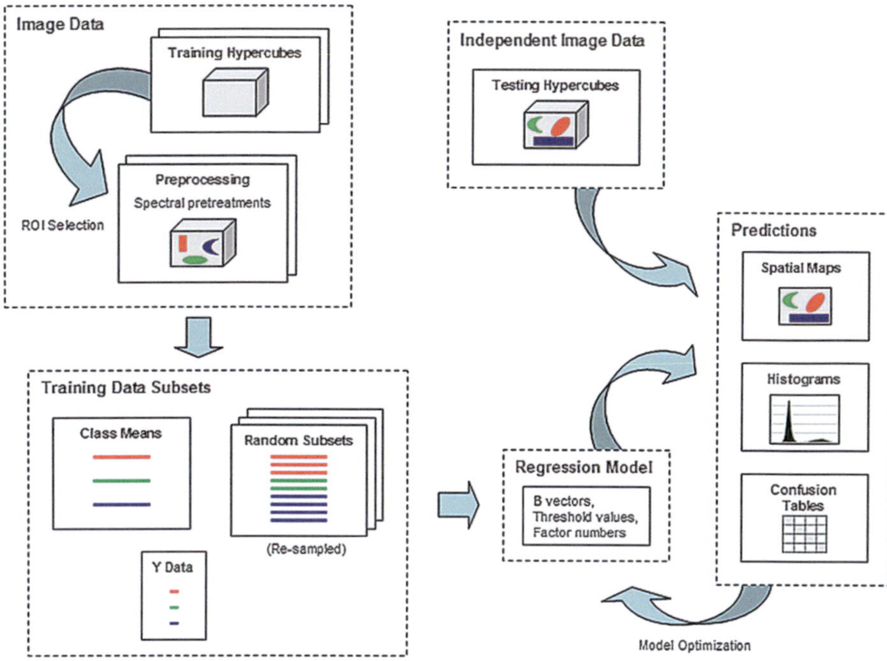

Fig. 5.6 Fundamental steps involved in developing classification models from hyperspectral chemical image (HCI) data

they treat the input data, their application to HSI data can be roughly generalized in a consistent manner. The development of the classification or prediction model generally begins with data pretreatment (spatial or spectral) followed by selection of regions of interest (ROIs). A data reduction step is then performed to create training set data, followed by calibration model development. Evaluation of independent test set is essential for model validation.

The typical steps in the building of a supervised classification model are shown in Fig. 5.6. The beginning phase of this modeling process provides the greatest opportunity for user input to impact the performance of the resulting model. First, spectral preprocessing options such as first or second derivatives, detrending, or normalizations, must be chosen to maximize intra-class variations while minimizing inter-class variations and the effects of systematic variances due to, for example, instrumentation instabilities. Such transformations are generally application specific and must simply be tested to explore which are most appropriate in any given situation. The second and probably most critical step of model construction involves the selection of proper spectra from the hyperspectral imaging data to adequately represent each class of interest. ROIs must be selected which capture as much as possible the variance of each desired class. This is especially important when a desired class contains spectral variations due to physical conditions such as sample height or shadow effects imposed by surface textures. ROIs may be selected from just one hyperspectral image, if all variations of all classes of interest are present in that image; however, it is preferable to select spectra from a number of hypercubes in order to include in the model additional potential sources of variability arising from images taken at different times (e.g., spectral differences arising from changes in the detector response or sample preparation and presentation methodology). This ROI selection process may result in spectral datasets containing thousands of spectra. One of the advantages of hyperspectral imaging is the availability of large collections of spectra which can be split into training and testing or model validation subsets. Further data reduction of model training sets can be achieved by computing either mean spectra, or by randomly selecting smaller subsets of spectra for each class. The implications of these approaches will be discussed further in following sections of this chapter.

For classification purposes a categorical variable must be maintained which is a vector of the same length as the spectral data matrix, containing information on the class that each spectrum belongs to. A classifier model must be computed for each individual class. A 'Y-block' reference vector is created for each class, typically containing values of one or zero, indicating whether each spectrum is or is not a class member. Once a suitable classifier has been trained it can be applied to the entire unfolded hypercube (or for classification of new hypercubes) providing a classification prediction for every spectrum. Because of the immense quantity of hyperspectral data, alternative representations of results should be explored:

1. Predictions may be refolded, resulting in a spatial image known as a prediction map. Typically each class is assigned a unique color or grayscale value for identification of classified pixels in the prediction image map.
2. Histograms can be created for examining the distribution of prediction values for an individual class.
3. Classification confusion tables can be created which summarize the counts of classified and mis-classified data. Interactive software which provides simultaneous displays of these results may lead to fully optimized classification models. For example, implications from modifications to the class threshold value, a limit which defines the boundary for class membership, can be immediately explored by examination of the spatial maps and confusion tables. These ideas are presented in the following case study using the BurgerMetrics *HyperSee*™ image analysis software.

5.2.3 Case Study Description

As a case study of the development of PLS-DA models from HCI data, we demonstrate the prediction of seven classes in the example image. As previously mentioned, the first step in developing a classification model is the selection of data to be included in the calibration set. We will present a variety of methods for doing this, and compare them in terms of the visual quality of prediction maps and classification accuracy of the resultant models. The selection methods we present

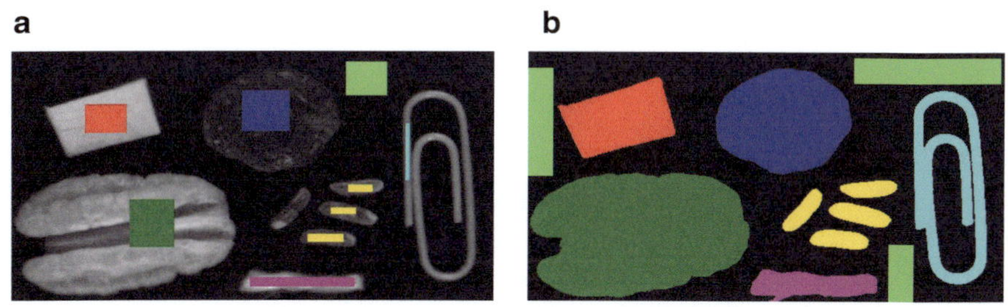

c

Number of Pixels	Wood	Raisin	Walnut	Rice	String	Paperclip	Background
Rectangle	1715	2800	3074	958	1500	350	2050
Full Object	8534	16038	31715	4848	3830	8557	13838

Fig. 5.7 Selecting regions of interest (ROIs) for development of classification models. (**a**) rectangle ROI, (**b**) full object ROI, and (**c**) number of pixels of each class in each ROI

can be broadly partitioned into two groups: in the first group, calibration spectra were selected from a rectangular ROI on each object, shown in Fig. 5.7a; in the second group, spectra were selected from spatial regions representing full objects (with the exception of the background class, where three ROIs were selected in order to keep the number of pixel spectra in each class comparable), shown in Fig. 5.7b). The ROIs were selected interactively using *HyperSee*™ software.

The various modeling strategies explored in this case study are listed in Table 5.2. The three sampling strategies employed were as follows:

1. *Mean spectra*: within each ROI, 200 pixel spectra were randomly selected. The mean of these 200 pixel spectra, matched to the categorical variable describing its class, was used for model building (cases 1 and 5 in Table 5.2)
2. *Pixel spectra*: within each ROI, 200 pixel spectra were randomly selected. Each was matched to the categorical variable corresponding to its class for model building (cases 2, 3, 6 and 7 in Table 5.2)
3. *Resampling*: within each ROI, 200 pixel spectra were randomly selected. Each was matched to the same categorical variable for model building. This random selection process was repeated 50 times (cases 4 and 8–13 in Table 5.2). The regression vectors from the 50 resulting calibration models were then averaged

It is often necessary to apply spectral pretreatments to HCI data. There is a wide variety to choose from, ranging from baseline to multiplicative scatter correction; however, in this case study we will consider only two: standard normal variate (SNV, case 12 in Table 5.2) and first derivative Savitsky Golay pretreatment (case 13 in Table 5.2).

As previously mentioned, we wish to build a PLS-DA model for each class present in the image. In order to do this for a given class, the calibration spectra representing that class are assigned a categorical value of 1. The remaining 'non-class' spectra in the calibration set are assigned a categorical value of 0. In the

Table 5.2 Training set details (in all cases, 200 spectra were randomly selected from each class ROI)

Case #	Mean	ROI	Threshold	Re-sample	Pre-process
1	Y	Rectangle	0.5	1	
2[a]		Rectangle	0.5	1	
3		Rectangle	Auto	1	
4		Rectangle	Auto	50	
5	Y	Full object	0.5	1	
6[b]		Full object	0.5	1	
7		Full object	Auto	1	
8		Full object	Auto	50	
9[c]		Full object	Auto	50	
10[d]		Full object	Conservative	50	
11[d]		Full object	Liberal	50	
12		Full object	Auto	50	SNV
13		Full object	Auto	50	FirstD

[a]Based on case 3 model, thresholds manually set to 0.5
[b]Based on case 7 model, thresholds manually set to 0.5
[c]Case 9 is a repeat of case 8
[d]Based on case 9 model, thresholds manually adjusted

development of a PLS-DA model it is first necessary to build a PLS regression (PLSR) model to predict the categorical variable. The number of latent variables to include in the classification model must be determined. This is by no means a trivial task—the inclusion of too many latent variables can lead to poor predictive performance on independent test sets due to overfitting. Numerous methods are available to aid in the selection of the number of latent variables in a PLSR model (Martens and Dardenne 1998; Wiklund et al. 2007; Gowen et al. 2011); however, it is outside of the scope of the current chapter to review them. In the presented case study, for comparative purposes the optimum number of latent variables for each model was systematically estimated as the number after which the % explained variance in the categorical variable became less than one.

After selecting the number of latent variables to include, the predicted class values for each sample may be investigated. Since PLSR is a regression method, the predicted class values are distributions of values around 0 and 1 rather than exact class values. In HCI, it is useful to visualize these predicted values as a histogram, as shown in Fig. 5.8. In order to step from PLSR to PLS-DA, it is necessary to select a threshold, such that samples with predicted values greater than the threshold will be classified as 'in class' or 1, and vice versa. In this case study, four threshold selection approaches were investigated:

1. *'0.5'* (cases 1, 2, 5 and 6 in Table 5.2): the threshold was set to 0.5. This threshold value is commonly selected, since it lies in the middle of the interval [0, 1].
2. *'Auto'* (cases 3, 4, 7, 8, 9, 12, 13 in Table 5.2): the threshold was set by an automatic procedure in *HyperSee*™ software, based on the means and

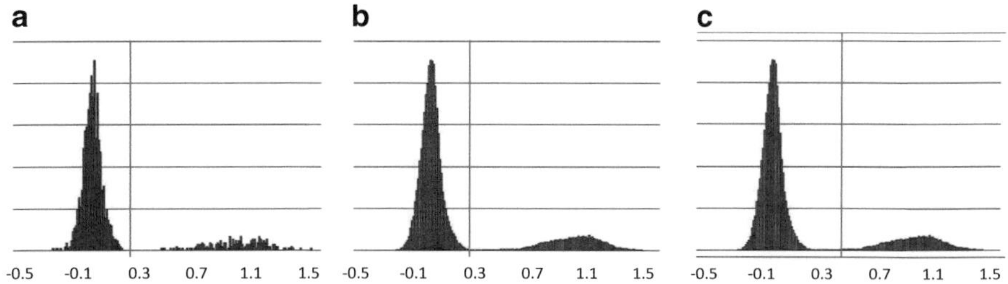

d

	Wood	Raisin	Walnut	Rice	String	Paperclip	Bkgnd	Total
Wood	99.50	0.00	0.00	0.50	0.00	0.00	0.00	100.00
Raisin	0.00	99.50	0.00	0.50	0.00	0.00	0.00	100.00
Walnut	0.00	0.00	100.00	0.00	0.00	0.00	0.00	100.00
Rice	0.50	0.00	0.00	94.00	2.50	2.50	0.00	99.50
String	3.50	0.00	0.00	1.00	94.50	1.00	0.00	100.00
Paperclip	0.50	0.00	0.00	1.00	0.00	98.50	0.00	100.00
Bkgnd	0.00	0.00	0.00	1.50	0.50	0.00	98.00	100.00
Total	104.00	99.50	100.00	98.50	97.50	102.00	98.00	99.93

Fig. 5.8 Setting thresholds in PLS-DA models of HCI data. The histograms shown here are for the raisin class and correspond to cases 9–11. (**a**) Histogram for calibration training set with automatic threshold displayed at 0.27, (**b**) histogram for test set with automatic threshold displayed at 0.27, (**c**) threshold manually adjusted to 0.46, and (**d**) confusion table for prediction of each class

standard deviations of the 'class' and 'non-class' samples. An example threshold estimated using this approach is shown in Fig. 5.8a.

3. *'Conservative'* (case 10 in Table 5.2): when the histogram of the test set is examined (Fig. 5.8b), the automatically selected threshold appears on the edge of the 'non class' distribution. Visually, a more satisfying threshold would lie somewhere in the valley between the two distributions. This is what we call the conservative approach. The threshold for each class model is selected in an interactive way by visualization of the histogram of predicted values (Fig. 5.8c).
4. *'Liberal'* (case 11): this method of threshold selection is highly interactive: it involves selecting the threshold by simultaneously considering the histogram and the prediction accuracy, viewing the display of the spatial class prediction map, and by selecting the threshold that maximizes the diagonal elements but minimizes the off diagonal elements of the confusion matrix (Fig. 5.8d).

5.2.4 Comparison of Prediction Model Results

The prediction maps for cases 1–8 (described in Table 5.2) are shown in Fig. 5.9a. The advantage of using pixel spectra (cases 2–4 and 6–8) over mean spectra (cases 1 and 5) is evident: the prediction maps arising from the models built with pixel spectra have far fewer misclassifications than those built with mean spectra. Comparing the models built with spectra selected from square ROIs (cases 1–4)

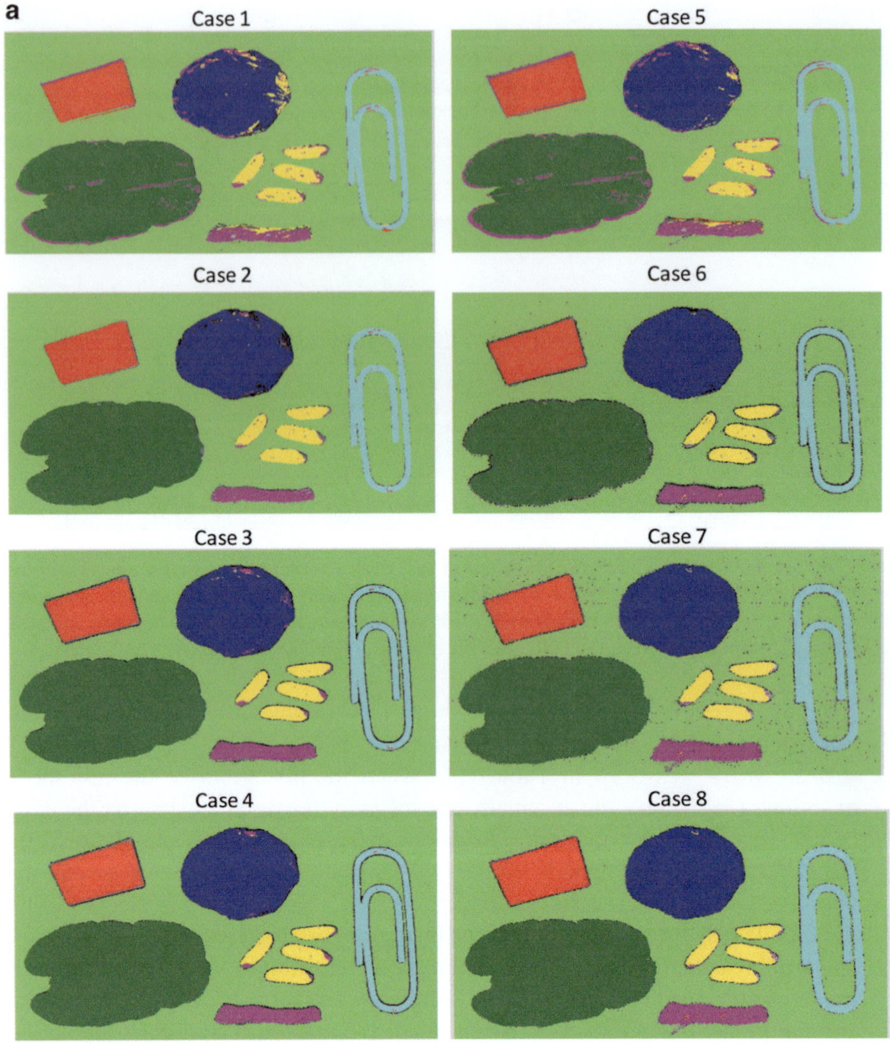

Fig. 5.9 Prediction maps for models built according to strategies described in Table 5.2

with those selected from full object ROIs (cases 5–8), there is obviously less misclassification when the full object ROIs are used. This is particularly evident for the raisin object, regions of which are misclassified as string or rice in a number of the prediction maps. How can we account for these differences? Imagine viewing the class spectra in a multi-dimensional spectral space scatter plot, similar to Fig. 5.5c. Basing a classification model on mean spectra from each class utilizes only the differences in distance between class centers. When variations within a class are included, improved discriminating class boundaries can be found which account for the variation in size and shape of each class. Additionally, when an

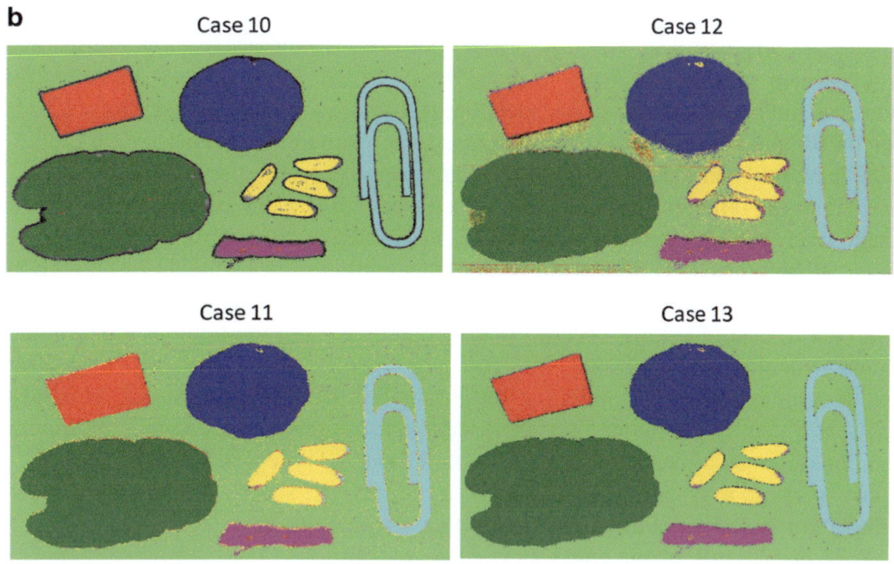

Fig. 5.9 (continued)

individual class contains excessive variations (e.g., peripheral sample shadow effects) it is important to include the full object, not just a subset rectangular ROI, to adequately define the complete class region.

There appears to be fewer misclassified pixels when the automatic thresholding method was applied (cases 3 and 7), as compared to setting the threshold to 0.5 (cases 2 and 6), although this advantage is more evident for the square ROIs (cases 3 and 2) than for the full object ROIs (cases 7 and 6). Some misclassification of background pixels becomes evident upon the use of automatic thresholding (case 7). There is a slight visible improvement upon the use of resampling for the models based on square ROIs (compare cases 3 and 4), in terms of fewer misclassifications on the raisin object, and fewer misclassified background pixels for the models based on full image ROIs (compare cases 7 and 8). The misclassification of edge regions of the rice grains persists throughout all prediction images. In this example case image, the effect of resampling is diminished by the fact that 200 randomly selected spectra were used to define the variation in each class. Resampling these 200 spectra provides little change in the variation of each class. Resampling is more effective when the availability of representative class spectra is reduced. Case 9 is simply a repeat of case 8 to check the repeatability of the resampling method. In this case the number of factors for each class model, the automatic threshold values, and the training and test set prediction counts were nearly identical to case 8 values. The random resampling method appears to create very repeatable robust models.

Comparing conservative thresholding (case 10) with automatic thresholding (case 8), it seems that by setting the threshold conservatively, there is a problem with edge classification, with most of the object edges being unclassified. Liberal thresholding (case 11) gives a better result, with a prediction map comparable to that obtained using the automatic threshold; however, a number of background pixels are misclassified. The repositioning of class thresholds is a very subjective process and must be carefully reconsidered with any application. Results should also be confirmed with truly independent test images, which for brevity, was not included here. The SNV pretreatment (case 12) resulted in a noisy looking prediction map; for example, many pixels in the background were misclassified. This may be due to the inflation of noise in the spectra upon using this pretreatment. Such a high level of noise was not as apparent on the application of the first derivative pretreatment (case 13). The application of this pretreatment resulted in a prediction map similar to that for resampled raw spectra selected from the full images (case 8). As with threshold values, spectral pretreatments are also application specific and must be applied with caution.

In conjunction with examining the prediction maps for each model, it is also useful to quantify the model performance by calculating the percentage of correct and mis-classifications associated with each model. These can be plotted for each class or on an average basis, as shown in Fig. 5.10. It should be noted that these charts are based on combined training and test set statistics, which represents a subset of the full image prediction maps of Fig. 5.9. The summary statistics displayed in Fig. 5.10 are very informative: immediately we can pick out the worst performing strategies—cases 1 and 5—which were based on mean spectra. This is in agreement with the visual inspection of full image prediction maps arising from these models. The advantage of using the automatic thresholding method instead of using a threshold of 0.5 is confirmed by comparing cases 6 and 7. The advantage of resampling was less clear; the percentage correct classification for models built using resampled spectra (cases 4 and 8) and those built without resampling (cases 3 and 7) were not obviously higher; however the % false positives was slightly lower for the resampled data. This is in agreement with visual analysis of the prediction maps. The conservative thresholding approach (case 10) resulted in the lowest % false positives; however, this came at the expense of lower % correctly classified spectra. The liberal thresholding (case 11) approach maintained a high % correct classification, but this came at the cost of a higher % false positives. The manual selection of thresholds used can be optimized according to application specific needs of total correct and false positive classifications. Figures 5.9 and 5.10 appear to show conflicting results for cases 12 and 13, i.e., prediction maps (Fig. 5.9b) for case 12 appear to contain more mis-classifications, whereas the summary statistics (Fig. 5.10) for case 12 indicated a slightly higher correct classification and lower false positive classification percentage than case 13. This can be explained because the summary statistics are based on only test plus training areas, which do not include for example, the highly mis-classified background regions seen in Fig. 5.9b.

In addition to examining the numbers of positive predictions or false positive predictions, sometimes it can be beneficial to create maps identifying the class members **not** correctly classified. Figure 5.11 presents these negative prediction maps for some of the processing cases. It is apparent that the majority of pixels not classified are along the edges of the featured objects, where dark shadows make

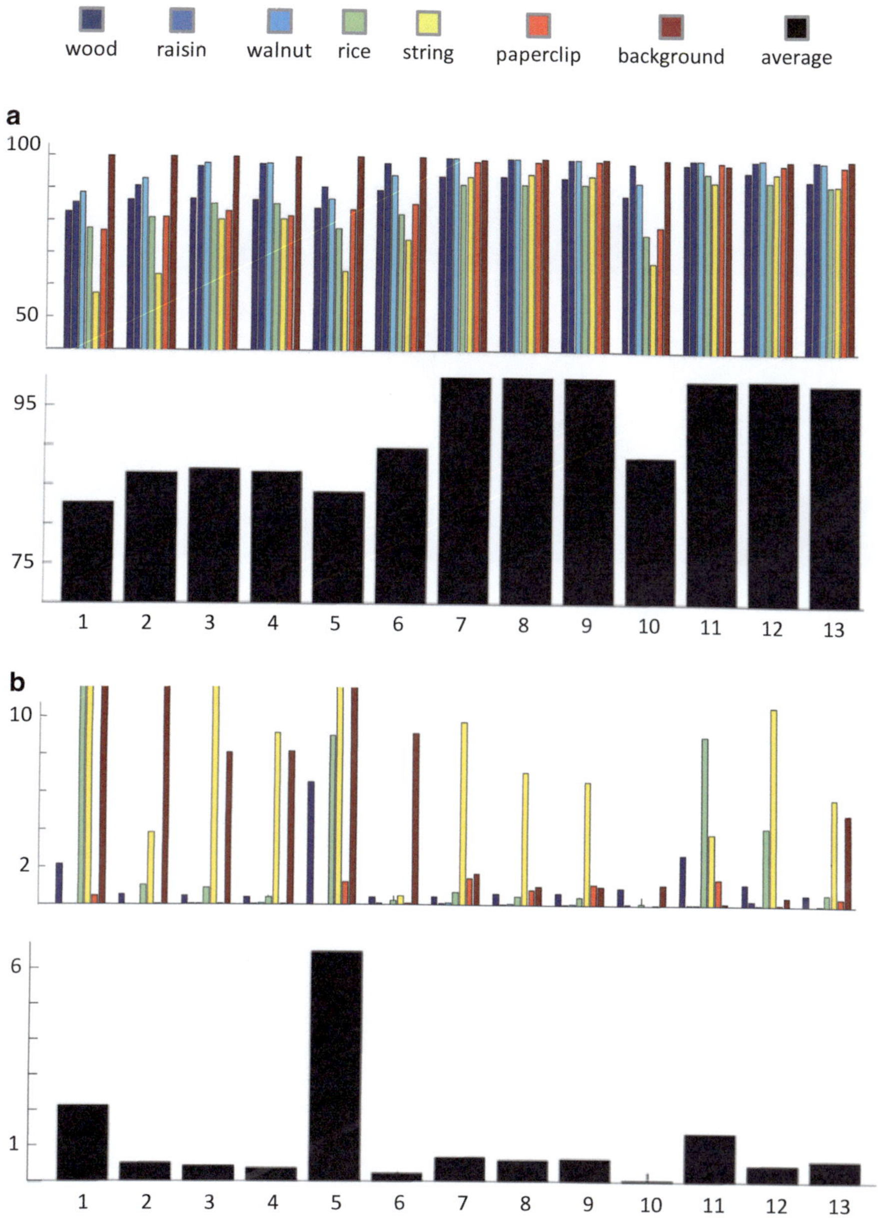

Fig. 5.10 (**a**) % Correct classification for each class/modeling method and (**b**) % false positives for each class/modeling method (see Table 5.2) (Average classification performance over different samples for each method is shown in *black* and *white* bar chart)

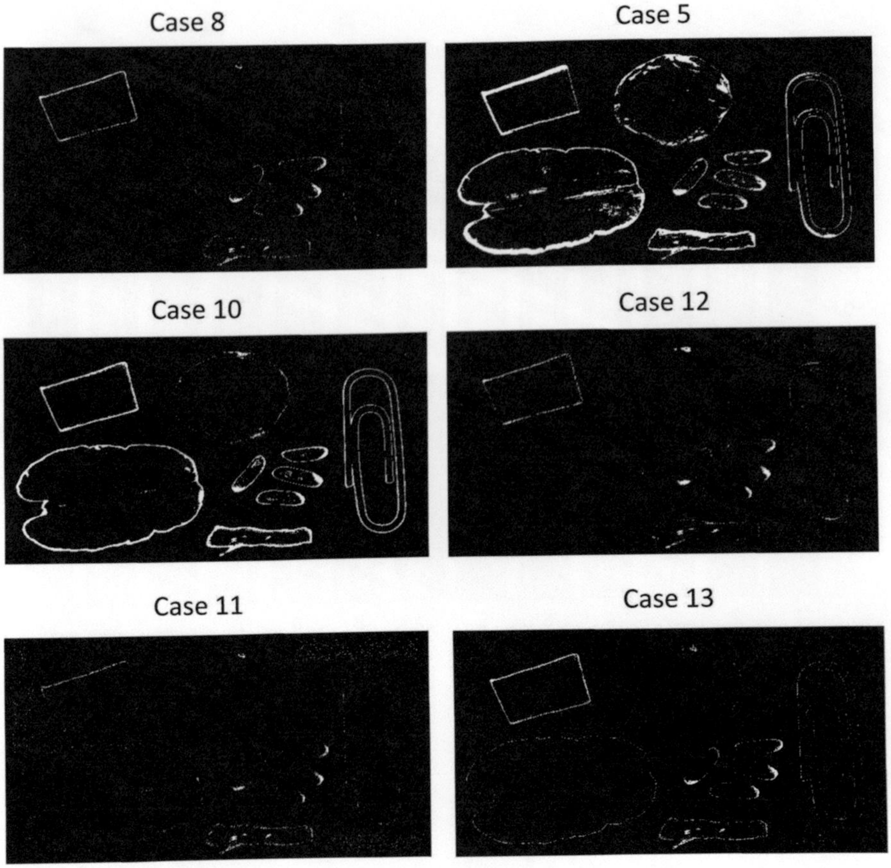

Fig. 5.11 Identification of training and test pixels which are **not** detected can also contribute to model optimizations

spectral measurements difficult. If these pixels are less important, then manually adjusting the threshold levels may be advantageous. The three cases displayed in the left column reflect this: the case 8 map shows the default 'automated' results. Increasing the threshold limits to minimize false positives (case 10) increased the non-classified edge pixels significantly. Whereas decreasing the thresholds enabled better classification of the wood and rice edge pixels, but increased the non-classification of string edge and background pixels. The case 5 map indicates the areas of the objects not classified because their variance was not captured properly by the mean spectra used for model building. The cases 12 and 13 maps provide additional information regarding the effect of spectral preprocessing on each specific class; here we can see that the SNV pretreatment (case 12) resulted in a lower number of incorrectly-classified pixels than the first derivative pretreatment (case 13).

5.3 Incorporating Spatial Information in Model Development

The methods for HSI classification discussed above take only information in the spectral domain into account, discarding the extra information available in the spatial domain. This is mainly due to the fact that these methods were developed for conventional NIR spectral data where spatial information is absent. Such methods do not exploit the fact that the spatial domain is also useful for model evaluation and development. Recently, a number of novel HSI classification methods have been presented that take the spatial information into account. One category of such approaches extends existing clustering methods by including spatial information in the feature vector (Camps-Valls et al. 2006; Noordam and van den Broek 2002). Another category of approaches considers the spectral and spatial domains in succession. A combined spectral-spatial method, applied to multispectral remote sensing data was presented by Marçal and Castro (2005). This method takes, as an input, a classification map obtained from the original multispectral image using an unsupervised classifier operating on the spectral domain. Hierarchical clustering is then applied to the classification map based on an aggregation index that simultaneously characterizes each class by its mean spectrum, size (relative to the image) and two spatial indices (boundary and compactness). The method is computationally efficient and has been demonstrated to work well on a test image. However, it suffers from the indeterminacy of tuning coefficients that weigh each element of the aggregation index. Moreover, the required prior classification step can result in different class maps depending on the choice of unsupervised classifier.

More recently, Gorretta et al. (2009) proposed an elegantly titled "butterfly approach" which combines topological concepts and chemometric analysis, using an iterative cross analysis of data in spectral and spatial domains. As a first step, spatial structures in an image are defined based on spectral structure; the second step is to extract a spectral structure based on spatial structure. PCA is employed for extracting spectral structures while spatial structures are extracted by applying region segmentation algorithms to the PC score images. This approach demonstrated promising results when applied to remote sensing hyperspectral images.

Another approach, proposed by Tarabalka et al. (2009), is based on partitional clustering. This method implies the joint use of an unsupervised classifier based on spectral data (e.g., ISODATA or expectation maximization) combined with a pixel-wise classification by SVM. The results from both methods are overlaid in the spatial domain and a maximum vote rule applied to classify pixels. An additional post-processing step is then applied that uses spatial information obtained from the classified image by assigning each pixel to the class *most represented* by its neighbors. The definition of *most represented* rests on the number of neighboring pixels required to have that class. Although the authors show that this method improved classification in a number of remotely sensed hyperspectral images, they also caution that such spatial post-processing might lead to the incorrect classification of pixels if a certain class is only present in one or a small number of disjoint pixels. This problem is potentially severe in remote sensing images where one pixel can represent a few square meters area, although it has its analogue in higher resolution imaging should the target of interest be small enough to span only one pixel. Therefore any prior knowledge on the expected size of object regions compared with pixel size should be incorporated into the analysis.

To demonstrate how spatial information might be incorporated in a classification model in a simple way, we present a spatial postprocessing of case 12 in our case study (Fig. 5.9b). Based on visual observation of the prediction maps, case 12 was a poor model, with numerous background pixels being misclassified. It is possible to improve the appearance of this classification map by spatial filtering. This consists of choosing a window around each pixel (e.g., a 3×3 window around a central pixel), and replacing each pixel in that window by the median or mode of the pixel values in the window. In Fig. 5.12 we compare the effects of changing the window size and the difference between using the median or mode value of the window pixels for filtering. It can be seen that these kinds of spatial operation are useful for the removal of isolated misclassified pixels. In this case, the mode filter was more effective than the median filter; for example, the median filter retained misclassified pixels on the edge of the wood sample, whereas the mode filter did not. In addition, the effect of increasing window size is apparent: this improved the classification of interior parts of the samples. However, this occurred at the expense of some 'smearing' where separate objects, e.g., rice grains, began to merge. Effects such as these depend greatly on the type of sample to be imaged, and should be taken into consideration when choosing a window size for spatial post processing.

5.4 Summary

In this chapter we have presented some specific hyperspectral image processing tools which allow extraction of information, focusing on data exploration (PCA) and classification (PLS-DA) of common objects identifiable in a representative image. We have demonstrated how various graphical data representations such as histograms and color coded prediction maps can be used to further the understanding of multivariate model building. In particular we have shown how interactive software tools can be used to assist with the optimization of model performance. Training and test data set selection is essential to capture the full variance of individual classes. The effects of model parameters such as the number of latent variables, classification thresholds, and spectral preprocessing treatments can be spatially visualized for comparison purposes. Additional prediction improvements may be achieved by applying spatial filters to prediction maps.

The selection of spectral pretreatment, calibration technique and model parameters (e.g., number of latent variables, class boundary thresholds) is generally

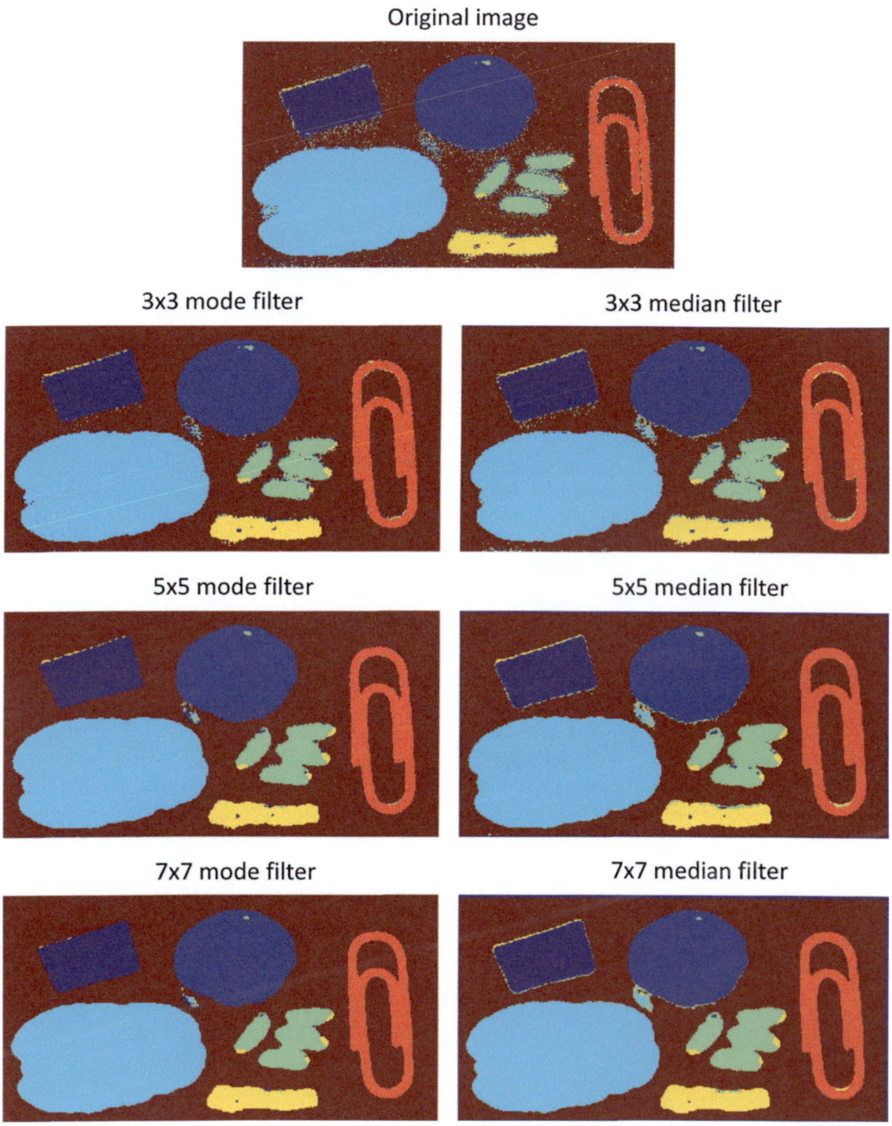

Fig. 5.12 Spatial filtering of case 12 prediction image using median and mode filters of different window sizes

application dependent. However, the graphical tools presented here: confusion matrices, histogram distributions and color coded maps of both positive and negative predictions, can always be used for model comparison, regardless of the modeling technique applied. In all cases, however, the results of model prediction should be validated with truly independent image data.

Acknowledgement The first author acknowledges funding from the EU FP7 under the European Research Council Starting Grant programme.

References

Brown SD, Tauler R, Walczak B (2009) Comprehensive chemometrics: chemical and biochemical data analysis. Elsevier B.V., Amsterdam. ISBN 978-0-444-52701-1

Camps-Valls G, Gomez-Chova L, Munoz-Mari J, Vila-Frances J, Calpe-Maravilla J (2006) Composite kernels for hyperspectral image classification. IEEE Geosci Remote Sens Lett 3(1):93–97

Gorretta N, Roger JM, Rabatel G, Bellon-Maurel V, Fiorio C, Lelong C (2009) Hyperspectral image segmentation: the butterfly approach. In: Hyperspectral image and signal processing: Evolution in remote sensing, 2009. WHISPERS '09. First Workshop on, 1(1):1,4. doi: 10.1109/WHISPERS.2009.5289062

Gowen AA, Downey G, Esquerre C, O'Donnell CP (2011) Preventing over-fitting in PLS calibration models of near-infrared (NIR) spectroscopy data using regression coefficients. J Chemometr 25(7):375–381. doi:10.1002/cem.1349

Marcal A, Castro L (2005) Hierarchical clustering of multispectral images using combined spectral and spatial criteria. IEEE Geosci Remote Sens Lett 2(1):59–63

Martens H, Dardenne P (1998) Validation and verification of regression in small data sets. Chemom Intell Lab Syst 44:99–121

Noordam JC, van den Broek WHAM (2002) Multivariate image segmentation based on geometrically guided fuzzy c-means clustering. J Chemometr 16:1–11

Otto M (1999) Chemometrics statistics and computer application in analytical chemistry. Wiley-VCH, Weinheim. ISBN 3-527-29628-X

Tarabalka Y, Benediktsson JA, Chanussot J (2009) Spectral-spatial classification of hyperspectral imagery based on partitional clustering techniques. IEEE Trans Geosci Remote Sens 47(8):2973–2987

Wiklund S, Nilsson D, Eriksson L, Sjöström M, Wold S, Faber K (2007) A randomisation test for PLS component selection. J Chemometr 21:427–439

Part II
Applications

Chapter 6
Safety Inspection of Plant Products

Haibo Yao, Zuzana Hruska, Robert L. Brown, Deepak Bhatnagar, and Thomas E. Cleveland

6.1 Introduction

Outbreaks of food-borne illness create major headlines. Some recent examples (Food Safety News 2011) are alfalfa and spicy sprouts contaminated with *Salmonella enteritidis* and cantaloupe contaminated with *Listeria monocytogenes* in the US, and the outbreak of *E. coli* O104:H4 in Germany due to contaminated fenugreek seeds. For consumers, the demand for safe food is paramount. The food industry and its associated research communities are always challenged to meet the safety requirements of the public and continually seek new technologies for safety inspection of food products and process control. In the past decade, one type of technology, hyperspectral imaging, has made significant progress for the food industry in addressing rapid and non-destructive inspection of food quality and safety issues (Kim et al. 2001, 2004; Park et al. 2002; Lawrence et al. 2003a; Lu 2003; Zavattini et al. 2004; Gowen et al. 2007; Chao et al. 2007a; Yoon et al. 2011). Hyperspectral imaging technology integrates both imaging and spectroscopy into unique imaging sensors producing hyperspectral images with exceptional spectral and spatial resolution. A single hyperspectral image has a

H. Yao (✉) • Z. Hruska
Geosystems Research Institute, Mississippi State University,
1021 Balch Blvd., Stennis Space Center, MS 39529, USA
e-mail: haibo@gri.msstate.edu; hruska@gri.msstate.edu

R.L. Brown
U.S. Department of Agriculture, Agricultural Research Service, Southern Regional Research Center, Room 2129, 1100 Robert E. Lee Blvd., New Orleans, LA 70124, USA
e-mail: Robert.Brown@ars.usda.gov

D. Bhatnagar • T.E. Cleveland
U.S. Department of Agriculture, Agricultural Research Service, Southern Regional Research Center, Room 2131, 1100 Robert E. Lee Blvd., New Orleans, LA 70124, USA
e-mail: Deepak.Bhatnagar@ars.usda.gov; Ed.Cleveland@ars.usda.gov

B. Park, R. Lu (eds.), *Hyperspectral Imaging Technology in Food and Agriculture*,
Food Engineering Series, DOI 10.1007/978-1-4939-2836-1_6

contiguous spectral resolution between one and several nanometers, with the number of bands ranging from tens to hundreds. Generally, high spectral resolution images can be used to study either the physical characteristics of an object at each pixel by looking at the shape of the spectral reflectance curves or the spectral/spatial relationships of different classes using pattern recognition and image processing methods. By investigating the physical, biological, and chemical properties of food products using both spatial and spectroscopic information, the hyperspectral technology offers an alternative and often superior method for assessment of food safety.

Hyperspectral imagery was traditionally used in earth remote sensing applications incorporating aerial or satellite image data. More recently, low cost portable hyperspectral sensing systems became available for laboratory-based research. Typically, hyperspectral imaging is used as a research vehicle with full wavelength analysis in the exploration phase of a research project. In most applications, the full wavelength spectrum of data can be reduced to several key wavelengths for specific subject matter. These wavelengths can then be implemented in a multispectral mode for faster data acquisition. This way, rapid inspection can be realized in real-time online applications. Many food-related research applications that used hyperspectral technology are reported in literature. An incomplete list of these applications includes *Fusarium* head blight (SCAB) detection in wheat (Delwiche and Kim 2000), fecal contamination on apples (Kim et al. 2002a, b), aflatoxin detection in corn kernels (Pearson et al. 2001; Yao et al. 2010a), identification of fecal contamination on poultry carcasses (Park et al. 2002; Lawrence et al. 2003a, b; Heitschmidt et al. 2007), detection of apple bruising (Lu 2003), on-line measurement of grain quality (Maertens et al. 2004), detection of pits in cherries (Qin and Lu 2005), detecting egg embryo development (Lawrence et al. 2006), apple firmness estimation (Peng and Lu 2006; Lu 2007), quality assessment of pickling cucumbers (Liu et al. 2006; Kavdir et al. 2007; Ariana and Lu 2008), differentiation of wholesome and systemically diseased chicken carcasses (Chao et al. 2007a), bone fragment detection in chicken breast fillets (Yoon et al. 2008), detection of insects in cherries (Xing et al. 2008), quality definition of sliced mushrooms (Gowen et al. 2008), quality evaluation of fresh pork (Hu et al. 2008), prediction of beef tenderness (Naganathan et al. 2008), detection of toxigenic fungi (Yao et al. 2008; Rasch et al. 2010), citrus canker detection (Qin et al. 2008), food-borne pathogen detection (Yoon et al. 2009), fecal contamination detection in vegetables (Yang et al. 2010), as well as contamination detection on processing equipment (Cho et al. 2007; Jun et al. 2009). While other chapters in this book discuss different issues and applications with hyperspectral technology, the present chapter focuses on food safety inspection of plant products.

The generic approach for applying hyperspectral technology in food related research includes experiment design, sample preparation, image acquisition, spectral pre-processing/calibration, sample ground truth characterization, data analysis and information extraction. Hyperspectral measurement can be point or image data, and this chapter will discuss research and applications using both types. A spectrometer is usually used for point data acquisition (Pearson et al. 2001;

Hu et al. 2008). For hyperspectral image data acquisition, two methods can be used. One approach is the frame-based method which acquires one band at a time in a sequential process. A liquid crystal tunable filter based (LCTF) system (Gat 2000; Peng and Lu 2006) follows the frame-based principal. The other approach is to use a line scanning method such as push-broom scan. The scan mechanism can be implemented by either moving the target (Kim et al. 2001) or moving the camera lens internally within the camera system (Mao 2000). For more details regarding hyperspectral imaging and data analysis, the readers are encouraged to visit the previous chapters.

In the following sections, the major food safety concerns that have drawn upon hyperspectral imaging research and applications will be discussed. The most common types of food contaminants including pathogens, physical and chemical contaminants, and their detection applications using sensing technology will be summarized. Next, hyperspectral technology will be briefly discussed. Since the technology has been covered in great detail in the previous chapters, the emphasis will be on the different aspects of hyperspectral data such as reflectance, fluorescence, and transmittance. The application feasibility of such data to plant product safety inspection will also be discussed. The third section will discuss different applications of using hyperspectral data. The main focus will be on grains, produce, nuts, and spices. Different application modes including lab research, online inspection, and in-field/remote monitoring will be presented after the applications. The last section will be a summary of the chapter.

6.2 Food Safety

Because presence of any pathogenic microorganism in food can lead to severe health consequences in both animals and humans, food safety is of major concern. Though the global incidence of food-borne disease is difficult to estimate, it was reported in 2005 that close to two million people died from gastrointestinal infections largely attributed to contaminated food and drinking water (WHO 2007). A high incidence of food-borne illness in developing countries is understandable; however, the numbers are high even in industrialized countries. Reports indicate that up to 30 % of the population is stricken with food-borne diseases each year in developed countries. For example, the Centers for Disease Control and Prevention (CDC 2011a) estimated that around 48 million cases of food-borne diseases, resulting in 128,000 hospitalizations and 3,000 deaths occur each year in the USA. Clearly, preventing food-borne illness remains a major public health challenge.

Due to a recent increase in health awareness worldwide, higher consumption of plant products including fresh produce, nuts and grains has increased the incidence of food-borne illness from contaminated foods precipitated partially by changes in food production (non-traditional farming practices spurred on by the "green" revolution) and supply (increase in international trade). According to the international trade statistics, the World Trade Organization (WTO 2007) reported that Europe is responsible for 46 % of world exports of agricultural products, where 80 % of the agricultural export represents food (WTO 2007). Trading of contaminated food between countries increases the potential for outbreaks and health risks posed by microbial pathogens present in the food. In 2007, the US Food and Drug Administration (FDA) developed a comprehensive 'Food Protection Plan', which implies that food must be considered as a potential vehicle for intentional contamination (FDA, Food Protection Plan 2007). Such intentional contamination of food could result in human or animal illnesses and deaths, as well as economic losses.

Leading causes of contamination worldwide are microbial toxins and chemicals from agriculture. Most prominent food-borne pathogens associated with plant products are certain species of *Salmonella*, *Escherichia coli* and *Listeria* (Arora et al. 2011) as well as mycotoxins. In the US the impact of international commerce connected to global incidence of food-borne illness was felt during the recent outbreak in Germany where infected fenugreek seeds originating from Egypt contributed to numerous deaths from *E. coli* O104:H4 shiga toxin infection, one of which was an American (Giordano 2011). In addition to the food-borne epidemic, the incident caused a diplomatic breach with Spain for mistakenly implicating Spanish cucumbers in the outbreak and causing massive losses to farmers in that country (Giordano 2011). Another example was the recent *Salmonella* outbreak in 23 US states resulting from contaminated papayas imported from Mexico. The 2008 *Salmonella saintpaul* infection caused serious losses for tomato producers in the US due to recall of tomatoes mistakenly implicated in spread of the infection. It was determined that the infection outbreak was caused by jalapeno peppers from a Texas farm and Serrano peppers from Mexico, distributed to Mexican restaurants in the US and used to prepare salsa (Behravesh et al. 2011). Another incident was reported in September 2011 when fresh cantaloupes from Jensen Farms, Colorado were linked to a *Listeria monocytogenes* outbreak in 20 states in the US resulting in an estimated 100 infections and 18 deaths (CDC 2011b). The most susceptible populations for contracting Listeriosis are pregnant women, the immunocompromised, and the elderly.

Conventional microbiology techniques for the detection of microorganisms include culture and colony counting (Allen et al. 2004), immunological assays (Van et al. 2001), polymerase chain reaction (PCR) (Burtscher and Wuertz 2003), imaging flow cytometry for complex food systems (Bisha and Brehm-Stecher 2009) and more recently, biosensors (Arora et al 2011; Velusamy et al. 2010; Lazcka et al. 2007). Current methods for detection of mycotoxins are largely dependent on chromatography methods (TLC—thin-layer chromatography, HPLC—high performance liquid chromatography, immunoaffinity column chromatography), and enzyme assays (e.g., ELISA—Enzyme-linked immunosorbent assay). While these methods can be sensitive, inexpensive and give both qualitative and quantitative information of the tested microorganisms, there are drawbacks associated when using them for detection of food-borne pathogens and toxins. The techniques are time-consuming as well as labor intensive. Some require elaborate instrumentation, and most must be performed by qualified personnel. Moreover, the results are not always accurate and each method requires destruction of samples. To overcome these limitations recent research has focused on the development of biological sensors for the detection of pathogens. Biosensors offer multiple, rapid, real time analyses from perishable or semi-perishable foods. However, applying biosensors also has some limitations in its sensitivity, cost, and the need for sample pre-treatment for the detection of pathogens (Arora et al. 2011). It is also an analytical approach and as such, subject to sampling error and sample destruction. Implementation of safe agricultural (Kay et al. 2008; Umali-Deininger and Sur 2007), and manufacturing practices (Mucchetti et al. 2008; Umali-Deininger and Sur 2007) as well as applying the hazard analysis and critical control point (HACCP) program (Jin et al. 2008) can significantly diminish pathogens in food. However, to more effectively address problems related to health and food safety a need remains for rapid, reliable, simple, specific and sensitive detection technology that is also suitable for real-time monitoring at a low cost. Hyperspectral imaging technology could prove to be an invaluable tool in reducing risk of food contamination by effectively detecting microbial pathogens such as *Salmonella* spp., *Escherichia coli* spp. (*E. coli*) and mycotoxins.

6.2.1 Contaminants and Detection

6.2.1.1 Pathogen Contamination

More than 250 different food-borne illnesses have been described in scientific literature. Most of these diseases are caused by a variety of bacteria, viruses, and parasites. Recently, a number of researchers have reported the potential of hyperspectral imaging for identification of microorganisms of concern in food. Dubois et al. (2005) demonstrated the potential application of NIR (near-infrared) hyperspectral imaging as a high throughput technique for the differentiation of bacteria based on their NIR spectra. NIR images of cards for specific foods containing both test and calibration bacteria samples were obtained in the spectral region 1,200–2,350 nm using an InSb focal-plane array detector. Some bacteria were identifiable from spectral differences observed at unique wavelengths; however, in situations where particular microorganisms of concern were sought, partial least square (PLS) classification was preferable to separate the genera of bacteria present.

A visible near-infrared (VNIR) hyperspectral technique was used for detecting and differentiating cultures of *Campylobacter* associated with human gastrointestinal infections from semi-raw poultry or unpasteurized milk, from non-*Campylobacter* cultures after 48 h (Yoon et al. 2009) and 24 h (Yoon et al 2010) incubations. The latter study found that a two-band ratio algorithm at 426 and 458 nm achieved up to 99 % detection accuracy for cultures grown on blood agar, determined to be an optimal culture medium for VNIR reflectance, after 24 h. Yao et al. (2008) conducted a similar VNIR study on fungal cultures. Their study reported that five fungal strains were differentiated with a classification accuracy of 97.7 %. In addition, all five fungi could be classified by using only three narrow bands (bandwidth = 2.43 nm) centered at 743 nm, 458 nm, and 541 nm. Jian et al. (2009) applied Principal Component Analysis (PCA) along with Support Vector Machine (SVM) for the classification of toxigenic and atoxigenic strains of *Aspergillus flavus* to VNIR hyperspectral data and concluded that although the technique was promising, broadening the spectrum range to include the ultraviolet and infrared regions would simplify classification.

The suitability of Raman hyperspectral imaging for the enumeration of water-borne pathogens has also been evaluated (Escoriza et al. 2006). Hyperspectral images in the range 3,200–3,700 nm were obtained from inoculated water samples using a Raman Chemical Imaging microscope containing a liquid crystal tunable filter. It was shown that Raman hyperspectral imaging can provide quantitative information for bacterial concentration in water samples. It was noted, however, that the Raman signal was poor for low bacteria concentration and requires the use of filters on dilute water samples prior to examination.

6.2.1.2 Chemical (Toxins) Contamination

Beside microbial contamination, certain food-borne illnesses are caused by chemical food contaminants from natural or industrial sources (Peshin et al. 2002). Natural toxins include compounds such as lectins and glycoalkaloids produced by plants (phytotoxins) and found in potatoes as well as legumes (Peshin et al. 2002; Rietjens and Alink 2003). Other sources of naturally occurring toxins are mycotoxins produced by some fungi and phycotoxins present in some marine algae (Peshin et al. 2002; Rietjens and Alink 2003). Under ambient environmental conditions certain fungi produce toxic secondary metabolites that contaminate agricultural products including grains, nuts, seeds, and various produce. Mycotoxins can enter the food chain in the field, during transport, in storage, or at any point during production and result in significant economic consequences for the producers (Bennett and Klich 2003). The major toxigenic fungi affecting agricultural products and of consequence to food inspectors include some species of *Aspergillus, Fusarium* and *Penicillium* (Peshin et al. 2002; Bennett and Klich 2003). Collectively, these fungi produce a variety of toxins, (e.g. aflatoxin, ochratoxin, fumonisin, among others), some of which have acute or chronic health effects (Peshin et al. 2002; Bennett and Klich 2003). Environmental contaminants including heavy metals and organic pollutants such as dioxins, and polychlorinated biphenyls, as well as pesticides and cleaning chemicals are found in the food chain by chance, as a result of transportation, storage or processing and are largely considered industrial contaminants (Peshin et al. 2002; Rietjens and Alink 2003; Schrenk 2004). Intentional food additives that enhance taste or appearance, such as sulfides and MSG are considered adulterants and may have deleterious effects on health of sensitive individuals (Peshin et al. 2002; Lipp 2011). A recent example of ill intended adulteration that resulted in a major heath consequence was the addition of scrap melamine to milk products (Lipp 2011). Unfortunately, globalization compounds the challenge of producing a safe food supply. Being able to rapidly verify the authenticity of a product could be a useful step in food production and manufacturing.

The development and application of nondestructive analytical techniques based on the optical properties of agricultural and biological products may be useful tools for quality control and evaluation of food products (Deshpande et al. 1984). Yao et al. (2010a) utilized fluorescence hyperspectral imaging techniques for detecting the mycotoxin aflatoxin in corn kernels inoculated with *Aspergillus flavus* under field condition. A Fluorescence Peak Shift (FPS) phenomenon was noted among different groups of kernels with different levels of aflatoxin contamination indicating the potential of fluorescence hyperspectral techniques for detecting mycotoxins in grains. Hiroaki et al. (2002) used Fourier transform infrared diffuse reflectance spectroscopy (FT-IR-DRS) to measure pesticide residue on heads of lettuce obtained directly from the field. The diffuse reflectance spectra ranged from 2,800 to 800 cm^{-1} and were transformed to spectra closely resembling the NIR transmission spectral features (Birth and Hecht 1987). The results of the study pointed to the potential of a more rapid (2 min) pesticide measurement system utilizing the PLS regression model for optimal calibration for measuring pesticides in agricultural products. Carrasco et al. (2003) tested hyperspectral systems with both reflectance and fluorescence outputs to assess the feasibility of remote sensing for food quality applications and found that hyperspectral imaging may be a valuable tool in detecting pesticide levels in agricultural products. To address the adulteration factor of food safety, September (2011) applied NIR hyperspectral imaging and multivariate image analysis (PCA, Partial least square discriminant analysis PLS-DA, and PLS regression) to determine presence and amount of foreign substances such as buckwheat and millet flour in ground black pepper samples. Overall the study results indicate that NIR hyperspectral imaging may be a promising technique for the identification of adulterants in ground black pepper which may have additional applications for identifying adulterants in other powdered foods.

6.2.1.3 Physical Contamination

Physical contamination of a food product can occur when objects such as glass, hair, dirt, paint chips, etc. get intermixed with food. In the tobacco industry, for example, a variety of products including plastics, paper, and cords, among other debris, get mixed with the leaves due to manual harvesting. Although tobacco is not exactly considered a food item, it is a plant product widely used in oral applications (e.g. chewing tobacco, cigars, etc.) and as such needs to be free of physical contaminants. Conde et al. (2006) developed a real-time spectral image-based system for the discrimination of tobacco leaves from unwanted debris as well as other non-tobacco plant materials. The authors utilized PCA analysis and ANN (Artificial Neural Network) classification for efficient discrimination of tobacco from debris. Furthermore, the technique appears applicable to other sorting needs as long as the ANN is properly trained based on total characterization of the spectra of the material in question.

Transmission hyperspectral imaging is potentially applicable for the online estimation of internal constituent concentrations and detection of internal defects within foods (Schmilovitch et al. 2004). Qin and Lu (2005) applied hyperspectral transmission imaging to detect pits, considered a choking hazard, in tart cherries. Light was transmitted through individual cherries from a light source placed below the sample holder, and recorded by an imaging spectrograph placed above the sample. Transmission images for four different sample orientations were tested, and it was shown that sample orientation and color did not significantly affect classification accuracy.

The shells in packaged nuts are also considered a potential choking hazard. Because of this, food processors in the pistachio industry have a very low or no tolerance policy for the presence of shells or shell fragments. Unfortunately, automated sorters are not that precise so a manual sorting must follow the machine sorting, incurring additional cost to the industry (Haff et al. 2010). A recently developed (Pearson 2009) relatively low cost, high speed, color sorter was used to sort small and large in-shell pistachio nuts from kernels. The system consisted of three cameras around the product stream to insure that all surfaces of each sample were inspected. The study employed two algorithms, DA and k-nearest neighbor (KNN), making use of the spatial information provided by the camera configuration. Both algorithms successfully (99 % accuracy) discriminated between regular in-shell and kernel pistachios. The accuracy declined when small in-shells and kernels were compared, with the KNN algorithm significantly outperforming the DA algorithm (Haff et al. 2010).

6.3 Hyperspectral Imaging

6.3.1 Reflectance

Reflectance measurement is the most common way of utilizing hyperspectral technology. Reflectance is the fraction of incident electromagnetic power reflected by a surface. The typical reflectance information is in the visible and near-infrared region of the electromagnetic spectrum. For plant product safety inspection, some examples of using reflectance measurement include mycotoxin detection in corn (Berardo et al. 2005), deoxynivalenol estimation in wheat (Beyer et al. 2010), classification of scab-damaged wheat kernels (Delwiche and Hareland 2004), detection of toxigenic fungi (Del Fiore et al. 2010; Jian et al. 2009; Yao et al. 2008), identification of fecal contamination on vegetable leaves (Yang et al. 2010), and assessment of *Fusarium* damage to wheat kernels (Shahin and Symons 2011).

In summary, hyperspectral reflectance imagery has proven to be a good tool for external inspection and evaluation for food quality and safety applications.

In processing reflectance data, the raw digital counts recorded by the camera are generally converted into reflectance as part of the image radiometric calibration. For example, the following equation can be used to convert raw digital counts of reflectance into percent reflectance:

$$Reflectance_{\lambda} = \frac{S_{\lambda} - D_{\lambda}}{R_{\lambda} - D_{\lambda}} \times 100\%, \tag{6.1}$$

where $Reflectance_{\lambda}$ is the reflectance at wavelength λ, S_{λ} is the sample intensity at wavelength λ, D_{λ} is the dark intensity at wavelength λ, and R_{λ} is the reference intensity at wavelength λ. Eventually, the calibrated reflectance value lies in the range from 0 to 100 %. More topics on hyperspectral image calibration can be found in Yao and Lewis (2010).

6.3.2 *Transmittance*

Another approach of using hyperspectral technology is through measurement of transmittance. Transmittance is the fraction of incident electromagnetic power that passes through a sample. Hyperspectral transmission measurement involves projecting light at one side of the target and recording light transmitted through the target at the opposite side with a hyperspectral imager. Consequently, hyperspectral images of transmittance can be useful for studying internal properties of food. It was reported that NIR spectroscopy in transmittance mode can penetrate deeper regions of fruit (>2 mm) compared to reflectance mode (McGlone and Martinsen 2004). The internal property of targets can be analyzed using light absorption within the detector's spectral range. One drawback of transmittance imaging is the low signal level from light attenuation due to light scattering and absorption.

Research activities related to using hyperspectral transmittance measurement for plant product safety inspection have been reported in detection of aflatoxin (Pearson et al. 2001) and fumonisin in corn (Dowell et al. 2002), corn kernel constituent analysis (Cogdill et al. 2004), detection of pits in cherries (Qin and Lu 2005), and detection of *Fusarium* in single wheat kernels (Polder et al. 2005). These studies demonstrated that hyperspectral transmittance data has potential for safety inspection of plant products.

Equation (6.1) used in reflectance calibration is also applicable for calculating the calibrated relative transmittance. Similarly, a dark current image and a reference transmittance image are needed in the calibration equation.

6.3.3 Fluorescence

In addition to reflectance and transmittance measurements, fluorescence hyperspectral imaging is also used for food and agriculture applications. Fluorescence is the emission of light by a sample that has absorbed electromagnetic radiation. Generally, the emitted light (emission) has a longer wavelength than the absorbed incident radiation (excitation). For example, certain organic and inorganic substances exhibit natural, intrinsic fluorescence when excited under an ultraviolet (UV) light source (<400 nm). Under UV excitation, plants can emit a fluorescence spectrum ranging from about 400 to 800 nm. Thus, fluorescence spectra are suitable for studying the properties of sample constituents and also chemical compositions related to safety inspections.

In the past decade, fluorescence hyperspectral imaging technology has been developed to enable the acquisition of fluorescence image data with both high spectral and spatial resolutions (Kim et al. 2001; Zavattini et al. 2004). A fluorescence hyperspectral imaging system is generally based on an imaging spectrometer or hyperspectral imager. Fluorescence emission can be excited by long wavelength UV radiation (Kim et al. 2001; Jun et al. 2009; Yao et al. 2010a), or induced by a laser source (Kim et al. 2003a, 2004; Lefcourt et al. 2004). Some research and applications in plant product safety inspection that found use for fluorescence hyperspectral imagery include detection of fecal contamination on apples (Kim et al. 2002a, 2005a) and on cantaloupes (Vargas et al. 2005), assessing bacterial biofilm on stainless steel surfaces (Jun et al. 2009), studying corn contaminated with aflatoxin (Yao et al. 2010a), and analyzing *Fusarium* influence on wheat ears (Bauriegel et al. 2011b).

6.4 Plant Products Safety Inspection Applications

6.4.1 Grains

6.4.1.1 Corn with Aflatoxin Contamination

Aflatoxins, secondary metabolites produced by certain *Aspergillus* species, are among the most toxic naturally occurring substances known (Bennett and Klich 2003). Repeated exposure to the most potent aflatoxin (B1) in the diet can lead to liver cancer, and prolonged inhalation to aspergillosis as well as lung cancer (Peshin et al. 2002; Bennett and Klich 2003; Wild and Turner 2002). Contaminated feed often results in deleterious health effects as well as eventual demise for several types of farm animals (Peshin et al. 2002; Bennett and Klich 2003). Thus, aflatoxins present a serious food safety issue when found in food and feed products. Figure 6.1 is an illustration of the chemical structure of aflatoxins B1, B2, G1, and G2. Because of their uneven distribution, particularly in grains, aflatoxins are difficult

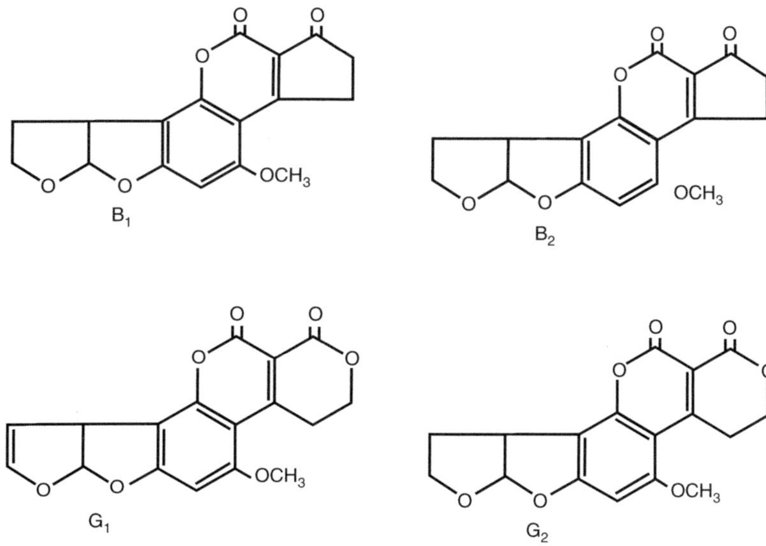

Fig. 6.1 Structure of aflatoxins B1, B2, G1, and G2

to isolate for elimination without destroying a large amount of product and incurring significant economic losses to the farmers. The estimated cost for the management of aflatoxin and other mycotoxins reported in 2003 was between \$500 M to \$1.5 M (Robens and Cardwell 2003). Aflatoxin is regarded as one of the most important food safety problems in the world (Robens 2008).

Aflatoxin contamination is a major concern, particularly in corn, as corn is one of the main crops in food and feed production. In preharvest corn plants, the problem usually starts when the toxigenic fungus infects the corn kernels in the field. Figure 6.2 illustrates (a) *Aspergillus flavus* fungus cultured on potato dextrose agar (PDA) medium and (b) a sporulating mycelia of *A. flavus* under Scanning Electron Microscope (SEM). Figure 6.3 is an illustration of corn ears exhibiting *Aspergillus flavus* infestation. The fungus begins to produce aflatoxin when the host corn plant is under heat and drought stress in the early dough stage. Aflatoxin levels in food and feed are regulated by the Food and Drug Administration (FDA) in the US as well as various agencies throughout the world. In the US, the regulation levels are 20 ppb (parts per billion) for human consumption and 100 ppb for feed. These levels (USDA 2002) allow farmers, the food industry, and the Federal Grain Inspection Service (FGIS) to take appropriate action when aflatoxin is found in food or feed. To screen for aflatoxin, conventional chemical-based analytical methods such as TLC and HPLC are used. These methods are time consuming, expensive (Collison et al. 1992; Brown et al. 2001) and require the destruction of samples.

It is thus important to be able to detect aflatoxin in corn in a rapid and non-destructive manner. Hyperspectral technology based detection provides a potential approach for this application. Pearson et al. (2001) used spectral reflectance (5,050–1,700 nm) and transmittance (500–950 nm) measurements for

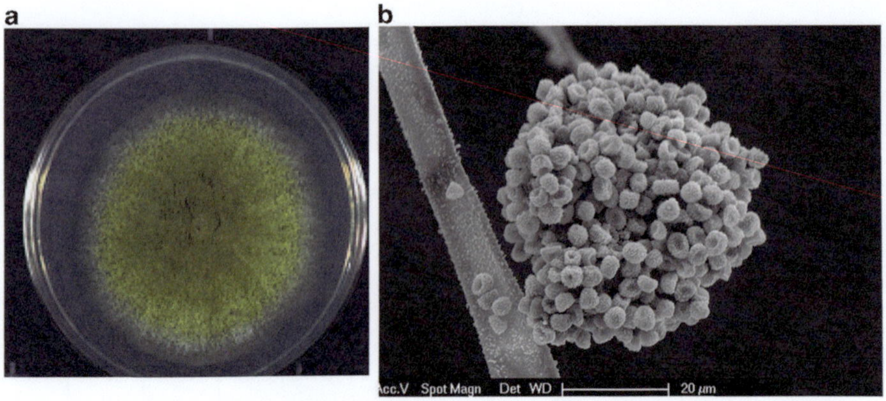

Fig. 6.2 *Aspergillus flavus* in culture and SEM image. (**a**) *Aspergillus flavus* culture and (**b**) *Aspergillus flavus* SEM Image

Fig. 6.3 Corn ears infected with *A. flavus* with artificial inoculation of the fungus

contamination detection in single corn kernels. The experiment utilized fiber optic spectrometers for the spectral measurements. A total of 200 kernels inoculated with *Aspergillus flavus* NRRL A-27837 and another 300 randomly selected kernels were used in the experiment. After spectral measurement, each single kernel was chemically analyzed with an affinity chromatography procedure for the actual aflatoxin concentration. The spectral data was treated with a 19-point Savitzky-Golay second order filtering for noise removal and was converted to absorbance (log (1/R). The average whole kernel reflectance is displayed in Fig. 6.4. In the plot, corn kernels were divided into three categories, i.e., <1 ppb, >100 ppb, and between 1 and 100 ppb. In the reflectance spectra the absorbance was higher below 850 nm and

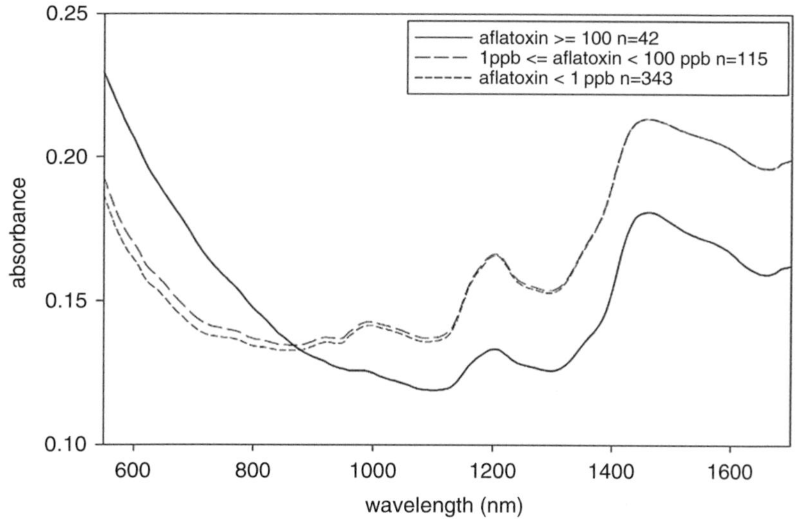

Fig. 6.4 Average whole kernel reflectance with different aflatoxin contamination levels (Pearson et al. 2001)

lower between 850 and 1,700 nm for contaminated kernels. In the transmittance spectra the absorbance was generally much higher for highly contaminated kernels. The authors speculated that the differences could be explained by the scattering and absorbance characteristics caused by the fungus in the kernel. With the fungal invasion, the kernel endosperm becomes powdery. Consequently, the scattering would cause less NIR (>750 nm) radiation to be absorbed in the reflectance mode. While in the transmission mode, more NIR radiation would be absorbed from the contaminated kernels.

The above corn kernel spectra and chemical data was analyzed using discriminant analysis and partial least squares regression. For the kernels with >100 ppb or <10 ppb levels of aflatoxin, the classification accuracy was over 95 %. For the kernels between 10 and 100 ppb the accuracy was 25 %. The study also pointed out that the two-feature discriminant analysis of the transmission data yielded the best result. This approach involved the use of a two band ratio for the analysis.

Another study (Fernández-Ibañez et al. 2009) also used a spectrometer (400–2,500 nm) and a Fourier transform near-infrared (FT-NIR) spectrophotometer (9,000–4,000 cm^{-1}) for aflatoxin (B1) detection in corn (66 samples) and barley (76 samples). The spectral data were acquired in reflectance mode and saved as log (1/R), where R is the reflectance. The grain samples were cultured at room temperature (20 ± 2 °C) for 3 months after natural infection. Each single grain was chemically analyzed to determine its aflatoxin contamination concentration, with above 20 ppb classified as positive and below 20 ppb as negative. Data processing was based on discriminant analysis with the partial least squares. The best model for aflatoxin detection based on reflectance data had $R^2 = 0.8$ for corn and $R^2 = 0.85$ for barley. When the FT-NIR data was used the R^2 numbers were 0.82 and 0.84, respectively. The study explored analysis with different chemometric models. The description of grain sample preparation did not reveal much detail.

Fluorescence measurement could provide another approach for screening aflatoxin contaminated corn kernels. Aflatoxin emits fluorescence when excited with UV light (Carnaghan et al. 1963; Goryacheva et al. 2008). *A. flavus*-infected grains also emit bright greenish yellow fluorescence under UV excitation. Marsh et al. (1969) pointed out that the ability of emitting fluorescence is a characteristic of living cells that exhibit peroxidase activity. For *A. flavus*, the fungus must infect the plant tissue and grow in it for some time, producing and transforming kojic acid, another known *A. flavus* metabolite, to one or more BGYF (bright greenish-yellow fluorescence) compounds in a peroxidase-type reaction. In other words, there is an apparent overlap between the BGYF compound and aflatoxin production. The BGYF phenomenon had been widely used in presumptive tests for the presence of aflatoxin in corn (Shotwell and Hesseltine 1981; Maupin et al. 2003) in order to determine if chemical analysis is required to measure the level of aflatoxin concentration of a given sample. The approach requires the use of 365 nm UV light for fluorescence identification. However, it only reveals the broad fluorescence response from the samples without identifying the source of fluorescence emission. As a consequence, the BGYF approach was not used as a quantitative or even qualitative measure of aflatoxin contamination.

The use of narrow band fluorescence spectra offers the potential for better aflatoxin detection in corn. Instead of using a spectrometer for single point data collection for each grain sample, a hyperspectral imaging system can capture hundreds of pixels over the same grain sample. The later approach can thus provide more optimal data collection, especially in the spatial domain. With this approach, one study (Yao et al. 2010a) used fluorescence hyperspectral imaging for aflatoxin detection in corn. The study focused on understanding the BGYF phenomenon in corn kernels infected with fungus, and the goal was to determine the relationship between fluorescence emission of corn kernels inoculated with *A. flavus* and aflatoxin contamination levels within the corn. The fluorescence hyperspectral imaging system was based on push-broom line scanning coupled with a spectrograph for spectral light dispersing. The imaging sensor was a 14 bit CCD camera. A long wave UV lamp with wavelength centered at 365 nm was used as a fluorescence excitation light source. The total sample was comprised of 504 individual kernels. After imaging, each kernel was chemically analyzed to ascertain aflatoxin levels. All the corn samples were obtained from a field where corn ears were artificially inoculated with toxigenic *A. flavus*.

The fluorescence hyperspectral image data were preprocessed and the region of interest (ROI) for each corn kernel was generated. The ROIs were used to extract fluorescence spectral information and to statistically compare it with the chemical measurement. A Fluorescence Peak Shift phenomenon was noted among different groups of kernels with different levels of aflatoxin contamination. It was found that the fluorescence emission peak shifted toward the longer wavelengths with highly contaminated corn kernels (Fig. 6.5). The fluorescence peak of highly contaminated

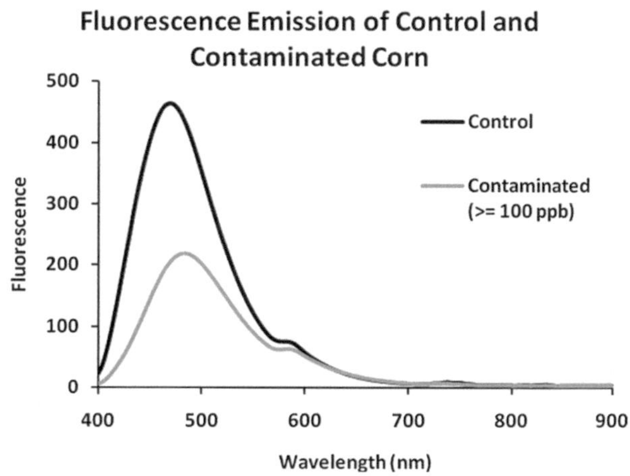

Fig. 6.5 Illustration of a fluorescence emission peak shift between control and corn kernels contaminated with aflatoxin

kernels was also found to have a lower level of intensity compared with less contaminated kernels. In addition, a general negative correlation existed between the measured aflatoxin and the fluorescence image bands in the blue and green regions. The correlation coefficient of determination, R^2, was 0.72 for the multiple linear regression model. The multivariate analysis of variance revealed that fluorescence means of the four aflatoxin groups, <1 ppb, 1–20 ppb, 20–100 ppb, and ≥ 100 ppb, were significantly different from each other at 0.01 level of α. Under a two-class schema, the classification accuracy ranged from 0.84 to 0.91 when either a threshold of 20 ppb or 100 ppb was used. Studies with other classification algorithms produced similar outcomes (Yao et al. 2010b, 2011b).

6.4.1.2 Corn Contaminated with Fumonisin by *Fusarium* and Other Mycotoxins

Another common mycotoxin in corn is fumonisin. Fumonisin is produced by the toxigenic fungi *Fusarium*. Similar to aflatoxin, fumonisin is considered to have cancer promoting ability. Based on FDA's regulations, the maximum level of fumonisin in corn and corn products is 2–4 ppm (parts per million) for human consumption. The maximum levels for animal consumption are 5–100 ppm based on different animal feeds (FDA/CFSAN 1978). Current sampling procedures account for up to 90 % of the variability in the results. Measurement of fumonisin is based on analytical methods. Corn samples are ground and chemically analyzed for accurate determination of fumonisin levels.

Visible and NIR reflectance and transmittance spectroscopy has provided the potential for rapid and non-invasive detection of fumonisin in corn kernels. Dowell et al. (2002) used spectral reflectance (400–1,700 nm) and transmittance (550–1,050 nm) measured with fiber optic spectrometers for fumonisin detection in single corn kernels. The total number of kernels was 330. The corn was artificially inoculated with *Fusarium verticillioides* NRRL 25457 in the late milk to early dough stage of kernel maturity. After spectral measurement, each single kernel was chemically analyzed for the actual fumonisin concentration. Both partial least square and discriminant analysis were used for data analysis. It was found that kernels with >100 ppm and <10 ppm could be accurately classified as fumonisin positive or negative, respectively. In another study, Berardo et al. (2005) acquired VNIR reflectance spectra of 280 corn kernels with *Fusarium* infection. The spectrometer was in the range from 400 to 2,500 nm and the data was recorded in log(1/R). Each kernel was chemically analyzed with HPLC for fumonisin concentration. A modified partial least-squares regression was used for data analysis. The model prediction $R^2 = 0.78$.

6.4.1.3 Corn Contaminated with Fungus

Generally, corn kernels damaged by fungi are low in quality with a higher probability of being contaminated with mycotoxins. If these kernels were removed, the overall corn quality could be greatly improved. This removal would also prevent such infected/contaminated materials from entering the food chain. Thus, in addition to mycotoxin detection, studies were also implemented on the detection of fungi infected corn kernels using spectroscopic methods.

Studies have shown that fungi could be spectrally quite different. For example, Yao et al. (2008) attempted to differentiate toxigenic fungi using hyperspectral imagery. In the study, five different types of fungi, *Penicillium chrysogenum, Fusarium verticillioides, Aspergillus parasiticus, Trichoderma viride,* and *Aspergillus flavus*, some of which are toxin producers, were imaged with a VNIR push-broom line-scanning hyperspectral camera. The resulting reflectance data indicated that these fungi are highly separable with a classification accuracy of 97.7 %. In another study, Jian et al. (2009) used hyperspectral images to classify one toxigenic *A. flavus* and three atoxigenic *A. flavi*. The data was collected with two different light sources, UV and halogen light. To process the image, Genetic Algorithm (GA) was first used to select the principal components based on Bhattacharya Distance. A support Vector Machine was then used for the classification of fungi. On average, the classification accuracy ranged from 0.67 to 0.85. The pair-wise classification achieved accuracy from 0.8 to 0.99.

Pearson and Wicklow (2006) found that NIR reflectance spectral bands centered at 715 and 965 nm correctly identified 96.6 % of corn kernels that showed extensive discoloration and were infected with *Aspergillus flavus, Aspergillus niger, Diplodia maydis, Fusarium graminearum, Fusarium verticillioides, or Trichoderma viride*. The spectral data was collected with a spectrometer. Other studies in this area also used hyperspectral images. Williams et al. (2010) focused on distinguishing between *Fusarium* infected and sound whole corn kernels. The hyperspectral images used were in the NIR an SWIR range (960–2,498 nm). It was pointed out that the infected and non-infected regions on individual kernels could be effectively identified with principal component analysis. Specifically, along principal component one

(PC1), there was a distinct difference between infected and sound kernels with two clusters. With partial least square discriminant analysis, the coefficient of determination was 0.73 and 0.86 for the two camera systems used in the study.

Del Fiore et al. (2010) worked to differentiate between kernels infected with fungi and healthy control corn kernels using hyperspectral reflectance data. The imaging system was based on push-broom line-scanning. For sample preparation, four types of *Aspergillus* strains, two *Fusarium* strains, and one *Penicillium* strains were used to inoculate 12 corn hybrids. For imaging, kernels were divided into 20 g samples. After imaging, the reflectance data was converted to apparent absorbance units expressed as log (1/R) and then transformed to principal components for statistical analysis. In this study, after inoculation the changes induced by fungal infection on the surface of the corn kernels were recorded through daily imaging. The results showed that the ANOVA and the following Fisher's LSD test could identify two wavelengths with high discrimination power for the presence of fungus and/or growth. Additionally, the study demonstrated an early detection of fungal infection on the corn, which appeared 48 h after inoculation of *A. flavus* and *A. niger*.

6.4.1.4 Wheat Contaminated with Mold-Scab and Other Fungi

One of the major problems in wheat is *Fusarium* head blight, or scab. The problem is a fungal disease caused by the *Fusarium* fungus in humid conditions when the wheat plant is in flowering or early kernel development stage. The symptoms of *Fusarium* head blight are shrunken kernels which are chalky white or pink in appearance (Fig. 6.6). The presence of scab can result in downgrading of wheat with eventual losses to the producers, as well as to the consumers. The current method used by USDA/GIPSA (Grain Inspection, Packers and Stockyards Administration) for scab inspection is based on manual visual inspection, which is labor intensive and subjective; therefore, new methods for rapid inspection are always sought. In this section, several spectroscopic methods will be reviewed, including the use of a fiber spectrometer, application of spectroscopic data in sorting, and hyperspectral imaging.

In one study, Delwiche (2003) grouped single wheat kernel samples into three categories, sound, mold-damaged, and scab-damaged. Each category had a minimum of 138 kernels. Reflectance of each kernel was measured with a spectrometer in the range from 940 to 1,700 nm. The reflectance data was then stored in log (1/R) format. The best combination of individual wavelengths, best difference of wavelengths, best ratio, and combinations of each were exhaustively searched. The best classification model, which was a combination of kernel mass and the difference of two wavelengths, 1,182 and 1,242 nm, achieved up to 95 % detection accuracy. If only two classes (sound vs. damaged) were used, the accuracy level was from 95 to 98 %. The study pointed out that the orientation of kernel placement also had an impact on the classification accuracy. In a later study, Delwiche and Hareland (2004) grouped hard red spring wheat kernels into sound and scab-damaged

Fig. 6.6 Fusarium Head Blight on Wheat. Courtesy of USDA ARS

kernels, with 868 kernels from each category for one trial and 1,790 for a second trial. The spectrometer instrument was the same as in the previous study with wavelength ranging from 1,000 to 1,700 nm. Statistical classification methods such as linear discriminant analysis and k-nearest-neighbor were used. It was found that the slope of the low-wavelength side of a broad carbohydrate absorption band (~1,200 m) was quite effective in separating the sound from the scab-damaged kernels (95 % accuracy).

Peiris et al. (2009) reported the NIR absorbance characteristics of sound and *Fusarium* damaged wheat kernels. Differences were found in the heights of the absorption peaks at 1,204, 1,365, and 1,700 nm. The differences could be caused by the changes in the levels of grain food reserves such as starches, proteins, lipids, and other structural compounds. There were also shifts in absorption peak positions between the two types of kernels at 1,425–1,400 nm and 1,915–1,930 nm. Finally, in a recent study (Rasch et al. 2010), it was found that data combined from different spectroscopic methods such as optimized excitation and emission wavelengths, fluorescence decay times and fluorescence quantum efficiency, and NIR spectroscopy is promising for the qualitative and quantitative identification of fungi and mycotoxin.

Efforts were also carried out to implement the related findings in the sorting of *Fusarium*-damaged wheat. In general, sorters are based on monochromatic or bichromatic light from broad visible or NIR wavelength regions. Delwiche and Gaines (2005a) aimed at locating the best monochromatic and bichromatic wavelengths in the visible and near-infrared regions for the sorting task. The results showed an accuracy of 94 % with 500 and 550 nm in the visible region, 97 % with 1,152 and 1,248 nm for the NIR region, and 86 % with 750 and 1,476 for the hybrid region. In the sorting process, one challenge for the sorters is to identify and remove the damaged kernels. Delwiche (2008) implemented a design with high-power pulsed LED (light emitting diode) lights combined with one silicon photo diode detector. In the sorting process, the kernel was free-falling and the LED lights were flashing at 2,000 Hz. In this setting the reflected light from the kernel would be captured by the detector for about 20 cycles of the pulsed light during the free-falling of the kernel. The resulting data was analyzed with linear discriminant analysis and a classification accuracy of 78 % for *Fusarium*-damaged kernels was reached. Even though the result was not comparable with the previous study, this was a step forward toward a real operation and was an improvement over the conventional bichromatic design (with 50 % efficiency). Further study (Yang et al. 2009) enhanced the detection accuracy of the same sorter system. The combined accuracy was 85 %. In another experiment (Wegulo and Dowell 2008), the performance of a single-kernel near-infrared sorting system (from Perten Instruments, Stockholm, Sweden) was compared with visual sorting of *Fusarium*-damaged wheat kernels. The conclusion was that the sorting system had better results due to its wider range of detection and greater consistency.

The above classification of scab-damaged wheat kernels was based on spectrometer readings. Each single kernel was measured with one spectral reading. A hyperspectral image, on the other hand, can provide hundreds of pixels (or hundreds of spectral readings) for one wheat kernel. Thus with more spatial and spectral information extraction capability, hyperspectral images were also used for detecting scab in wheat. Image acquisition was usually achieved with push-broom line scanning techniques. In one study, Delwiche and Kim (2000) imaged three wheat sample varieties, each with 32 normal and 32 scab-damaged kernels. The spectral range of the reflectance image was from 424 to 858 nm. The wheat kernels were arranged in an 8 by 8 matrix on a black velvet cloth. The rows of normal and scab kernels alternated in the arrangement. After image calibration and preprocessing, the reflectance of each kernel was averaged into a single reflectance number. It was found that scab-damaged kernels generally exhibited higher reflectance than normal kernels. Statistical analysis included step-wise discriminant analysis and discriminant analysis. The first process selected the best 22 bands for classification. The best two-band combination from the 22 bands was then selected for the discriminant analysis. The misclassification rate was related to sample varieties ranging from 2 to 17 %. Alternatively, Shahin and Symons (2011) used reflectance images (400–1,000 nm) for the detection of *Fusarium* damaged kernels in Canada Western Red Spring wheat. In this work, 800 kernels were separated into three classes of samples: sound, mildly damaged, and severely damaged. Images were analyzed with principal component analysis and linear discriminant analysis. In a two-class (sound and damaged) scenario, an overall accuracy of 92 % was achieved. This work also pointed out that the performance of the reduced image space (6 wavelengths) and the full-spectrum image was comparable.

In order to exploit a wider range of spectral data, Polder et al. (2005) used both VIS-NIR (430–900 nm) and SWIR (900–1,750 nm) hyperspectral images for the detection of *Fusarium* in single wheat kernels. The imaging data recorded was transmission spectra. In order to capture transmittance data, the two imaging systems required different physical configurations. For sample preparation, 96 kernels were selected from samples artificially inoculated with *Fusarium culmorum*, including both damaged and healthy looking kernels. The actual infestation with *Fusarium* (i.e., the amount of *F. culmorum* DNA) was determined for each kernel through TaqMan real-time PCR analysis. The spectral absorption data was analyzed through unsupervised fuzzy c-means clustering and supervised partial least squares regression. The results indicated that the SWIR data had better performance than the VNIR data. The analysis could clearly identify kernels with more than 6,000 pg (picograms) *Fusarium* DNA.

In addition, wheat kernels infected with other fungi were also studied. Singh et al. (2007) used near-infrared hyperspectral images in the wavelength range 1,000–1,600 nm for fungal detection in wheat. Fungal infection was introduced with *Penicillium, Aspergillus glaucus*, and *Aspergillus niger*. PCA was applied to reduce the image dimension. Classification was then implemented with K-means clustering and discriminant analysis. An average of 97.8 % accuracy was obtained to detect the infected kernels in a 2-class scenario. When four classes were used, 95 % of the *Penicillium* infected and 91.7 % of the healthy kernels could be correctly identified. Misclassification happened to the kernels infected with *Aspergillus* glaucus and *Aspergillus niger*. With the same data, Zhang et al. (2007) used support vector machine for the analysis. Accuracy for the above 4 classes, *Penicillium., A. glaucus, A. niger*, and healthy kernels, was 99.3 %, 87.2 %, 92.9 %, 100 %, respectively. Between the *A. glaucus* and *A. niger* classes, the misclassification rate was 10 %.

6.4.1.5 Wheat/Barley Contaminated with Don

In addition to *Fusarium* head blight, the *Fusarium* fungi can produce a metabolite called deoxynivalenol (DON), also known as vomitoxin. In the US, the FDA's regulation on the levels of DON in finished wheat products is 1 ppm for human consumption and 5–10 ppm for livestock and poultry feed. Conventional methods of detecting DON are chemical-based analytical methods such as HPLC or immunoassays (ELISA). Generally, rapid and non-invasive approaches are desired for the detection, and ideally, quantification of the wheat samples.

Earlier research focused on using spectroscopic methods on single wheat kernels. For example, Dowell et al. (1999) used a spectrometer (400–1,700) to measure absorbance of wheat kernels infected with DON. The PLS model had a prediction $R^2 = 0.64$ for DON. The model was developed with kernels when DON > 5 ppm. The relationship between actual DON and spectral measurements of bulk samples (more than one kernel) was also investigated. Ruan et al. (2002) utilized spectrometer absorbance data (400–2,500 nm) for non-destructive determination of deoxynivalenol levels in barley. The focus was to develop neural-network models for the prediction. The model R^2 was 0.933 when the full spectrum NIR data was used. The R^2 was 0.912 when data from 700 to 1,000 nm was used. It was concluded that the neural network approach performed better than PLS analysis. Pettersson and Aberg (2003) used NIR transmittance data (570–1,100 nm) for the determination of DON in wheat kernels. Both PCA and PLS methods were used in the analysis. The best regression model was developed in the wavelength range 670–1,100 nm with a correlation coefficient of 0.984. In a more thorough approach, Beyer et al. (2010) first separated the *Fusarium* damaged wheat kernels into 6 levels of damage, 0, 20, 40, 60, 80, and 100 %, based on visual inspection. Each group had more than 120 kernels. Spectral reflectance (350–2,500 nm) was collected with an ASD spectrometer and DON was determined chemically (HPLC) for each group. Both partial least square regression and LDA were used in the analysis. The estimation R^2 was 0.84. However, it was concluded that this approach alone was not reliable enough to separate grain samples with DON contents at the legal limit of 1.25 ppm (1.25 mg/kg, EU standard).

Work was also extended to automated detection and sorting of wheat kernels with DON contamination. Delwiche et al. (2005b) worked on high-speed optical sorting of soft wheat for reduction of deoxynivalenol. A commercial high-speed bichromatic sorter was modified with two wavelengths, 675 and 1,480 nm. The wavelength selection was based on prior studies on single wheat kernels. The sorter had a capacity of 0.33 kg/ (channel-min) with a rejection rate of 10 %. After the first run of sorting, the fraction of DON contamination level in the sorted wheat to that in the original wheat ranged from 18 to 112 %. The mean fraction was 51 %. If a multiple sorting approach was used, in which the samples are re-sorted, the fraction could be reduced to 16 to 69 % of the original level. Peiris et al. (2010) evaluated an automated single kernel new-infrared spectroscopic sorter for the estimating of DON levels. The results indicated that single kernels could be predicted as having low (<60 ppm) or high (>60 ppm) DON with ~96 % accuracy. For the kernels with high DON concentration, the single kernel DON levels could be estimated with $R^2 = 0.87$. The sorter could run at 1 kernel/s and could be helpful for breeders with more information on the single seed.

6.4.1.6 Soybean Contaminated with Fungus

Wang et al. (2004) used NIR reflectance (400–1,700 nm) to classify single fungal damaged soybean seeds. The reflectance was converted to absorbance for PLS and NN analysis. The samples were 800 fungal-damaged and 500 healthy soybean seeds manually selected by trained grain inspectors. The two-class PLS model in the wavelength region of 490–1,690 nm had an accuracy of more than 99 %. The study further worked on a five class model when the fungal-damaged seeds were divided into 4 different categories based on the damage. In this case, the NN model had better classification accuracy, which was 94.6 % for the validation sample set.

6.4.2 Produce

In addition to its application in the safety inspection of grains, hyperspectral imaging has viable applications for implementation at critical control points of food processing when inspecting for potential contaminants of produce. One of the main food safety issues for fresh produce such as fruits and vegetables is fecal contamination on the surfaces of the products. Thus the detection of such contamination is of importance for the producers and processors.

6.4.2.1 Apple with Fluorescence Detection

Several early studies used apples to demonstrate the feasibility of using spectroscopic and imaging techniques to detect surface contaminants (Schatzki et al. 1997; Wen and Tao 2000; Leemans and Destain 1998). In the 1990s, outbreaks of *Escherichia coli* O157:H7, a bacterial strain that often leads to hemolytic uremic syndrome, were found to be associated with unpasteurized apple juice and cider (Steele et al. 1982; Besser et al. 1993; CDC 1996, 1997; Cody et al. 1999). Animal droppings from cattle and deer were thought to be the main source of contamination found in the fruit juices (Riordan et al. 2001; Uljas and Lngham 2000). Because this presented a threat to the general US population and children in particular, the FDA came out with a mandate in an attempt to reduce fecal contamination of apples in the food chain (FDA 2001). The FDA also indicated a need for rapid and non-invasive technology for the detection of fecal contaminants on apples. In response to the federal mandate and request for an automated, non-destructive, imaging system for quality and safety inspection of produce, researchers at the USDA, Beltsville, Maryland developed lab-based (Kim et al. 2001, 2003b; Lefcourt et al. 2005a) and in-line (Kim et al. 2008b) hyperspectral and multispectral imaging systems for potential commercial application. Apples from different cultivars were used as the test model for the evaluation of the various systems and techniques.

An early study (Kim et al. 2001) compared fluorescence and reflectance images of a normal apple and an apple with fungal contamination and bruises. The images were acquired with a lab-based hyperspectral reflectance and fluorescence imaging system specifically developed for food quality and safety research. With a spectral range between 430 and 930 nm, spectral resolution of 10 nm and spatial resolution of 1 mm^2, the system was equipped with halogen as well as UV-A illumination sources to measure both reflectance and fluorescence emission. The results of the study provided baseline spectral features of unblemished Golden Delicious apples and the effects due to contamination and defects from reflectance and fluorescence emission images (Kim et al. 2001). In two follow up experiments, hyperspectral reflectance (Kim et al. 2002a) and fluorescence (Kim et al. 2002b) data, acquired with the above mentioned imaging system, was used to determine optimal wavelengths that may be used in an online multispectral system. Principal component analysis was used to aid in visualizing the hyperspectral data and to develop criteria for multispectral detection. Four different cultivars of apples were selected (Red Delicious, Fuji, Gala and Golden Delicious). Feces from a dairy farm were applied onto apples in both a patch and a transparent invisible smear. Reflectance results identified three VIS-NIR and two NIR wavelengths that could potentially be implemented in a multispectral imaging system for fecal detection on apples. Fecal contamination on apples for the different cultivars was classified with the use of a single threshold method which worked well for the thick fecal smears, but not as well for the transparent ones (Kim et al. 2002a).

In contrast, results from the fluorescence imaging experiment were more promising. Four multispectral bands (450, 530, 685, and 735 nm) were identified by PCA and examination of fluorescence emission maxima as being optimal for discriminating contaminated apple surfaces. Additionally, a simple two-band ratio (685 to 450 nm) reduced variation on normal apple surfaces and emphasized the differences between contaminated and uncontaminated areas (Kim et al. 2002b). The authors also noted that because auto-fluorescence from fecal matter is low, enhancement of the fluorescence signal as well as using an excitation wavelength more appropriate for fecal detection (410–420 nm, Kim et al. 2003b) would be required in an in-line commercial application.

A larger scale experiment was conducted where 96 apples from the Golden Delicious cultivar were treated with diluted (1:2, 1:20, 1:200) fecal matter prior to fluorescence imaging (Kim et al. 2004). Fluorescence hyperspectral images were acquired with an updated version of the original system (Kim et al. 2001) developed for food safety research. The main difference in the imaging system was the inclusion of a thermo-electrically cooled electron multiplying charge coupled device (EMCCD) camera, which was able to attenuate the low fluorescence signals resulting in better signal to noise ratio, which replaced the original CCD camera. Following hyperspectral imaging, PCA was used to determine several wavelengths for the classification of feces on the apples. A two-band fluorescence ratio using combinations of the PCA-selected wavelengths determined that the ratio images at 556 and 663 nm provided automated detection of several fecal dilutions on the surfaces of Golden Delicious apples with a minimum of false positives. The results were reproduced regardless of color variation of individual apples (Kim et al. 2004).

All the previously acquired information was utilized in the subsequent development of multispectral imaging systems. The individual steps including algorithm development and testing were outlined by Lefcout et al. (2006).

One of the multispectral imaging systems developed was a portable multispectral fluorescence system which was used to isolate the optimal red bands for detecting feces on Red Delicious apples (Kim et al. 2005a). The system consisted of a UV light source, an intensified camera (ICCD) with a six-position filter wheel, and software for controlling the system and automatic analysis of images (Kim et al. 2005a; Lefcourt et al. 2005a). It was found that the best contrast between contaminated and uncontaminated sites on apple surfaces was provided by fluorescence emission bands at 670 nm. It was further demonstrated that the use of an unsupervised automated threshold algorithm along with a two-band ratio multispectral fusion increased the power of detection of the system to 100 % (Kim et al. 2005a). The system was exhibited in a demonstration at a public domain for the purpose of increasing public awareness to the presence of fecal contamination that may not be readily visible. In fact, the system was able to detect fecal remnants after the apples were washed as long as the feces remained on the apple for at least 5 h before washing (Lefcourt et al. 2005a).

6.4.2.2 Apple with Laser Induced Fluorescence Detection

The next step in the quest for automated detecting device for in-line inspection of fecal contamination of apples was the introduction of the multispectral laser-induced fluorescence (LIF) imaging system. The system was developed by Kim et al. (2003a) and was intended for capturing multispectral fluorescence emission images, at bands 450, 550, 678 and 730 nm in the blue, green, red, and far-red regions, from large biological samples including meat and produce. The system consisted of a pulse laser, a beam expander, a lens, a common aperture adapter and a fast-gated intensified camera (Kim et al. 2003a; Lefcourt et al. 2003).

One of the applications for the multispectral laser-induced fluorescence imaging system was to test the ability and sensitivity of laser-induced fluorescence in detecting fecal contaminants on Red Delicious apples. Serial dilutions of fecal matter from dairy cows, deer, and an abandoned cow pasture, were applied to the apples. Images were acquired 1 day and 7 days after apple surface treatment and after washing as well as after washing and brushing. Because a wide and steep gradient in fluorescence surrounded the treated areas, a gradient method for detection was applied. The results revealed close to 100 % detection for the 1:2 and 1:20 dilutions, and over 80 % for 1:200 (<15 ng) dilutions 1 day after application. The lowest detection was noted on apples contaminated with feces from the pasture that were washed and brushed with detection for 1:2, 1:20 and 1:200 corresponding to 100 %, 30 % and 0 %, respectively. Analysis of the ratios of red band to blue band images were found to improve detection of 1:2 dilutions, where as detection of 1:200 dilutions was better when only the red band images were used for analysis. The results were encouraging and indicated that laser-induced fluorescence imaging was a sensitive method for fecal contamination detection on apples including obscure areas like the calyx (Lefcourt et al. 2005b), however, there were several practical issues that needed to be addressed before a viable in-line inspection system could be developed (Lefcourt et al. 2003).

The next few studies concentrated on perfecting detection methodology when using the multispectral laser-induced fluorescence system for the specific application of discriminating feces-contaminated from non-contaminated apples from various cultivars. Lefcourt et al. (2005c) analyzed potential methods for increasing selectivity and specificity of the LIF system using the previously determined optimum excitation wavelength for feces of 417 nm (Kim et al. 2003b), and a pulsed laser with a parametric oscillator for tuning to a particular wavelength. The study examined the role of timing of the fluorescence responses to the pulsed laser excitation when detecting fecal contamination on apples. Results of the study found that consideration of the timing was an important factor in enhancing the signal from the feces-contaminated areas. Utilizing the time-dependence of the fluorescence signal, Kim et al. (2005b) examined the nano-scale fluorescence emission decay characteristics of fecal matter on apples in order to find the optimal red band and gate-delay time for detecting the contamination. The results revealed that the optimum emission band providing the best difference in time-dependent fluorescence between treated and untreated apples was 670 nm and a gate-delay of 4 nm from the laser excitation peak (Kim et al. 2005b). Ultimately, the optimal detection was achieved after images underwent universal power transformation followed by edge detection (Lefcourt et al. 2005b). The typical intensity variation between apples was reduced by the transformation and contrast between contaminated and uncontaminated areas was improved (Lefcourt and Kim 2006).

An updated time-resolved multispectral laser induced imaging system was developed that incorporated tunable wavelength capability for excitation band selection and nanosecond-scale characterization of fluorescence responses (lifetime imaging). The system was tested on Red Delicious apples contaminated with cow manure. Several excitation-emission wavelengths were employed and it was confirmed that 670 nm emission and 418 nm excitation bands provide the best separation between clean and treated spots on apples (Kim et al. 2008a). Another benefit of the system is a large field of view (13×13 cm^2) for the fluorescence lifetime imaging that can accommodate relatively large biological samples (Kim et al. 2008a).

6.4.2.3 Apple with Reflectance Detection

Although fluorescence hyperspectral and multispectral imaging was considered more sensitive when applied to fecal detection on apples (Kim et al. 2001), studies continued to explore reflectance image properties for potential food safety applications.

An experiment with 153 samples was conducted by Mehl et al. (2002) where the hyperspectral system (Kim et al. 2001) in reflectance mode, was used to isolate three bands for the discrimination of normal and contaminated Golden Delicious, Red Delicious, and Gala apples. The PCA and chlorophyll absorption peak methods were used to elucidate the optimal spectral bands that were implemented in a multispectral imaging system consisting of a three channel (RGB) prism-based camera with specific band pass filters placed between the prisms and the three CCDs to attain spectral specificity for the detection of various apple anomalies including defects, disease and contamination. Good separation between normal and contaminated apples was obtained for the Gala (95 %) and Golden Delicious (85 %) samples; however, separation of the Red Delicious (76 %) apples was less satisfying (Mehl et al. 2002). Although PCA is the most frequently utilized multivariate method when analyzing hyperspectral data, other approaches may be developed depending on the need presented by the data set. Mehl et al. (2004) compared several methods for image analysis of hyperspectral reflectance data for the detection of defects and contaminants on the surfaces of Red Delicious, Golden Delicious, Gala, and Fuji apples. The authors also presented an asymmetric second difference method which applied a chlorophyll absorption waveband at 685 nm and two bands in the near-infrared region. The method provided detection of defects and contamination on apples independent of apple color or cultivar with results comparable to PC image analysis. Because the method required only three spectral bands and considerably less processing time, it was considered suitable for application in an in-line multispectral imaging system (Mehl et al. 2004).

Further experiments using hyperspectral reflectance (Liu et al. 2007) revealed that the 675–950 nm VNIR region exhibited the most obvious spectral differences between clean and fecal contaminated surfaces of Red Delicious and Golden Delicious apples. The authors applied several image analyses methods and determined that the dual band ratio (725/811 nm) algorithm provided the best detection of the contaminated areas on both cultivars. Because the two bands (725 and 811 nm) are away from the natural pigments, there is no interference from color variations among apples of different varieties. In addition, the algorithm was effective in identifying bruises on surfaces of the fruit. The development of the potentially universal two-band algorithm advanced the possibility of using a reflectance based application in an in-line detection system.

Results from the combined fluorescence and reflectance studies led to the testing of an in-line system. The hyperspectral line-scan imaging system (Kim et al. 2004) was integrated with a commercial apple sorting machine for the simultaneous imaging of apples for both quality (cuts and bruises) and safety (fecal contamination). It was demonstrated that fluorescence imaging with two-band ratio analysis achieved a 100 % detection rate of apples artificially contaminated with cow feces and no false positives. A NIR two-band reflectance ratio coupled with a simple classification method based on the heterogeneity of the ratio values achieved a 99.5 % apple defect classification with a false positive rate of 2 %. The system in the hyperspectral mode achieved a speed of approximately three apples/s (Kim et al. 2007). A subsequent "multitask" multispectral imaging system was developed with an updated EMCCD camera designed to operate with a commercial apple sorting machine. Details of the system are described in Kim et al. (2008b). Test line-scan images were acquired with 200 s exposure time with processing speed exceeding four apples/s. A total of 333 lines per second and 40 pixels per apple resulted in sufficient spatial resolution for image based online inspection (Kim et al. 2008b).

One problem that has not been adequately addressed is the ability of the camera to see all of the surfaces of individual fruits on an automated inspection system. Reese et al. (2009) proposed the use of concave parabolic mirrors, an idea that has merit but will require some logistical adjustments to be of practical use. Software for the automated online inspection multitask system is under development. The system has the potential of incorporating several sorting components as well as the added benefit of detecting fecal contaminants, thus making it suitable for both safety and quality inspection of other food products besides apples (Kim et al. 2011).

6.4.2.4 Other Fruit (Cantaloupes, Strawberries)

The hyperspectral system (Kim et al. 2001) in the fluorescence mode was also utilized in the detection of fecal contamination of cantaloupes (Vargas et al. 2004, 2005) and strawberries (Vargas et al. 2004). Image analysis of the cantaloupe data indicated that due to the natural variation of the fruit surfaces, both single band and two-band ratio methods showed a number of false positives. Subjecting the entire hyperspectral data set to the PCA method was more useful for discriminating between fecal contamination and false positives as well as isolating specific bands for potential multispectral application. The strawberry imaging was less promising, assumedly, due to the combination of high moisture content of the fruit and less than ideal lighting conditions (Vargas et al. 2004).

6.4.2.5 Vegetables (Leafy Greens, Cucumbers)

At least 26 outbreaks of *E. coli* infection reported in the US since the early1990s were traced to contaminated leafy greens including spinach and lettuce (Maki 2009). In spite of great efforts on the part of the various food safety agencies to prevent the spread of these infections, the fact remains that every year in the US thousands of people acquire *E. coli* infections, and many of them do not survive (Maki 2006). Because of its non-invasive and adaptive nature, remote sensing may be ideally suited for contamination monitoring of leafy greens as well as other vegetables.

To test this theory, hyperspectral technology was employed for classification of fecal contamination on leafy greens, where fluorescence hyperspectral imaging (421–700 nm) and a two-band ratio (666 nm/680 nm) image analysis method successfully classified fecal contamination on spinach and lettuce leaves (Yang et al. 2010). The imaging system was based on push-broom line-scanning with an EMCCD camera. The samples were excited with 365 nm UV light. The leaf samples were romaine lettuce and baby spinach leafs. Fecal contamination was artificially introduced onto the surface of each leaf. The emission peaks of the leaf were between 660 and 690 nm, which was related to the intensity of chlorophyll emissions. Compared to the leaf peaks, the emission peak for fecal contamination spots had a shift toward the shorter waveband, named blue shift. Chlorophyll in general emits red and far red fluorescence under long wavelength UVA excitation. This was in agreement with the observation of the study. An exhaustive search was implemented over all possible two-band combinations and the bands with the highest correlation between the two-band ratio (666/680 nm) and fecal contamination were selected to develop a classification algorithm. The result showed that the classification accurately detected all of the contaminated spots, but not all the contaminated pixels. A different study (Siripatrawana et al. 2011) conducted on packaged fresh spinach contaminated with *E. coli* used hyperspectral reflectance to develop a rapid detection method for quantifying *E. coli* infection with a combined PCA and artificial neural network (ANN) chemometric analysis.

Lu and Ariana (2011) applied reflectance and transmittance hyperspectral techniques in order to detect fruit-fly infestation in pickling cucumbers. Partial least square discriminant analysis was employed for the classification of infested and un-infested samples. It was found that the hyperspectral imaging system outperformed manual inspection accuracy (75 %) in either mode. In the reflectance mode classification was between 82 and 88 %, and in the transmittance mode it was 88–93 %.

6.4.3 *Tree Nuts*

6.4.3.1 Almonds

Internally damaged almonds that brown after cooking are rejected by both the almond industry and the consumer, mainly because of their unpleasant, bitter taste and dark color. The damage is not readily apparent and it is difficult to distinguish the affected nuts from normal ones. In 1999, Pearson found that he could distinguish raw, internally damaged almonds from undamaged nuts with the use of full spectrum (700–1,400 nm) transmission hyperspectral imaging. Although promising, the methodology proved to be cumbersome due to the large data set produced by the full spectrum of the hyperspectral system. A new feature selection algorithm was recently developed for internally damaged almond detection that requires only two sets of ratio features for classification. Results demonstrate that the method gives a higher classification rate than using the best feature selection subset of separate wavebands or when using a feature extraction algorithm where all wavelength data is included (Nakariyakul and Casasent 2011). The method can potentially be used in real-time practical multispectral sensor systems for other applications including insect damage assessment in almonds, an occurrence linked to aflatoxin contamination (Schatzki and Ong 2001).

6.4.3.2 Walnuts

Another specialty crop subjected to hyperspectral image sorting is the black walnut. Because of risks associated with swallowing shell pieces, the application of automated discrimination between walnut shell and pulp has become a safety concern in the walnut postharvest processing industry in the US. Jiang et al. (2007) employed hyperspectral fluorescence imaging along with a Gaussian-kernel based support vector machine (SVM) approach to classify the walnut shells and pulp. Results of their experiments showed an overall 90.3 % recognition rate based on 6,257 samples. Zhu et al. (2007) used fluorescence hyperspectral imagery with ICA-kNN optimal wavelength selection, to avoid data redundancy of hyperspectral data, for classification of walnut shells and pulp. Results of both studies show the feasibility of using hyperspectral fluorescence for classification of black walnut shells and pulp in order to prevent a possible choking hazard or an oral injury.

6.4.3.3 Pistachios

Because of their pleasing taste and many health benefits, pistachios are a popular snack throughout the world and of economic importance in several countries. Unfortunately, like other nuts subject to variable storage conditions, pistachios are also prone to aflatoxin contamination associated with *Aspergillus* spp. fungal infestation (Farsaie et al. 1978; Pearson and Schatzki 1998; RahaieI et al. 2010). The acceptable export limits for pistachios worldwide are 10 ppb for aflatoxin B1 and 15 ppb for total aflatoxin (RahaieI et al. 2010). Early attempts at using spectroscopic methods for differentiating between aflatoxin contaminated and non-contaminated pistachios were made by Farsaie et al. (1978). Excitation and emission spectra were derived from Iranian pistachios in order to develop an index for sorting aflatoxin contaminated and uncontaminated nuts based on the natural BGY fluorescence. Six indices were tested with excitation wavelengths at 360 and 420 nm. It was determined that the BGY fluorescence emission peak was at 490 nm and was most probably due to a pure substance and was obviously different from the other emission peaks. Thus the feasibility of using machine vision was demonstrated. In a later study, Pearson and Schatzki (1998) used machine vision for automated detection of pistachios also contaminated with aflatoxin, based on a particular staining pattern of contaminated nuts. In their study the authors reported that the automated image-based system they used outperformed hand sorting. There were some unaddressed issues with their method. One obvious problem is that not all pistachio varieties have the particular marking associated with aflatoxin contamination described in the present study. There do not appear to be recent imaging studies that address the aflatoxin problem in pistachios, yet the problem persists (Rahaie et al. 2010).

6.4.3.4 Other Nuts (Hazelnuts, Peanuts)

Due to differences in allowable limits in exported goods, care must be taken when exporting to certain countries. Europe, in particular, has very stringent aflatoxin limits. Because aflatoxin contamination is highly heterogeneous and contaminated seeds or nuts are often unevenly distributed, it would be useful to be able to non-invasively prescreen goods meant for export and remove the contaminated nuts or seeds instead of having to deal with the consequences of a rejected shipment. In a recent study, Kalkan et al. (2011) applied multispectral imaging techniques to detect aflatoxin contaminated hazelnuts. The authors used a two-dimensional, local-discriminate bases algorithm that identifies the optimal band-pass width and center frequencies of optical filters to be used in a multispectral imaging system. The classification accuracy for aflatoxin contaminated and uncontaminated hazelnuts was 92.3 % with a resulting decrease of aflatoxin concentration from 608 to 0.84 ppb. Hirano et al. (1998) used the transmittance ratio of (T700/T1100 nm) when attempting to differentiate aflatoxin contaminated from uncontaminated shelled peanuts.

6.4.4 *Spices*

The increased use of spices and herbal ingredients obtained through international commerce brings to attention possible microbiological contamination of products during prolonged storage or those that may have not received a proper processing treatment. Outbreaks of salmonella linked to contaminated spices present a problem especially when herbs and spices are added to foods ready for immediate consumption (Sagoo et al. 2009). Other contaminants in spices may be in the form of mycotoxins (Hernandez-Hierro et al. 2008). Although there exist established guidelines for sampling and testing of dry herbs and spices (EC 2004; ICMSF 2005), practical application and monitoring or enforcing the guidelines may be problematic. Currently, prevention of microbial contamination of spices and herbs largely depends on good hygiene practices during all stages of production from farm to the table. Reliance on end product testing for compliance with costly analytical methods (e.g., HPLC) may be avoided if an in-step monitoring system was employed to ascertain product integrity. Spectroscopy based systems may present a viable rapid and cost-effective option. A recent study in Turkey (Atas et al. 2010) used hyperspectral image data to classify aflatoxin-contaminated chili peppers with the application of ANN for detecting the mycotoxin aflatoxin. Their results led them to design a machine vision system based on hyperspectral imaging and machine learning (Ataş et al. 2011).

Another study (Kalkan et al. 2011) used multispectral imaging for detecting aflatoxin in red chili pepper flakes. With their two-dimensional, local-discriminant bases algorithm, they achieved classification accuracy of 80 %. Another popular spice susceptible to mycotoxins is the red paprika. Hernandez-Hierro et al. (2008) found that using NIR spectroscopy for detecting aflatoxin and ochratoxin in paprika was an alternative method offering lower cost and higher speed than conventional chemical analysis methods.

6.5 Inspection Mode

6.5.1 *In Research*

Using hyperspectral imagery for food quality and safety inspections is a natural extension from using this type of data in space or terrestrial remote sensing. Different from traditional earth based hyperspectral remote sensing applications where solar radiation is the sole source for target illumination, the aforementioned research activities all utilized artificial light. The artificial light can be fiber light (Armstrong 2006; Cho et al. 2007; Kim et al. 2001; Lawrence et al. 2003b; Lu 2003; Pearson and Wicklow 2006), tungsten halogen light (Haff and Pearson 2006; Yao et al. 2008), tungsten halogen light in a diffuse lighting chamber (Naganathan et al. 2008), or light emitting diode (Chao et al. 2007b; Lawrence et al. 2007). Typically, this type of experiment is implemented in an indoor environment at a close distance.

Among the applications utilizing hyperspectral imaging for safety inspection in plant products, almost all follow a similar research to implementation track. The research step would examine the full wavelength response from the samples under different imaging modes (reflectance, transmittance, fluorescence) and lighting conditions. Once sufficient knowledge is obtained regarding a certain subject, key wavelengths can be selected to fit in specially designed inspection devices to speed up the inspection process. The former task is generally carried out in lab conditions, and the latter is applied in an online inspection setting.

In the lab, all of the imaging factors are well-controlled in order to collect the best image data from the samples. Generally, the instrumentation used are spectrometers or hyperspectral cameras (imaging spectrometers). Since inspection speed is not of major concern in lab based research, full wavelength data is generally collected. The artificial light is tuned to provide the best lighting conditions for the imaging experiment. The samples are treated under controlled conditions set by the experimenter. Natural samples are carefully selected in order to provide a representative sample of the real world conditions. Most of the above discussed applications fall into the "research" category. A typical hyperspectral imaging camera used for research purposes is presented in Fig. 6.7. The camera is based on push-broom line-scanning.

Fig. 6.7 Hyperspectral camera system showing the CCD detector, the housing for the prism-grating-prism spectrograph, and the motor-controlled lens positioning assembly (Lawrence et al. 2003b)

6.5.2 Online Inspection

Online inspection uses spectral information for rapid safety inspection of large quantities of a specific product. The application environment is generally in an industrial or simulated industrial situation. In this case, the lighting condition is pre-defined and well controlled. Large quantities of sample move at high speeds on a conveyor-belt. The spectral information is generally narrowed to several key wavelengths for rapid data acquisition. To collect spectral information at those wavelengths, narrow band filters (Kise et al. 2008) or a filter wheel (Kim et al. 2005a), wavelength switching devices such as LCTF (Gat 2000) or AOTF (Acousto-optic tunable filters, Park et al. 2011), or wavelength addressing on a line scanner (Yoon et al. 2011) can be used.

For online grain inspection, Fig. 6.8 provides a conceptual view of a corn inspection instrument for aflatoxin detection. The narrow band fluorescence wavelength was selected with a LCTF mounted in front of the camera. In a study (Yao et al. 2011a) that simulated the operation of the inspection instrument and tested the feasibility of rapid detection, both 25 and 1,000 g corn samples were used. The detection accuracies were generally in the range from 80 to 90 %.

One of the main applications in online inspection is sorting. In a test (Pearson et al. 2004) using high-speed dual-wavelength sorting for the reduction of aflatoxin and fumonisin contamination in corn, the absorbance at 750 and 1,200 nm were used. The two wavelengths were selected from a discriminant analysis process using full wavelength data. Filtering of the two wavelengths was implemented through a dual-peak filter mounted in front of the silicon and indium-gallium-arsenide (InGaAs) detectors equipped with a high-volume optic sorter. It was reported that the sorter was able to reduce aflatoxin levels by 81 % from an initial average of 53 ppb. For fumonisin the levels were reduced by 85 % from an initial level of 17 ppm. In other work, Delwiche et al. (2005) modified a commercial high-speed bichromatic sorter with two wavelengths 675 and 1,480 nm in order to implement high-speed sorting of soft wheat for the reduction of Deoxynivalenol.

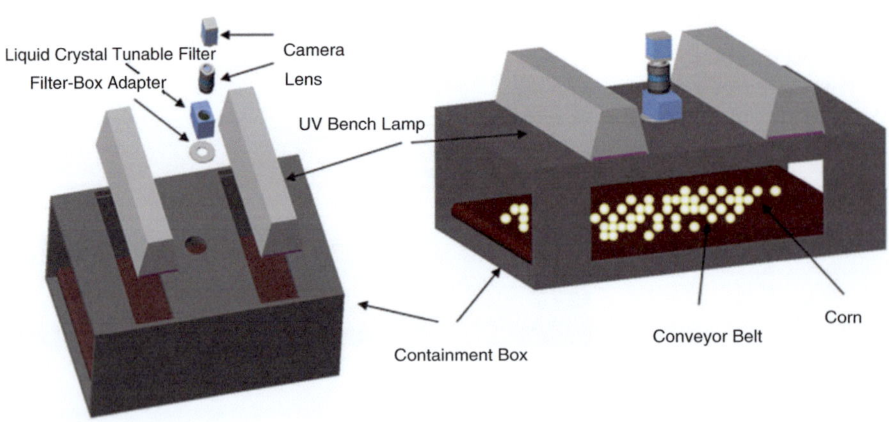

Fig. 6.8 Schematic of an automated instrument for aflatoxin inspection in corn

Pasikatan and Dowell (2001) reviewed optical-based sorting systems for detecting and removing seeds infested internally by insects or fungi. The focus of the review was on sorting indices based on selected wavelengths for different grains and applications. The subjects included *A. flavus* and aflatoxin, *Fusarium*, and internal insects in grains. It was concluded that both wavelength identification and proper selection of sorting criteria are important for high sorting accuracy. The selection of sorting criteria is closely related to the nature of the products because acceptable and unacceptable samples can have spectral characteristics ranging from very pronounced to very subtle. The consequence is that products are not clearly classified as "accept" and "reject". Thus it is difficult to define the cutoff point to reduce both false-positive and false-negative rate. False-positive is defined as a good product mistakenly marked as contaminated, and vice versa for false-negative. When the decision is for economic reasons, the commercial sorters would provide the option to allow the operator to adjust the thresholds.

On line inspection of apples is an important application for the apple industry. For apple defects detection, Cheng et al. (2003) used a dual-camera approach. The method included the use of a near-infrared and a mid-infrared (MIR) camera. The results demonstrated that the MIR camera could identify only the stem-end/calyx parts of the apple, while the NIR camera could identify both the stem-end/calyx portions and the true defects. When the two cameras worked together a 100 % recognition rate for good apples and a 92 % recognition rate for defective apples was achieved. Kim et al. (2008a) developed a line-scan imaging system to simultaneously acquire a combination of multispectral reflectance and fluorescence for the inspection of quality and safety attributes of apples. The system was capable of working on a sorting line with a speed of three to four apples per second. In Figure 6.9 is described this

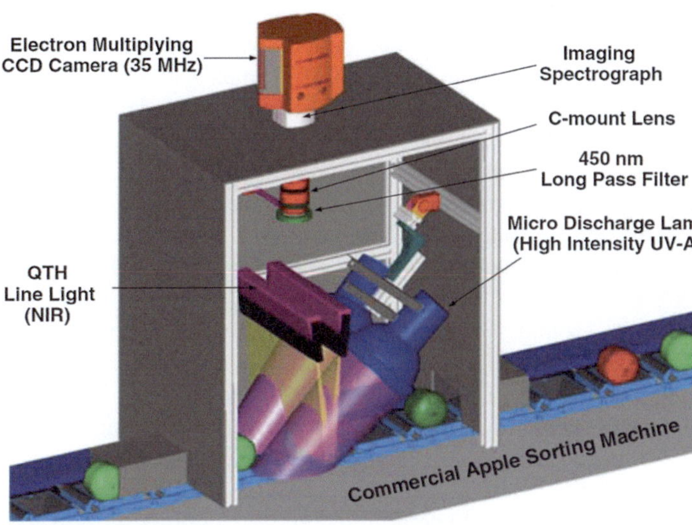

Fig. 6.9 Multispectral line-scan imaging inspection instrument for quality and safety applications of apples (Kim et al 2008b)

inspection system. Two light sources, a UV-A (320–400 nm) lamp and a tungsten halogen lamp, were used for fluorescence excitation and reflectance illumination. When in operation, the system could collect fluorescence images in the visible light range and reflectance data in the NIR range. The imaging system incorporated an EMCCD camera and a spectrograph for line-scan imaging. The EMCCD camera could be configured for wavelength addressing, indicating that a few selected wavelengths could be acquired rather than imaging the full wavelength. In the paper (Kim et al. 2008a) the fluorescence ratio from 665 to 530 nm was used for fecal contamination detection. The reflectance ratio from 750 to 800 nm was used for defect detection. This online system would be a good validation instrument for testing results obtained from full wavelength hyperspectral imagery for many food inspection applications.

A similar concept regarding online inspection is the Process Analytical Technology (PAT) initiative for the pharmaceutical industry. As described by Federal Drug Administration (Kourti 2006), PAT are "*systems for the analysis and control of manufacturing processes based on timely measurements during processes of critical quality parameters and performance attributes of raw and in-process materials and processes to assure acceptable end product quality at the completion of the process*". The essential components of PAT including process analysis, real time measurement and monitoring, multivariate statistical analysis, and in-situ control are also suitable for the online inspection of agricultural products. Gowen et al. (2007) reviewed the use of hyperspectral imaging as PAT implementation for food quality and safety control. Similar to the pharmaceutical industry, it is expected that hyperspectral imaging will be increasingly adopted as a PAT by the food industry. For example, in food processing industry, it is important to maintain cleanliness of the work environment and keep the food processing equipment free from contaminants. Specifically, bacteria can establish a community of biofilm on food processing equipment surface materials such as stainless steel. The biofilm can cause cross-contamination of the food being processed. Although many methods are available for the detection of biofilms, technology for rapid and non-invasive detection of biofilms on large equipment surfaces is always in demand. One study (Jun et al. 2009) examined the feasibility of using hyperspectral fluorescence imaging to inspect bacterial biofilms on the surface of stainless steel material. In this work two biopathogen samples, *E. coli* and *Salmonella*, were applied to the stainless steel surface. Under long wavelength UV excitation (320–400 nm), both biopathogen films exhibited blue emission peaks at around 480 nm. The emission peaks had the highest contrast between the biopathogen films and the stainless steel background. The results showed that the second principal component image from the hyperspectral fluorescence data had the most distinguishable morphological differences between the concentrated biofilm formations of *E. coli* and *Salmonella*. The authors suggested this method could be used to pre-screen surfaces of food processing equipment and could compliment the use of other biosensors designed for specific microbial targets.

6.5.3 Field/Remote Monitoring

Although previously discussed research was focused on controlled environments such as those that exist indoors and under lab conditions, work has also been conducted in out-door environments. Identifying or detecting food safety issues on plant products with hyperspectral imaging techniques is more challenging in this situation. This type of research typically used canopy reflectance collected from in-field or remote platforms to correlate with the safety problems in question. Thus many remote sensing methods were used in the research. In this section, studies ranging from leaf level spectrometer reflectance data to field level airborne hyperspectral data will be discussed.

Typically, vegetation indices are used for the analysis of vegetation reflectance. Vegetation indices have been used widely in remote sensing. The most important vegetation index is the normalized difference vegetation index (NDVI) calculated by using the red and near-infrared wavelengths. The use of hyperspectral images makes it possible to build more refined vegetation indices by using distinct narrow-bands and improving the indices for the correction of the effects of soil background. Many hyperspectral vegetation indices have been developed for different applications, with the simplest vegetation index being based on individual bands.

Some research used point measurement of spectral data. Muhammed and Larsolle (2003) and Muhammed (2005) collected canopy reflectance with a spectrometer for fungal disease severity assessment in wheat. The wheat was naturally infected with *Drechslera tritici-repenti* that can cause tan spot disease. Several analysis methods including independent component analysis, PCA, and a nearest-neighbor classification were used in the above research. There were two effects on the spectral curve observed with increased disease severity. The first was a flattening of the green reflectance peak together with a general decrease in reflectance in the near-infrared region. The second was a decrease of the shoulder of the near-infrared reflectance plateau together with a general increase in the visible region between 550 and 750 nm. Mahlein et al. (2010) also used a spectrometer to measure reflectance spectra (400–1,050 nm) of leaves infected with the sugar beet fungal pathogens *Cercospora beticola, Erysiphe betae,* and *Uromyces betae* causing *Cercospora* leaf spot, powdery mildew and rust. The purpose of the study was to examine the potential of hyperspectral sensor systems for the nondestructive detection and differentiation of plant diseases. Among the vegetation indices evaluated, the spectral vegetation indices NDVI, Anthocyanin Reflectance Index (ARI) and modified Chlorophyll Absorption Integral (mCAI) differed in their ability to assess the different diseases at an early stage of disease development, or even before first symptoms became visible. The conclusion was that the use of spectral vegetation indices for the differentiation of the three sugar beet diseases is possible when using the combination of two or more indices.

Other research that took advantage of hyperspectral imaging used hyperspectral images for the detection of *Fusarium* infection (head blight) in wheat Bauriegel et al. (2011a, b). The hyperspectral images were acquired with a push-broom line-scanning imaging system (400–1,000 nm). Instead of collecting canopy

Fig. 6.10 Sample classification of *Fusarium* infected wheat ears (**a**) RGB image, (**b**) *grey-scale* image according to head blight index, (**c**) classification result (*dark grey/red*: diseased, *light grey/green*: healthy) (Bauriegel et al. 2011a)

reflectance, single wheat ears were imaged (Fig. 6.10). The fungal infection was introduced artificially and time series hyperspectral images were obtained. It was found that the head blight could be readily detected with the imaging method 7 days after inoculation up to a disease severity of 50 %. The healthy and diseased ears were hardly distinguishable after the beginning of ripening. The best time for the detection was at the beginning of the medium milk stage. A head blight index (HBI), which uses spectral differences in the ranges of 665–675 nm and 550–560 nm, was developed for the detection of head blight. The above results provided useful information for possible selective harvesting operations.

Zhang et al. (2003) also applied hyperspectral remote sensing for disease detection on a large scale in the field. The application was the detection of stress in tomatoes induced by late blight disease. An airborne visible infrared imaging spectrometer (AVIRIS) image with 224 bands within the wavelength range of 0.4–2.5 m acquired during the growing season, together with field data, was used. It was found, based on the spectral reflectance of the field samples (measured with a spectrometer), that the near-infrared region, especially 700–1,300 nm, was much more valuable than the visible region for detecting the crop disease. The disease level was divided into four levels from light symptoms to severe damage. The classification results based on SAM (Spectral Angle Mapper) showed that the late blight diseased tomatoes at stage 3 or above could be separated from the healthy plants while the less infected plants (at stage 1 or 2) were difficult to separate from the healthy plants.

6.6 Summary/Conclusions

Advances in hyperspectral imaging technology have provided enormous opportunity for the food industry and research community to develop rapid and non-invasive inspection methods for food safety inspection. This chapter reviewed and discussed different aspects of using this technology in safety inspection of plant products. The global demand for fresh plant products is on the rise due to improved living standards and increased health awareness. Among issues related to food production, food safety is always of major concern. Three types of major contaminants including pathogens, chemical contaminants, and physical contaminants related to food safety, are discussed in this chapter. Pathogen contamination in food is caused by certain microorganisms such as bacteria, viruses, fungi, and parasites. Consumption of pathogen contaminated food can easily lead to sickness and even be fatal. Chemical contamination can occur either in natural or industrial situations. Similar to pathogen contamination, the ingestion of food with chemical contaminants, for example toxins, can also be deadly. Physical contamination of a food product occurs when foreign objects are intermixed with food. Different from the other two contamination sources, physical contaminants generally do not cause acute biological reactions, but the contamination is still a very important food safety problem. Current safety inspection methods may not adequately address all of the aforementioned safety issues; therefore, there is a need for novel, preferably rapid, non invasive and cost effective technology applicable to food safety. Research and studies in safety inspection have demonstrated that hyperspectral technology is a viable method to address the above food contaminants.

The following steps are generally used in application-based research using hyperspectral technology: experiment design, sample preparation, image acquisition, spectral pre-processing/calibration, sample ground truth characterization, data analysis and information extraction. Within this framework, the full wavelength response from the samples is normally studied first. When sufficient knowledge is obtained for a certain subject, key wavelengths can be chosen for specially designed inspection devices to speed up the inspection process. The hyperspectral technology is used as a research tool which transfers knowledge learned from the laboratory to real world conditions. Hyperspectral data in the format of reflectance, transmittance, and fluorescence are generally used in applying the technology for different plant safety inspection applications. The reflectance data can be used for external inspection and evaluation of surface contaminants, while the transmittance hyperspectral imagery can be useful for studying internal safety issues of food. The fluorescence hyperspectral data is suitable for studying the properties of sample constituents and also for chemical composition related to safety inspection. In some cases the combination of different spectral data can be used. For example, reflectance and fluorescence were used for online inspection of apples (Kim et al. 2008b). In this application, defects and diseases were detected with the reflectance data and fluorescence data was used for detecting fecal contamination.

This chapter discusses utilizing hyperspectral imaging for potential safety inspection of plant products such as grain, produce, nuts, and spices. The products discussed are corn, wheat, barley, soybean, apple, vegetables, almond, walnut, pistachio, peanut, spice, etc. The main contaminants discussed are aflatoxin, fumonisin, fungal infection, and fecal contamination.

With the development of innovative detection technology achieved by fusion of different hyperspectral data, or the combination of sensors, the detection accuracy of hyperspectral applications is expected to continue improving. It is therefore expected that hyperspectral technology will be adapted and extended to more food safety applications in the near future.

References

Allen MJ, Edberg SC, Reasoner DJ (2004) Heterotrophic plate count bacteria—what is their significance in drinking water? Int J Food Microbiol 92(3):265–274

Ariana DP, Lu R (2008) Quality evaluation of pickling cucumbers using hyperspectral reflectance and transmittance imaging: part I. Development of a prototype. Sens Instrumen Food Qual Saf 2(3):144–151

Armstrong PR (2006) Rapid single-kernel NIR measurement of grain and oil-seed attributes. Appl Eng Agric 22(5):767–772

Arora P, Sindhu A, Dilbaghi N, Chaudhury A (2011) Biosensors as innovative tools for the detection of food borne pathogens. Biosens Bioelectron 28(1):1–12

Atas M, Temizel A, Yardimci Y (2010) Using hyperspectral imaging and artificial neural network for classification of aflatoxin contaminated chili pepper. In: IEEE 18th signal processing, communication and applications conference

Ataş M, Yardimci Y, Temizel A (2011) Aflatoxin contaminated chili pepper detection by hyperspectral imaging and machine learning. SPIE defense, security, and sensing (Sensing for Agriculture and Food Quality and Safety III)

Bauriegel E, Giebel A, Geyer M, Schmidt U, Herppich WB (2011a) Early detection of Fusarium infection in wheat using hyper-spectral imaging. Comput Electron Agr 75(2):304–312

Bauriegel E, Giebel A, Herppich WB (2011b) Hyperspectral and chlorophyll fluorescence imaging to analyse the impact of *Fusarium culmorum* on the photosynthetic integrity of infected wheat ears. Sensors 11(4):3765–3779

Behravesh CB, Mody RK, Jungk J, Gaul L, Redd JT, Chen S (2011) 2008 outbreak of *Salmonella saintpaul* infections associated with raw produce. N Engl J Med 364(10):918–927

Bennett JW, Klich M (2003) Mycotoxins. Clin Microbiol Rev 16(3):497–516

Besser RE, Lett SM, Weber JT, Doyle MP, Barret TJ, Wells JG, Griffin PM (1993) An outbreak of diarrhea and hemolytic uremic syndrome from *Escherichia coli* O157:H7 in fresh pressed apple cider. J Am Med Assoc 269(17):2217–2220

Berardo N, Pisacane N, Battilani P, Scandolara A, Pietri A, Marocco A (2005) Rapid detection of kernel rots and mycotoxins in maize by near-infrared reflectance spectroscopy. J Agric Food Chem 53(21):8128–8134

Beyer M, Pogoda F, Ronellenfitsch FK, Hoffmann L, Udelhoven T (2010) Estimating deoxynivalenol contents of wheat samples containing different levels of Fusarium-damaged kernels by diffuse reflectance spectrometry and partial least square regression. Int J Food Microbiol 142(3):370–374

Bisha B, Brehm-Stecher BF (2009) Flow-through imaging cytometry for characterization of *Salmonella* subpopulations in alfalfa sprouts, a microbiologically complex food system. Biotechnol J 4:880–887

Birth GS, Hecht HG (1987) The physics of near-infrared reflectance. In: Near-infrared technology in the agricultural and food industries. American Association of Cereal Chemists, St. Paul, pp 1–16

Brown RL, Chen ZY, Menkir A, Cleveland TE, Cardwell K, Kling J, White DG (2001) Resistance to aflatoxin accumulation in kernels of maize inbreds selected for ear rot resistance in west and central Africa. J Food Prot 64(3):396–400

Burtscher C, Wuertz S (2003) Evaluation of the use of PCR and reverse transcriptase PCR for detection of pathogenic bacteria in biosolids from anaerobic digestors and aerobic composters. Appl Environ Microbiol 69(8):4618–4627

Carnaghan RBA, Hartley RD, O'kelly J (1963) Toxicity and fluorescence properties of the aflatoxins (*Aspergillus flavus*). Nature 200:1101

Carrasco O, Roper WE, Gomez RB, Chainani A (2003) Hyperspectral imaging applied to medical diagnoses and food safety. Proc SPIE 5097:215–221

CDC (1996) Outbreak of *E. coli* O157:H7 infections associated with drinking unpasteurized commercial apple juice-October 1996. Morbid Mortal Weekly Rep 45:975–982

CDC (1997) Outbreaks of *Escherichia* coli O157:H7 infection and cryptosporidiosis associated drinking unpasteurized apple cider-Connecticut and New York, October 1996. Morbid Mortal Weekly Rep 46(1):4–8

CDC (2011a) Increase in Salmonella infections, according to CDC report 2011. Food Safety Magazine 17(4): 10, 79

CDC (2011b) Multistate foodborne outbreaks. http://www.cdc.gov/outbreaknet/outbreaks.html

Chao K, Yang CC, Chen YR, Kim MS, Chan DE (2007a) Hyperspectral-multispectral line-scan imaging system for automated poultry carcass inspection applications for food safety. Poult Sci 86(11):2450–2460

Chao K, Yang CC, Chen YR, Kim MS, Chan DE (2007b) Fast line-scan imaging system for broiler carcass inspection. Sens Instrumen Food Qual Saf 1(2):62–71

Cheng X, Tao Y, Chen YR, Luo Y (2003) NIR/MIR dual-sensor machine vision system for online apple stem-end/calyx recognition. Trans ASAE 46(2):551–558

Cho BK, Chen YR, Kim MS (2007) Multispectral detection of organic residues on poultry processing plant equipment based on hyperspectral reflectance imaging technique. Comput Electron Agr 57:177–180

Cody SH, Glynn MK, Farrar JA, Cairns KL, Griffin PM, Kobayashi J, Fyfe M, Hoffman R, King AS, Lewis JH, Swaminathan B, Bryant RG, Vugia DJ (1999) An outbreak of *Escherichia coli* O157:H7 infection from unpasteurized commercial apple juice. Ann Internal Med 130 (3):202–209

Cogdill RP, Hurburgh CR Jr, Rippke GR, Bajic SJ, Jones RW, McClelland JF, Jensen TC, Liu J (2004) Single-kernel maize analysis by near-infrared hyperspectral imaging. Trans ASAE 47 (1):311–320

Collison E, Ohaeri G, Wadul-Mian M, Nkama I, Negbenebor C, Igene J (1992) Fungi associated with stored unprocessed cowpea and groundnut varieties available in Borno State, Nigeria. J Hyg Epidemiol Microbiol Immunol 36(4):338–345

Conde OM, Garcia-Allende PB, Cubillas AM, Gonzalez DW, Madruga FJ, Lopez-Higuera JM (2006) Industrial defects discrimination applying imaging spectroscopy and neural networks. ECNDT; Poster 74

Del Fiore A, Reverberi M, Ricelli A, Pinzari F, Serranti S, Fabbri AA, Bonifazi G, Fanelli C (2010) Early detection of toxigenic fungi on maize by hyperspectral imaging analysis. Int J Food Microbiol 144(1):64–71

Delwiche SR, Kim SM (2000) Hyperspectral imaging for detection of scab in wheat. Proc SPIE 4203:13–20

Delwiche SR (2003) Classification of scab and other mold-damaged wheat kernels by near-infrared reflectance spectroscopy. Trans ASAE 46(3):731–738

Delwiche SR, Hareland GA (2004) Detection of scab-damaged hard red spring wheat kernels by near-infrared reflectance. Cereal Chem 81(5):643–649

Delwiche SR, Gaines CS (2005) Wavelength selection for mono chromatic and bichromatic sorting of Fusarium-damaged wheat. Appl Eng Agric 21(4):681–688

Delwiche SR, Pearson TC, Brabec DL (2005) High-speed optical sorting of soft wheat for reduction of deoxynivalenol. Plant Dis 89(11):1214–1219

Delwiche SR (2008) High-speed bichromatic inspection of wheat kernels for mold and color class using high-power pulsed LEDs. Sens Instrumen Food Qual Saf 2(2):103–110

Deshpande SS, Cheryan M, Gunasekaran S, Paulsen MR, Salunkhe DK (1984) Nondestructive optical methods of food quality evaluation. Crit Rev Food Sci Nutr 21(4):323–379
Dowell FE, Ram MS, Seitz LM (1999) Predicting scab, vomitoxin, and ergosterol in single wheat kernels using near-infrared spectroscopy. Cereal Chem 76(4):573–576
Dowell FE, Pearson TC, Maghirang EB, Xie F, Wicklow DT (2002) Reflectance and transmittance spectroscopy applied to detecting Fumonisin in single corn kernels infected with *Fusarium verticillioides*. Cereal Chem 79(2):222–226
Dubois J, Lewis EN, Fry FS Jr, Calvey EM (2005) Bacterial identification by near-infrared chemical imaging of food-specific cards. Food Microbiol 22(6):577–583
Escoriza M, VanBriesen J, Stewart S, Maier J, Treado P (2006) Raman spectroscopy and chemical imaging for quantification of filtered waterborne bacteria. J Microbiol Meth 66(1):63–72
EC (2004) European Commission. Commission recommendation of 19 December 2003 concerning a coordinated programme for the official control of food stuffs for 2004. Off J Eur Union L6:29–37
Farsaie AL, McClure WF, Monroe RJ (1978) Development of indices for sorting Iranian pistachio nuts according to fluorescence. J Food Sci 43(5):1550–1552
FDA/CFSAN (1978) Guidance for industry: fumonisin levels in human foods and animal feeds. Final guidance. US FDA, Washington, DC, 2001
FDA (2001) Hazard analysis and critical control point (HAACP); procedures for the safe and sanitary processing and importing of juices. Fed Regist 66:6137–6202
FDA (2007) Food protection plan. Department of Health and Human Services. U.S. Food and Drug Administration, College Park
Fernández-Ibañez V, Soldado A, Martínez-Fernández A, de la Roza-Delgado B (2009) Application of near infrared spectroscopy for rapid detection of aflatoxin B1 in maize and barley as analytical quality assessment. Food Chem 113(2):629–634
Food Safety News (2011) http://www.foodsafetynews.com/sections/foodborne-illness-outbreaks/
Gat N (2000) Imaging spectroscopy using tunable filters: a review. Proc SPIE 4056(1):50
Giordano G (2011) Germany's E. Coli nightmare. Food Qual 18(4):12–13, 21
Goryacheva IY, Rusanova TY, Pankin KE (2008) Fluorescent properties of aflatoxins in organized media based on surfactants, cyclodextrins, and calixresorcinarenes. J Anal Chem 3(8):751–755
Gowen AA, O'Donnell CP, Cullen PJ, Downey G, Frias JM (2007) Hyperspectral imaging-an emerging process analytical tool for food quality and safety control. Trends Food Sci Technol 18(12):590–598
Gowen AA, O'Donnell CP, Taghizadeh M, Gaston E, O'Gorman A, Cullen PJ, Frias JM, Esquerre C, Downey G (2008) Hyperspectral imaging for the investigation of quality deterioration in sliced mushrooms (*Agaricus bisporus*) during storage. Sens Instrumen Food Qual Saf 2(3):133–143
Haff RP, Pearson T (2006) Spectral band selection for optical sorting of pistachio nut defects. Trans ASABE 49(4):1105–1113
Haff RP, Pearson TC, Toyofuku N (2010) Sorting of In-shell pistachio nuts from kernels using color imaging. Appl Eng Agr 26(4):633–638
Heitschmidt GW, Park B, Lawrence KC, Windham WR, Smith DP (2007) Improved hyperspectral imaging system for fecal detection on poultry carcasses. Trans ASABE 50(4):1427–1432
Hernandez-Hierro JM, Garcia-Villanova RJ, Rodríguez Torrero P, Toruño Fonseca IM (2008) Aflatoxins and ochratoxin A in red paprika for retail sale in Spain: occurrence and evaluation of a simultaneous analytical method. J Agric Food Chem 56(3):751–756
Hirano S, Okawara N, Narazaki S (1998) Near infra red detection of internally moldy nuts. Biosci Biotechnol Biochem 62(1):102–107
Hiroaki I, Toyonori N, Eiji T (2002) Measurement of pesticide residues in food based on diffuse reflectance IR spectroscopy. IEEE Trans Instrum Meas 51(5):886–890
Hu Y, Guo K, Suzuki T, Noguchi G, Satake T (2008) Quality evaluation of fresh pork using visible and near-infrared spectroscopy with fiber optics in interactance mode. Trans ASABE 51(3):1029–1033

International Commission on Microbiological Specifications for Foods (ICMSF) (2005) Spices, herbs, and dry vegetable seasonings. In: ICMSF (ed) Microorganisms in foods 6, microbial ecology of food commodities, 2nd edn. Kluwer Academic/Plenum, London

Jian J, Tang L, Hruska Z, Yao H (2009) Classification of toxigenic and atoxigenic strains of *Aspergillus flavus* with hyperspectral imaging. Comput Electron Agric 69(2):158–164

Jiang L, Zhu B, Rao X, Gerald B, Tao Y (2007) Discrimination of black walnut shell and pulp in hyperspectral fluorescence imagery using Gaussian kernel function approach. J Food Eng 81(1):108–117

Jin SS, Zhou J, Ye J (2008) Adoption of HACCP system in the Chinese food industry: a comparative analysis. Food Control 19:823–828

Jun W, Kim MS, Lee K, Millner P, Chao K (2009) Assessment of bacterial biofilm on stainless steel by hyperspectral fluorescence imaging. Sens Instrumen Food Qual Saf 3(1):41–48

Kalkan H, Beriat P, Yardimci Y, Pearson TC (2011) Detection of contaminated hazelnuts and ground red chili pepper flakes by multispectral imaging. Comput Electron Agr 77(1):28–34

Kavdir I, Lu R, Ariana D, Ngouajio M (2007) Visible and near-infrared spectroscopy for nondestructive quality assessment of pickling cucumbers. Postharvest Bio Technol 44 (2):165–174

Kay D, Crowther J, Fewtrell L, Francis CA, Hopkins M, Kay C (2008) Quantification and control of microbial pollution from agriculture: a new policy challenge? Environ Sci Policy 11(2):171–184

Kim MS, Chen YR, Mehl PM (2001) Hyperspectral reflectance and fluorescence imaging system for food quality and safety. Trans ASAE 44(3):721–729

Kim MS, Lefcourt AM, Chao K, Chen YR, Kim I, Chan DE (2002a) Multispectral detection of fecal contamination on apples based on hyperspectral imagery: part I. Application of visible and near-infrared reflectance imaging. Trans ASAE 45(6):2027–2037

Kim MS, Lefcourt AM, Chen YR, Kim I, Chan DE, Chao K (2002b) Multispectral detection of fecal contamination on apples based on hyperspectral imagery: part II. Application of hyperspectral fluorescence imaging. Trans ASAE 45(6):2039–2047

Kim MS, Lefcourt AM, Chen YR (2003a) Multispectral laser-induced fluorescence imaging system for large biological samples. Appl Opt 42(19):3927–3933

Kim MS, Lefcourt AM, Chen YR (2003b) Optimal fluorescence excitation and emission bands for detection of fecal contamination. J Food Prot 66(7):1198–1207

Kim MS, Lefcourt AM, Chen YR, Kang S (2004) Uses of hyperspectral and multispectral laser induced fluorescence imaging techniques for food safety inspection. Key Eng Mat 270–273:1055–1063

Kim MS, Lefcourt AM, Chen YR, Yang T (2005a) Automated detection of fecal contamination of apples based on multispectral fluorescence image fusion. J Food Eng 71(1):85–91

Kim MS, Lefcourt AM, Chen YR (2005b) Multispectral laser induced fluorescence imaging techniques for nondestructive assessment of postharvest food quality and safety. In: Proceedings of 5th international postharvest symposium, vol 682, pp 1379–1386

Kim MS, Chen YR, Cho BK, Chao K, Yang CC, Lefcourt AM, Chan D (2007) Hyperspectral reflectance and fluorescence line-scan imaging for online defect and fecal contamination inspection of apples. Sens Instrumen Food Qual Saf 1(3):151–159

Kim MS, Cho BK, Lefcourt AM, Chen YR, Kang S (2008a) Multispectral fluorescence lifetime imaging of feces-contaminated apples by time-resolved laser-induced fluorescence imaging system with tunable excitation wavelengths. Appl Opt 47(10):1608–1616

Kim MS, Lee K, Chao K, Lefcourt AM, Jun W, Chan DE (2008b) Multispectral line-scan imaging system for simultaneous fluorescence and reflectance measurements of apples: multitask apple inspection system. Sens Instrumen Food Qual Saf 2(2):123–129

Kim MS, Chao K, Chan DE, Yang C, Lefcourt AM, Delwiche SR (2011) Hyperspectral and multispectral imaging technique for food quality and safety inspection. In: Cho Y, Kang S (eds) Emerging technologies for food quality and food safety inspection. CRC, New York, pp 207–234

Kise M, Park B, Lawrence KC, Windham WR (2008) Development of handheld two-band spectral imaging system for food safety inspection. Biol Eng 1(2):145–157

Kourti T (2006) The process analytical technology initiative and multivariate process analysis, monitoring, and control. Anal Bioanal Chem 384(5):1043–1048

Lawrence KC, Park B, Heitschmidt GW, Windham WR, Mao C (2003a) Calibration of a pushbroom hyperspectral imaging system for agricultural inspection. Trans ASABE 46(2):513–521

Lawrence KC, Windham WR, Park B, Jeff Buhr R (2003b) A hyperspectral imaging system for identification of fecal and ingesta contamination on poultry carcasses. J Near Infrared Spectrosc 11:269–281

Lawrence KC, Smith DP, Windham WR, Heitschmidt GW, Park B (2006) Egg embryo development detection with hyperspectral imaging. Int J Poult Sci 5(10):964–969

Lawrence KC, Park B, Heitschmidt GW, Windham WR, Thai CN (2007) Evaluation of LED and tungsten-halogen lighting for fecal contaminant detection. Appl Eng Agric 23(6):811–818

Lazcka O, Campo FJD, Muñoz FX (2007) Pathogen detection: a perspective of traditional methods and biosensors. Biosens Bioelectron 22(7):1205–1217

Leemans V, Destain MF (1998) Defect segmentation in 'golden delicious' apples by using colour machine vision. Comput Elec Agric 20(2):117–130

Lefcourt AM, Kim MS, Chen YR (2003) Automated detection of fecal contamination of apples by multispectral laser-induced fluorescence imaging. Appl Opt 42(19):3935–3943

Lefcourt AM, Kim MS, Chen YR (2004) Portable multispectral fluorescence imaging system for food safety applications. Proc SPIE 5271:73–84

Lefcourt AM, Kim MS, Chen YR (2005a) A transportable fluorescence imagining system for detecting fecal contaminants. Comput Electron Agric 48(1):63–74

Lefcourt M, Kim MS, Chen YR (2005b) Detection of fecal contamination in apple calyx by multispectral laser-induced fluorescence. Trans ASAE 48(4):1587–1593

Lefcourt AM, Kim MS, Chen YR (2005c) Detection of fecal contamination on apples with nanosecond-scale time-resolved imaging of laser-induced fluorescence. Appl Opt 44(7):1160–1170

Lefcout AM, Kim MS, Chen YR, Kang S (2006) Systematic approach for using hyperspectral imaging data to develop multispectral imagining systems: detection of feces on apples. Comput Electron Agr 54(a):22–35

Lefcourt AM, Kim MS (2006) Technique for normalizing intensity histograms of images when the approximate size of the target is known: detection of feces on apples using fluorescence imaging. Comput Electon Agr 50(2):135–147

Lipp M (2011) A closer look at chemical contamination. Food Safety Magazine pp 28–31

Liu Y, Chen YR, Wang CY, Chan DE, Kim MS (2006) Development of hyperspectral imaging technique for the detection of chilling injury in cucumbers spectral and image analysis. Appl Eng Agric 22(1):101–111

Liu Y, Chen YR, Kim MS, Chan DE, Lefcourt AM (2007) Development of simple algorithms for the detection of fecal contaminants on apples from visible/near infrared hyperspectral reflectance imaging. J Food Eng 81(2):412–418

Lu R (2003) Detection of bruises on apples using near-infrared hyperspectral imaging. Trans ASAE 46(2):523–530

Lu R (2007) Nondestructive measurement of firmness and soluble solids content for apple fruit using hyperspectral scattering images. Sens Instrumen Food Qual Saf 1(1):19–27

Lu R, Ariana DP (2011) Detection of fruit fly infestation in pickling cucumbers using hyperspectral imaging. In: Proceedings of SPIE, vol 8027: 80270K

Maertens K, Reyns P, De Baerdemaeker J (2004) On-line measurement of grain quality with NIR technology. Trans ASABE 47(4):1135–1140

Mahlein AK, Steiner U, Dehne HW, Oerke EC (2010) Spectral signatures of sugar beet leaves for the detection and differentiation of diseases. Precision Agric 11(4):413–431

Maki DG (2009) Coming to grips with foodborne infection—peanut butter, peppers, and nationwide salmonella outbreaks. N Engl J Med 360:949–953

Maki DG (2006) Don't eat the spinach—controlling foodborne infectious disease. N Engl J Med 355:1952–1955

Mao C (2000) Focal plane scanner with reciprocating spatial window. U.S. Patent No. 6,166,373

Marsh PB, Simpson ME, Ferretti RJ, Merola GV, Donoso J, Craig GO (1969) Mechanism of formation of a fluorescence in cotton fiber associated with aflatoxin in the seeds at harvest. J Agric Food Chem 17(3):468–472

Maupin LM, Clements MJ, White DG (2003) Evaluation of the MI82 Corn Line as a source of resistance to aflatoxin in grain and use of BGYF as a selection tool. Plant Dis 87(9):1059–1066

McGlone VA, Martinsen PJ (2004) Transmission measurements on intact apples moving at high speed. J Near Infrared Spectrosc 12(1):37–42

Mehl PM, Chao K, Kim M, Chen YR (2002) Detection of defects on selected apple cultivars using hyperspectral and multispectral image analysis. Appl Eng Agr 18(2):219–226

Mehl PM, Chen YR, Kim MS, Chan DE (2004) Development of hyperspectral imaging technique for the detection of apple surface defects and contaminations. J Food Eng 61(1):67–81

Mucchetti G, Bonvini B, Francolino S, Neviani E, Carminati D (2008) Effect of washing with a high pressure water spray on removal of *Listeria innocua* from Gorgonzola cheese rind. Food Control 19(5):521–525

Muhammed HH, Larsolle A (2003) Feature vector based analysis of hyperspectral crop reflectance data for discrimination and quantification of fungal disease severity in wheat. Biosyst Eng 86(2):125–134

Muhammed HH (2005) Hyperspectral crop reflectance data for characterising and estimating fungal disease severity in wheat. Biosyst Eng 91(1):9–20

Naganathan GK, Grimes LM, Subbiah J, Calkins CR, Samal A, Meyer GE (2008) Partial least squares analysis of near-infrared hyperspectral images for beef tenderness prediction. Sens Instrumen Food Qual Saf 2(3):178–188

Nakariyakul S, Casasent DP (2011) Classification of internally damaged almond nuts using hyperspectral imagery. J Food Eng 103(1):62–67

Park B, Lawrence KC, Windham WR, Buhr RJ (2002) Hyperspectral imaging for detecting fecal and ingesta contaminants on poultry carcasses. Trans ASAE 45(6):2017–2026

Park B, Lee S, Yoon SC, Sundaram J, Windham WR, Hinton A Jr, Lawrence KC (2011) AOTF hyperspectral microscope imaging for foodborne pathogenic bacteria detection. SPIE Proc 8027:1–11

Pasikatan MC, Dowell FE (2001) Sorting systems based on optical methods for detecting and removing seeds infested internally by insects or fungi: a review. Appl Spectrosc Rev 36(4):399–416

Pearson TC, Schatzki TF (1998) Machine vision system for automated detection of aflatoxin-contaminated pistachios. J Agric Food Chem 46(6):2248–2252

Pearson TC (1999) Spectral properties and effect of drying temperature on almonds with concealed damage. Food Sci Technol-Leb 32(2):67–72

Pearson TC, Wicklow DT, Maghirang EB, Xie F, Dowell FE (2001) Detecting aflatoxin in single corn kernels by transmittance and reflectance spectroscopy. Trans ASAE 44(5):1247–1254

Pearson TC, Wicklow DT, Pasikatan MC (2004) Reduction of aflatoxin and fumonisin contamination in yellow corn by high-speed dual-wavelength sorting. Cereal Chem 8(4):490–498

Pearson TC, Wicklow DT (2006) Detection of corn kernels infected by fungi. Trans ASABE 49(4):1235–1245

Pearson TC (2009) Hardware-based image processing for high-speed inspection of grains. Comput Electron Agric 69(1):12–18

Peiris KHS, Pumphrey MO, Dowell FE (2009) NIR absorbance characteristics of deoxynivalenol and of sound and Fusarium damaged wheat kernels. J Near Infrared Spectrosc 17:213–221

Peiris KHS, Pumphrey MO, Dong Y, Maghirang EB, Berzonsky W, Dowell FE (2010) Near-infrared spectroscopic method for identification of Fusarium head blight damage and prediction of Deoxynivalenol in single wheat kernels. Cereal Chem 87(6):511–517

Peng Y, Lu R (2006) An LCTF-based multispectral imaging system for estimation of apple fruit firmness: part I. Acquisition and characterization of scattering images. Trans ASABE 49 (1):259–267

Peshin SS, Lall SB, Gupta SK (2002) Potential food contaminants and associated health risk. Acta Pharmacol Sin 23(3):193–202

Pettersson H, Aberg L (2003) Near infrared spectroscopy for determination of mycotoxins in cereals. Food Control 14(4):229–232

Polder G, van der Heijden GWAM, Waalwijk C, Young IT (2005) Detection of Fusarium in single wheat kernels using spectral imaging. Seed Sci Technol 33(3):655–668

Qin J, Lu R (2005) Detection of pits in tart cherries by hyperspectral transmission imaging. Trans ASAE 48(5):1963–1970

Qin J, Burks TF, Kim MS, Chao K, Ritenour MA (2008) Citrus canker detection using hyperspectral reflectance imaging and PCA-based image classification method. Sens Instrumen Food Qual Saf 2(3):168–177

Rahaie S, Emam-Djomehl Z, RazaviI SH, Mazaheri M (2010) Immobilized Saccharomyces cerevisiae as a potential aflatoxin decontaminating agent in pistachio nuts. Braz J Microbiol 41(1):82–90

Rasch C, Kumke M, Löhmannsröben HG (2010) Sensing of mycotoxin producing fungi in the processing of grains. Food Bioprocess Technol 3:908–916

Reese D, Lefcourt AM, Kim MS, Lo YM (2009) Using parabolic mirrors for complete imaging of apple surfaces. Bioresour Technol 100(19):4499–4506

Rietjens IM, Alink GM (2003) Nutrition and health-toxic substances in food. Ned Tijdschr Geneeskd 147(48):2365–2370

Riordan OCR, Sapers GM, Hankinson TI, Magee M, Mattrazzo AM, Annous BA (2001) A study of U. S. orchards to identify potential sources of *Escherichia coli* 01 57:H7. J Food Prot 64 (9):1320–1327

Robens J, Cardwell K (2003) The costs of mycotoxin management to the USA: management of aflatoxins in the United States. Toxin Rev 22(2–3):139–152

Robens J (2008) Aflatoxin—recognition, understanding, and control with particular emphasis on the role of the agricultural research service. Toxin Rev 27(3–4):143–269

Ruan R, Li Y, Lin X, Chen P (2002) Non-destructive determination of deoxynivalenol levels in barley using near-infrared spectroscopy. Appl Eng Agric 18(5):549–553

Sagoo SK, Little CL, Greenwood M, Mithani V, Grant KA, McLauchlin J, de Pinna E, Threlfall EJ (2009) Assessment of the microbiological safety of dried spices and herbs from production and retail premises in the United Kingdom. Food Microbiol 26(1):39–43

Schatzki TF, Haff RP, Young R, Can I, Le LC, Toyofuku N (1997) Defect detection in apples by means of x-ray imaging. Trans ASAE 40(5):1407–1415

Schatzki TF, Ong MS (2001) Dependence of aflatoxin in almonds on the type and amount of insect damage. J Agric Food Chem 49(9):4513–4519

Schrenk D (2004) Chemical food contaminants. Bundesgesundheitsblatt Gesundheitsforschung Gesundheitsschutz 47(9):841–847

Schmilovitch Z, Shenderey C, Shmulevich I, Alchanatis V, Egozi H, Hoffman A (2004) NIRS detection of mouldy core in apples. In: 2004 CIGR international conference, Beijing

September DJF (2011) Detection and quantification of spice adulteration by near infrared hyperspectral imaging. Master Thesis. Stellenbosch University

Shahin MA, Symons SJ (2011) Detection of Fusarium damaged kernels in Canada Western Red Spring wheat using visible/near-infrared hyperspectral imaging and principal component analysis. Comput Electron Agr 75(1):107–112

Shotwell OL, Hesseltine CW (1981) Use of bright greenish yellow fluorescence as a presumptive test for aflatoxin in corn. Cereal Chem 58:124–127

Singh CB, Jayas DS, Paliwal J, White NDG (2007) Fungal detection in wheat using near-infrared hyperspectral imaging. Trans ASABE 50(6):2171–2176
Siripatrawana U, Makinob Y, Kawagoeb Y, Oshitab S (2011) Rapid detection of *Escherichia coli* contamination in packaged fresh spinach using hyperspectral imaging. Talanta 85(1):276–281
Steele BT, Murphy N, Arbus GS, Rance CP (1982) An outbreak of hemolytic uremic syndrome associated with the ingest ion of fresh apple juice. J Pediatr 101(6):963–965
Umali-Deininger D, Sur M (2007) Food safety in a globalizing world: opportunities and challenges for India. Agric Econ 37:135–147
Uljas HE, Lngham SC (2000) Survey of apple growing, harvesting, and cider manufacturing practices in Wisconsin: implications for safety. J Food Saf 20(2):85–100
USDA (2002) USDA aflatoxin handbook. Grain Inspection Service Publication, Washington, DC
Van DE, Ieven M, Pattyn S, Van Damme L, Laga MJC (2001) Detection of *Chlamydia trachomatis* and *Neisseria gonorrhoeae* by enzyme immunoassay, culture, and three nucleic acid amplification tests. J Clin Microbiol 39:1751–1756
Vargas AM, Kim MS, Tao Y, Lefcourt AM, Chen YR (2004) Safety inspection of cantaloupes and strawberries using multispectral fluorescence techniques. ASAE paper no. 043056. ASAE, St. Joseph
Vargas AM, Kim MS, Tao Y, Lefcourt AM (2005) Detection of fecal contamination on cantaloupes using hypersepctral fluorescence imagery. J Food Sci 70(8):471–476
Velusamy V, Arshak K, Korostynska O, Oliwa K, Adley C (2010) An overview of foodborne pathogen detection: in the perspective of biosensors. Biotechnol Adv 28(2):232–254
Wang D, Dowell FE, Ram MS, Schapaugh WT (2004) Classification of fungal-damaged soybean seeds using near-infrared spectroscopy. Int J Food Prop 7(1):75–82
Wegulo SN, Dowell FE (2008) Near-infrared versus visual sorting of Fusarium-damaged kernels in winter wheat. Can J Plant Sci 88:1087–1089
Wen Z, Tao Y (2000) Dual-camera NIR/MIR imaging for stem-end/calyx identification in apple defect sorting. Trans ASAE 43(2):449–452
WHO (2007) Food safety & food-borne illness. Fact sheet no. 237 (reviewed March 2007). World Health Organization, Geneva
Wild CP, Turner PC (2002) The toxicology of aflatoxins as a basis for public health decisions. Mutagensis 17(6):471–481
Williams P, Manley M, Fox G, Geladi P (2010) Indirect detection of *Fusarium verticillioides* in maize (Zea mays L.) kernels by near infrared hyperspectral imaging. J Near Infrared Spectrosc 18(1):49–58
WTO (2007) International trade statistics. World Trade Organization, Geneva
Xing J, Guyer D, Ariana D, Lu R (2008) Determining optimal wavebands using genetic algorithm for detection of internal insect infestation in tart cherry. Sens Instrumen Food Qual Saf 2(3):161–167
Yang CC, Jun W, Kim MS, Chao K, Kang S, Chan DE, Lefcourt A (2010) Classification of fecal contamination on leafy greens by hyperspectral imaging. SPIE Proc 7676:76760F
Yang IC, Delwiche SR, Chen S, Lo YM (2009) Enhancement of Fusarium head blight detection in free-falling wheat kernels using a bichromatic pulsed LED design. Opt Eng 48(2):1
Yao H, Hruska Z, Kincaid R, Brown RL, Cleveland TE (2008) Differentiation of toxigenic fungi using hyperspectral imagery. Sens Instrumen Food Qual Saf 2(2):215–224
Yao H, Hruska Z, Kincaid R, Brown RL, Cleveland TE, Bhatnagar D (2010a) Correlation and classification of single kernel fluorescence hyperspectral data with aflatoxin concentration in corn kernels inoculated with *Aspergillus flavus* spores. Food Addit Contam 27(5):701–709
Yao H, Hruska Z, Kincaid R, Ononye A, Brown RL, Cleveland TE (2010b) Spectral angle mapper classification of fluorescence hyperspectral image for aflatoxin contaminated corn. In: Proceedings of IEEE 2nd workshop on hyperspectral image and signal processing: evolution in remote sensing, Iceland

Yao H, Lewis D (2010) Spectral pre-processing and calibration techniques. In: Sun D-W (ed) Hyperspectral imaging for food quality analysis and control. Academic, San Diego, Chapter 2

Yao H, Hruska Z, Kincaid R, Ononye A, Brown RL, Bhatnagar D, Cleveland TE (2011a) Development of narrow-band fluorescence indices for the detection of aflatoxin contaminated corn. In: Proceedings of SPIE conference, "Sensing for Agriculture and Food Quality and Safety III", 8027-12, April 26–27, Orlando, FL

Yao H, Hruska Z, Kincaid R, Ononye A, Brown RL, Bhatnagar D, Cleveland TE (2011b) Selective principal component regression analysis of fluorescence hyperspectral image to assess aflatoxin contamination in corn. In: Proceedings of IEEE 3rd workshop on hyperspectral image and signal processing: evolution in remote sensing conference, Lisbon

Yoon SC, Lawrence KC, Smith DP, Park B, Windham WR (2008) Bone fragment detection in chicken breast fillets using transmittance image enhancement. Trans ASABE 50(4):1433–1442

Yoon SC, Lawrence KC, Line JE, Siragusa GR, Feldner PW, Park B, Windham WR (2009) Hyperspectral reflectance imaging for detecting a foodborne pathogen: campylobacter. Trans ASABE 52(2):651–662

Yoon SC, Lawrence KC, Siragusa GR, Line JE, Park B, Feldner PW (2010) Detection of Campylobacter colonies using hyperspectral imaging. Sens Instrumen Food Qual Saf 4(1):35–49

Yoon SC, Park B, Lawrence KC, Windham WR, Heitschmidt GW (2011) Line-scan hyperspectral imaging system for real-time inspection of poultry carcasses with fecal material and ingesta. Comput Electron Agr 79:159–168

Zavattini G, Vecchi S, Leahy RM, Smith DJ, Cherry SR (2004) A hyperspectral fluorescence imaging system for biological applications, 2003. IEEE Nucl Sci Symp Conf Record 2:942–946

Zhang H, Paliwal J, Jayas DS, White NDG (2007) Classification of fungal infected wheat kernels using near-infrared reflectance hyperspectral imaging and support vector machine. Trans ASABE 50(5):1779–1785

Zhang M, Zhang Q, Liu X, Ustin SL (2003) Detection of stress in tomatoes induced by late blight disease in California, USA, using hyperspectral remote sensing. Int J Appl Earth Obs 4(4):295–310

Zhu B, Jiang L, Jin F, Qin L, Vogel A, Tao Y (2007) Walnut shell and meat differentiation using fluorescence hyperspectral imagery with ICA-kNN optimal wavelength selection. Sens Instrumen Food Qual Saf 1(3):123–131

Chapter 7
Foodborne Pathogen Detection

Seung-Chul Yoon

7.1 Introduction

Foodborne pathogens can cause various diseases and even death when humans consume foods contaminated with microbial pathogens. Although foodborne pathogens can be originated from various sources, such as animals, environments, and even humans, foods of animal origin are the primary source of many foodborne pathogens, such as Salmonella (poultry, eggs, meat, and produce), Campylobacter (poultry), Escherichia coli (ground beef, leafy green, and raw milk), Yersina (pork), Vibrio (fish) and Listeria (deli meats, unpasteurized soft cheese, and produce). According to the 2011 FoodNet report of the Centers for Disease Control and Prevention (CDC), the aforementioned six pathogens were responsible for the most foodborne diseases, in terms of prevalence rates (Centers for Disease Control and Prevention).

Rapid detection and identification of food-borne pathogens is increasingly important for development of intervention and verification strategies for the food industry and regulatory agencies. Traditional culture-based direct plating methods are still the "gold standard" for presumptive-positive pathogen screening (Dwivedi and Jaykus 2011; Beauchamp and Sofos 2009; Meng and Doyle 1998; Gracias and McKillip 2004). Direct plating methods provide good specificity, sensitivity, and information about the number of viable cells in food samples, broth cultures or liquid media at relatively low costs. However, a major challenge is that competitive microflora often grow together with target microorganisms on agar media and can appear morphologically similar. In practice, highly skilled technicians visually choose presumptive-positive colonies by trial and error for microscopic, biochemical, serological and molecular confirmation tests.

S.-C. Yoon (✉)
U.S. Department of Agriculture, Agricultural Research Service, US National Poultry Research Center, 950 College Station Road, Athens, GA 30605, USA
e-mail: seungchul.yoon@ars.usda.gov

B. Park, R. Lu (eds.), *Hyperspectral Imaging Technology in Food and Agriculture*, Food Engineering Series, DOI 10.1007/978-1-4939-2836-1_7

Although considerable research has been devoted to development and use of optical, biochemical, serological, and molecular methods for confirmation of presumptive-positive colonies, such as latex agglutination and polymerase chain reaction (PCR), little research has been done to develop methods and techniques for non-invasive screening of presumptive-positive colonies, while keeping them on agar-filled Petri dishes (Gracias and McKillip 2004; Lazcka et al. 2007; Velusamy et al. 2010; Mandal et al. 2011). Previous research on non-invasive detection of presumptive-positive colonies was concentrated on measuring light scattering (Bayraktar et al. 2006; Hirleman et al. 2008; Banada et al. 2009) and absorption (Yoon et al. 2009, 2010, 2013a, b; Windham et al. 2012) characteristics of colonies. Hyperspectral imaging is a non-destructive and non-contact optical imaging technique that combines aspects of conventional imaging and vibrational spectroscopy so that data can provide two-dimensional spatial information on colony shapes and spectral information at every pixel in each colony under test. The spectral "fingerprints" of bacteria provided by hyperspectral imaging can be used for detection and identification of pathogens. Researchers at the United States Department of Agriculture (USDA) Agricultural Research Service (ARS) have demonstrated the potential and efficacy of hyperspectral imaging for detection and identification of pathogenic colonies including Campylobacter (Yoon et al. 2009, 2010) and Escherichia coli (E. coli) (Windham et al. 2012; Yoon et al. 2013a, b).

This chapter covers research on the development of pathogen colony detection and classification using visible and near-infrared (VNIR) hyperspectral imaging. The first part of this chapter describes the important issues frequently encountered in hyperspectral imaging study for pathogen colony detection, including bacterial cultures, VNIR hyperspectral imaging system, and image acquisition and preprocessing. The second part of the chapter then presents two case studies for Campylobacter and E. coli colony detection.

7.2 Hyperspectral Imaging for Pathogen Detection

This section provides an overview of pathogens and agar media in a more general context and the experiment protocol that was used and modified, as needed, for the two case studies. The generic protocol included the sample preparation, imaging system, image acquisition and preprocessing.

7.2.1 Pathogens and Agar Media

Campylobacter and *E. coli* are commonly found bacteria in poultry and ground beef, respectively. *Campylobacter* is widespread in warm-blooded food-producing animals, and the presence of *Campylobacter* in foods of animal origin has been the most common cause of bacterially induced human gastroenteritis (diarrheal illness) in the United States (U.S.) and other developed countries. The most common species among more than a dozen of *Campylobacter* species causing human illness is *Campylobacter jejuni* (*C. jejuni*), followed by *Campylobacter coli* (*C. coli*) and other species (National advisory committee on microbiological criteria for foods (NACMCF) 2007). *E. coli* are bacteria living in the intestine of warm-blooded animals and humans. Many *E. coli* strains do not cause human disease; there is however, a pathogenic group of *E. coli* that produces Shiga toxin. Symptoms of human illnesses caused by the consumption of Shiga toxin-producing Escherichia coli (STEC) are diarrhea, stomach cramps, vomiting, and a potentially lethal kidney complication called hemolytic uremic syndrome (HUS). The most prevalent and commonly recognized STEC serotype is *E. coli* O157:H7; non-O157 STEC serogroups such as O26, O45, and O103 are also increasingly recognized (Bosilevac and Koohmaraie 2011; Griffin 1998). The U.S. Center for Disease Control (CDC) estimated that as many as 265,000 STEC infections occur every year in the U.S. alone, with about 64 % non-O157 STEC infections (Scallan et al. 2011). According to a previous study, from 1983 to 2002, about 70 % of non-O157 STEC infections were caused by six major serogroups, including O26, O45, O103, O111, O121, and O145 (called "Big Six") (Brooks et al. 2005).

Detection and isolation of bacteria are typically done in laboratories by growing them in liquid or solid culture media. Liquid culture (or called growth) media are typically used to grow a large amount of cells suspended in a nutrient broth. Some of the cells grown in a nutrient broth are transferred to solid growth media such as an agar plate (a Petri dish) that is typically used to detect, isolate and enumerate pure cultures of bacteria that appear as individual colonies on the plate. Common types of agar media are selective, differential, and non-selective. Selective media refer to a type of growth media with specific antibiotics or nutrients for supporting the growth of only certain types of microorganisms while inhibiting the growth of others. The selective media are widely used to determine whether specific organisms are present or not in a specimen. Differential media are used to distinguish certain types of organisms from others with characteristic changes in appearance or growth patterns, produced by adding dyes or chemicals, e.g. to change pH levels. Non-selective media are also popular to harvest many different types of bacteria by allowing the un-restricted growth of all microorganisms in the specimen that was plated. A chromogenic agar medium is a differential medium that produces the different coloration of certain bacteria colonies. A blood agar (BA) medium is a non-selective growth medium that is mainly used to grow many types of bacteria. The selection of appropriate agar media is usually determined by what types of pathogens are used for a study.

Table 7.1 summarizes the types of common agar media used for detection of Campylobacter and STEC and colony colors on each agar medium. Although most agar media are both selective and differential, the degree of selectivity and differentiation varies greatly among agar types. For example, Campy Line agar is more selective than Campy-Cefex agar. Campy-Cefex agar is one of commonly used agar media to isolate Campylobacter species from food sources. Campy-Cefex is more selective than blood-supplemented agars. STEC O157:H7 colonies on both Sorbitol

Table 7.1 Types of agar media and typical colony morphologies

Pathogen	Media	Characteristics	Colony color
Campylobacter	Campy-Cefex agar	Selective and differential	Colorless to grayish or light cream
Campylobacter	Campy Line agar	Selective and differential	Pink
Campylobacter	Blood agar	Non-selective and differential	Gray
STEC O157: H7	Sorbitol MacConkey agar	Selective and differential	Clear
STEC O157: H7	Rainbow agar O157	Selective and differential	Black or gray
STEC non-O157	Rainbow agar O157	Selective and differential	Pink, magenta, gray, or black

MacConkey agar and Rainbow agar are very selective and differential and, therefore, easy to discern STEC O157:H7 colonies even when background microflora are present on the same plate. However, in general, agar media for Campylobacter and STEC non-O157 are not very effective in their selectivity and differentiation of the targeted pathogens, especially when background microflora are present together at the same plate.

7.2.2 Hyperspectral Imaging System and Image Acquisition

A pushbroom line-scan VNIR hyperspectral imaging system used in a laboratory setting typically consists of a hyperspectral camera covering a spectral range of 400 to 1,000 nm, active illumination, and a translational motion control to move either a stage or a lens. The VNIR hyperspectral imaging system used for the studies described in this chapter had a moving lens set up (ITD, Stennis Space Center, MS, USA). The system was equipped with a copy stand to attach a hyperspectral camera, a computer for motion control and image acquisition, tungsten halogen lamps, and a Petri dish holder. The specifications of the VNIR image system shown in Fig. 7.1 are as follows:

- Prism-grating-prism spectrograph (ImSpector V10E, Specim, Oulu, Finland) for pushbroom line scanning
- 12-bit camera (SensiCam QE SVGA, Cooke Corporation, Auburn Hills, MI, USA)
- C-mount objective lens
- Spectral range between 400 and 1,000 nm with a nominal spectral resolution of 2.8 nm and a bandpass of 2.95 nm
- Detector- 17 mm (2/3-in.) silicon-based CCD with 1280 × 1024 pixel resolution
- A translation stage (STGA-10, Newmark Systems, Mission Viejo, CA, USA) attached to the slit end of the spectrograph. The translation stage moves by a

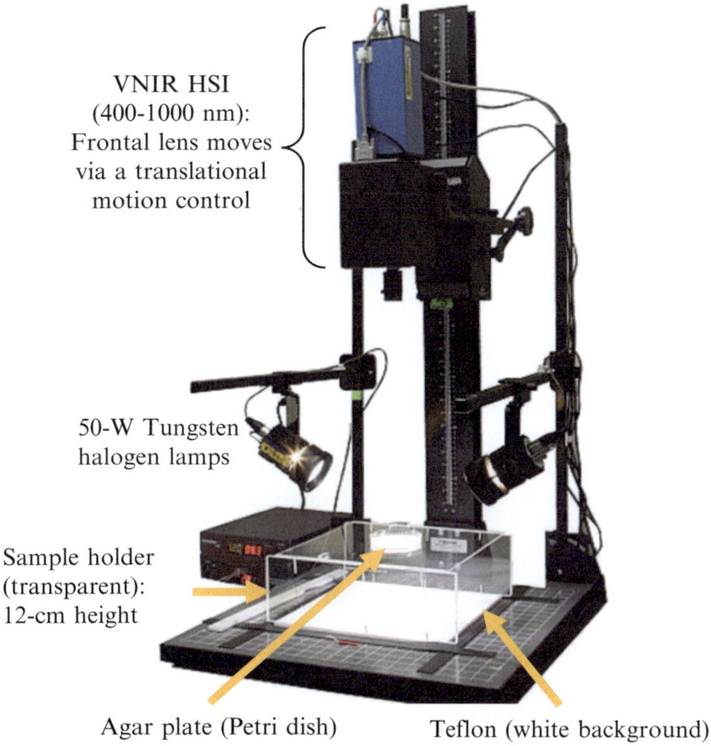

Fig. 7.1 Imaging system designed for imaging spread plates

motion controller (NCS-1S, Newmark Systems, Mission Viejo, CA, USA). Thus, the motorized translation stage moves the lens assembly so that successive lines of the Petri dish are scanned while the Petri dish remains stationary

For reflectance imaging, a Petri dish was laterally illuminated at about 45° pointing down from the left and right sides with two 50 W tungsten halogen lamps having a color temperature of 4,700°K. A white Teflon plate was placed under the Petri dish holder to increase the apparent reflectance of thin layered colonies on semi-transparent agar.

7.2.3 Hyperspectral Image Preprocessing

The acquired hyperspectral images were pre-processed to reduce the image size and to suppress spectral noise. All images were binned down spatially and spectrally. A Savitzky-Golay smoothing filter (window size: 25; order of moment: 4) was applied to the spectrum at each pixel position independently to reduce the random noise.

The reflectance intensity value at each image pixel was calibrated to relative reflectance value R with a 75 % reflectance Spectralon® target (13 × 13 cm, SRT-75-050, Labsphere, North Sutton, NH, USA). The relative reflectance value R at each pixel and each wavelength band was obtained by a reflectance calibration model below

$$R(x, y, z) = \frac{I_m(x, y, z) - I_d(x, y, z)}{I_r(x, y, z) - I_d(x, y, z)} \times C$$

where I_m is a measured raw value, I_r is a raw value measured with a reference target, I_d is a dark current value measured with a lens cap covered, and C is the nominal average reflectivity of the target material that was set to 75 % for the case studies. Percent reflectance calibration, spatial and spectral cropping and Savitzky-Golay smoothing filtering are conducted within with HyperVisual software (ITD, Stennis Space Center, MS, USA).

Calibrated images were arranged into a single image mosaic. An image mosaic approach was adopted to facilitate data analysis and algorithm development because a mosaic can be treated as a single hyperspectral image. In the mosaic, images measured during the same date were vertically stacked. The stacked images were then added to the mosaic in chronological order (latest right). The duplicated plates were arranged in the adjacent rows.

A binary mask was made in order to suppress the background noise outside and around the rim of each Petri dish. These binary masks were served as areas valid for testing classification algorithms (i.e., cross-validation of classification algorithms) and for facilitating other tasks, including image processing and analysis. In addition, ground-truth regions-of-interest (ROIs) were prepared in such a way that only pure organisms could be selected in the ROIs. Glints and rim shadows were excluded in the ROIs. Glints were usually observed around colony edges along the direction of the lateral illumination. If possible, pixels that might have contained mixed spectra of agar media and organisms were also excluded.

7.3 Detection of *Campylobacter*

This section describes the development of hyperspectral imaging techniques for differentiating *Campylobacter* and non-*Campylobacter* contaminants (i.e. background microflora) on spot plates, incubated for 48 and 24 h, respectively. The first study was done with the 48 h cultures and the second study was done with 24 h cultures for early detection.

For both 48 and 24 h cultures, a total of 17 different bacteria were used, including 11 *Campylobacter* strains (Campy.) and six non-*Campylobacter* bacteria (non-Campy.). The *Campylobacter* species used were *C. jejuni* (5 strains), *C. coli* (5 strains), and *C. lari* (1 strain). The non-*Campylobacter* species were the bacteria frequently found in poultry, such as *Sphingomonas paucimobilis*, *Acinetobacter baumannii*, *Brevundimonas diminuta*, *Ochrobacterium* sp., and *Flavobacterium odoratum*. With the exception of the American Type Culture Collection (ATCC) strains for 3 *Campylobacter* strains (1 *C. jenuni*, 1 *C. coli*, and 1 *C. lari*), all 14 other bacterial cultures were isolated from poultry samples consisting of either whole-carcass rinses or fecal specimens from conventionally reared broiler chickens or processing plants (Line 2001; Siragusa et al. 2004; Stern et al. 2001). The strains were maintained in frozen culture stock and propagated in culture tubes in 9 mL of enrichment broth. Tubes were incubated at 42 °C for up to 72 h in a Campy-gas atmosphere (85 % N_2, 10 % CO_2, 5 % O_2) created by placing the tubes within a re-sealable plastic bag, gas flushing three times, and refilling with the gas mixture for dense liquid growth. Following the initial liquid culturing step for 3 days, 5 μL spots (10 μL spots at the early stage of the study) were inoculated to the surface of the respective agar plate and incubated in Campy-gas atmosphere at 42 °C for a total of 48 h for the first study and 24 h for the second study. The recommended incubation time for growing *Campylobacter* is 48 h (National advisory committee on microbiological criteria for foods (NACMCF) 2007). Thus, the second study was designed to investigate the possibility of early detection of *Campylobacter* using hyperspectral imaging. Agars used were blood agar (BA; 5 % sheep's blood agar, Remel, Inc., Lenexa, Kans.), Campy-Cefex (Cefex) (Stern et al. 2001), and Campy-Line agar (CLA) (Line 2001). Agar plates were inoculated with spots at known and well-spaced locations on the agar surface. The primary reason for the use of spot plating was to make it simpler to build a spectral library of pure spectral signatures and use it for detection algorithm development, compared to spread plating that will be described in the *E. coli* detection study later in this chapter. Spot plating also made it easier to build ground-truth information of each pixel's identity by avoiding confluent growth and cross-contamination between colonies. The 17 spots were inoculated on two separate plates with nine and eight spots, respectively. Figure 7.2 shows an example picture of a spot plate (Campy-Cefex agar) with full grown microorganisms at eight different spots (*Campylobacter*

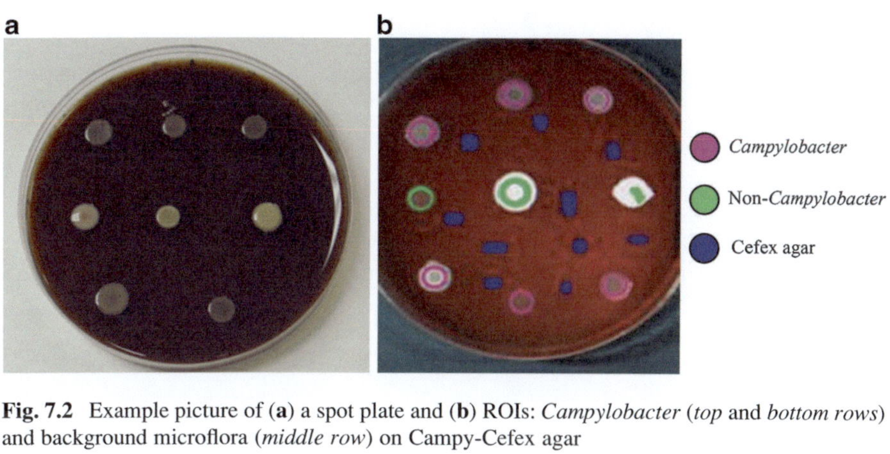

Fig. 7.2 Example picture of (**a**) a spot plate and (**b**) ROIs: *Campylobacter* (*top* and *bottom rows*) and background microflora (*middle row*) on Campy-Cefex agar

spots at the top and bottom rows and non-*Campylobacter* spots at the middle row) and an example of ROIs.

The hyperspectral imaging system used for both 48 and 24 h *Campylobacter* culture studies was essentially the same as the one in Fig. 7.1 (Yoon et al. 2009, 2010). Two-dimensional spectral images (i.e., line-scan images) were captured by 2 (spatial) × 4 (spectral) binning and 90 ms exposure time. The resolution of one hyperspectral image data was 640 (spatial) × 475 (scan lines) × 256 (spectral). Then, the images were further cropped down to 421 (width: spatial) × 475 (height: scan lines) × 194 (spectral). The 193 spectral bands in the range of 400–900 nm were kept. After the reflectance calibration and spectral de-nosing, the images were put into an image mosaic.

7.3.1 *Detection of 48 h* Campylobacter *Cultures on Spot Plates*

The first hyperspectral imaging study for discriminating *Campylobacter* and background microflora was designed with pure cultures grown on spot plates for 48 h. Eight experiments were repeated over 4 months, where one experiment took 5 days from sample preparation (72 h + 48 h) to plate imaging. A total of 108 spot plates (for all of three agar types: BA, Cefex, and CLA) were prepared for the four-month period. Figure 7.3 shows an image mosaic of Cefex cultures, where color composite images were created for a display purpose.

A spectral library consisting of mean reflectance spectra and their standard deviations was obtained from the pixels confined in the ground-truth ROI masks of all 17 organisms grown on the three types of agar media over the 400–900 nm range, as shown in Fig. 7.4. First, there was no significant difference in spectral

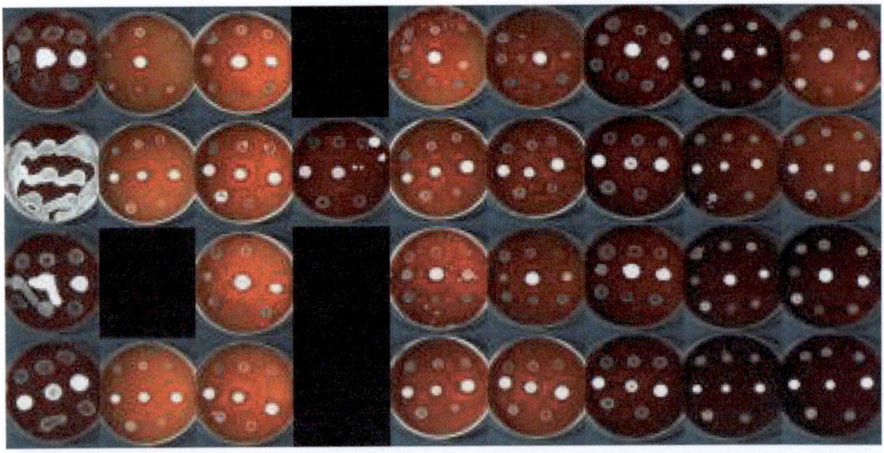

Fig. 7.3 Hyperspectral image mosaic (Cefex cultures). Color-composite images are displayed

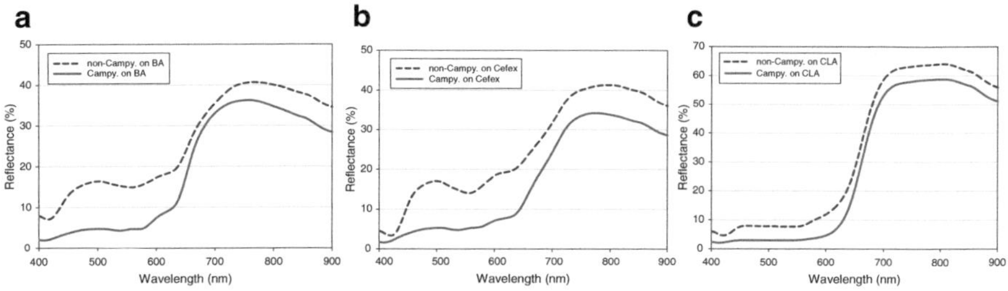

Fig. 7.4 Mean reflectance spectra of Campylobacter and background microflora on (**a**) BA, (**b**) Cefex, and (**c**) CLA spot plates

responses between Campylobacter strains. Second, except CLA, there was no significant difference in spectral responses between agar types as well. In the cases of BA and Cefex, *Campylobacter* showed low reflectance responses (~5 %) in the range below 650 nm and high reflectance (~35 %) in the near-infrared (NIR) range beyond 700 nm. Third, non-*Campylobacter* showed a characteristic reflectance feature at around 500 nm and their reflectivity was much higher than *Campylobacter* in the 500-nm region. In the case of Cefex, the mean spectrum of a particular non-*Campylobacter* organism *Sphingomonas paucimobilis* was very similar to that of *Campylobacter*. The mean spectra of non-*Campylobacter* on CLA were much different in the visible spectral range between 400 and 650 nm than on BA and Cefex. From the examination of the spectral library, it was obvious that Campy. and non-Campy. on both BA and Cefex plates could be well separable at around 500 nm, whereas the spectra obtained from CLA may not provide the same level of separability. Hence, it was hypothesized that the target organisms on BA and Cefex plates would be statistically separable in the approximated range of 450–550 nm. This hypothesis was evaluated by estimating the Bhattacharyya distances at a few bands determined by the principal component analysis (PCA) band weight analysis.

From the PCA-band weight analysis, two bands at 503 and 578 nm (the BA case) and three bands at 501, 606, and 827 nm (the Cefex case) were found as local peaks. Class separability in each of these bands was measured by the Bhattacharyya distance which measures a statistical distance between two populations. In the BA case, the band at 503 nm showed the largest statistical separability between *Campylobacter* and non-*Campylobacter*. In the Cefex case, the band at 501 nm was the best. In the CLA case, the overall separability was smaller than the other agars, as predicted from the spectral analysis. Hence, it was confirmed that *Campylobacter* and non-*Campylobacter* was statistically well separable at 503 nm for BA cultures and 501 nm for Cefex culture. Based on this statistical analysis, the development of classification algorithms was done for discriminating *Campylobacter* and background microflora on BA and Cefex plates only. The detection methods described next were designed as three-class classification algorithms that categorize the label of each pixel in the ROIs into one of three classes: Campy., non-Campy. and agar.

For BA cultures, a classification algorithm using a single wavelength band was designed. The algorithm was applied to a 503-nm band for the 3-class classification, and evaluated with the pre-defined ROI pixels. The algorithm was based on the following classification rule: (1) BA if a reflectance value is below 3 % ($T_1 = 3$), (2) background microflora (i.e. non-*Campylobacter*) if the value is greater than 7 % ($T_2 = 7$), and (3) *Campylobacter* otherwise. The total number of the ground-truth ROI pixels was 136,370, and the overall classification accuracy was 98.07 % (133,740 pixels/136,370 pixels). The Kappa coefficient was 0.9703. The detection accuracy of agar pixels was 100 %. When the agar pixels were excluded in the accuracy calculation, the classification accuracy only with *Campylobacter* and non-*Campylobacter* was 96.60 % (74,723 pixels/77,353 pixels). The classification results with pseudo-color were mapped in the image domain.

For Cefex cultures, a two-step detection algorithm was developed by performing (1) the aforementioned single band algorithm for 3-class classification with $T_1 = 3$ and $T_1 = 10$ and (2) a two-class minimum distance classifier using the Mahalanobis distance measure, which used all 193 spectral bands. The Mahalanobis distance classifier was applied to only the pixels predicted as *Campylobacter*, in order to find a non-Campy strain (*Sphingomonas paucimobilis*). The resulting classification results were again quantitatively evaluated against the ground-truth ROI pixels and predicted on the image space. The total number of the ground-truth ROI pixels was 264,984. The classification accuracy of the single band algorithm (first step) was 96.81 % (256,529 pixels/264,984 pixels) when agar pixels were included and 94.99 % (160,157 pixels/168,612 pixels) when agar pixels were excluded in the accuracy calculation. After the second step, the overall classification accuracy of the two-step algorithm became 99.29 % (263,104 pixels/264,984 pixels) when agar pixels were included and 98.58 % (166,221 pixels/168,612 pixels) when agar pixels were excluded in the accuracy calculation. The Kappa coefficient was 0.9893. Figure 7.5 shows a mosaic of classification images obtained by the two-step algorithm for the Cefex cultures. At the spot level, all spots except one whose prediction was mixed were correctly classified. At the pixel level, commonly misclassified pixels, although they were not many errors, were observed either on the *Campylobacter* and *Sphingomonas paucimobilis* spots. In conclusion, the classification results showed that hyperspectral imaging had a potential to be expanded for early detection of Campylobacter and for detection of other pathogens grown on agar media.

7.3.2 24 h Campylobacter *Cultures on Spot Plates for Early Detection Study*

Although the 48-h incubation time was often used for growing *Campylobacter* cultures in various studies such as the Food Safety and Inspection Service (FSIS) baseline study (Food Safety and Inspection Service (FSIS) 2006) and the International

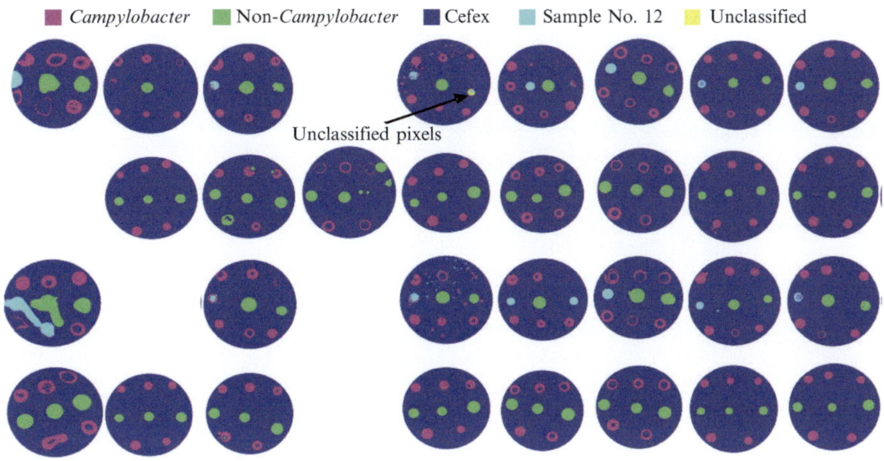

Fig. 7.5 Classification result to detect Campylobacter and background microflora on Cefex agar. Sample No. 12 refers to a non-Campy. contaminant, *Sphingomonas paucimobilis*

Organization for Standardization (ISO) standard for detecting *Campylobacter* (International Organization for Standardization (ISO) 2006), there was a need for a hyperspectral imaging study with the cultures incubated for less than 48 h for more rapid detection of *Campylobacter*.

Five replicates of experiments were carried out over the period of 6 months. The imaging system and experimental protocol for the sample preparation and imaging remained the same as the 48 h culture study throughout the experiments. The two duplicate plates were prepared per agar type and per experiment. Therefore, a total of 40 Petri dishes were prepared for imaging. Agars used were BA and Cefex. Figure 7.6 shows the hyperspectral image mosaics in RGB color composites for a display purpose.

A spectral library was also built for spectral analysis. The mean reflectance spectra of both 24-h and 48-h cultures from both blood agar and Cefex agar are presented in Fig. 7.7 for comparison. Although the locations of local maxima and minima were not much different between the 24-h and 48-h cultures, the overall reflectivity tended to decrease as the incubation time reduced from 48 to 24 h. In the visible spectral range from 400 to 650 nm especially at around 425 and 460 nm, there were pronounced local minimum and maximum features. In the red spectral range from 650 to 700 nm, reflectance responses increased sharply. Finally, in the near-infrared range from 700 to 900 nm, no prominent features were noticeable except that 24-h cultures on blood agar showed a weak absorption at around 760 nm. This weak absorption feature disappeared in the mean spectra of the 48-h cultures nonetheless.

The hyperspectral imaging study with 48-h *Campylobacter* cultures found a wavelength showing the statistical separability large enough to differentiate *Campylobacter* species from non-*Campylobacter* microorganisms. According to the

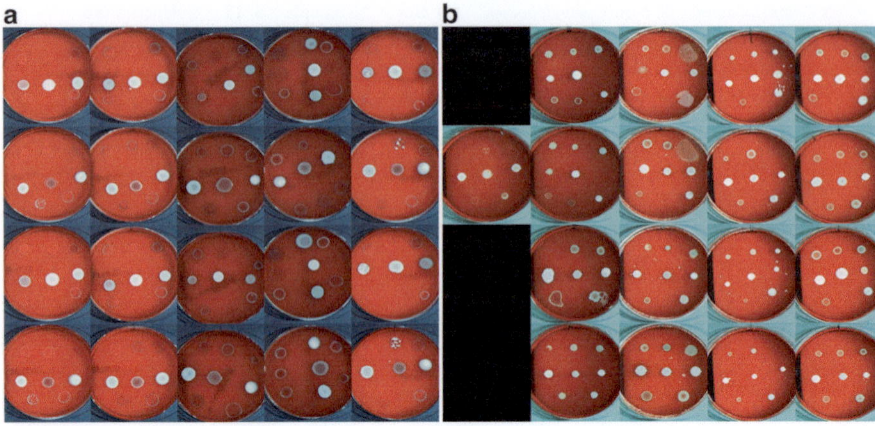

Fig. 7.6 Image mosaic showing 24-h spot cultures on (**a**) Blood agar and (**b**) Cefex agar. RGB color composite images are shown for a display purpose

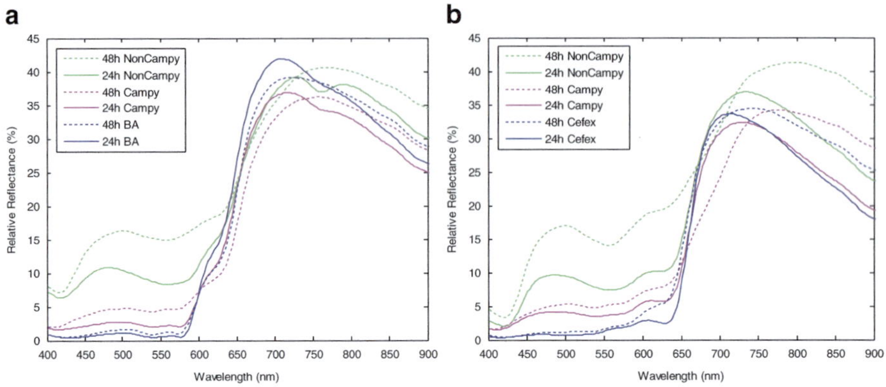

Fig. 7.7 Mean reflectance spectra of *Campylobacter* and non-*Campylobacter* spots grown on (**a**) Blood agar and (**b**) Cefex agar

study, the wavelength was at 503 nm for blood agar and 501 nm for Cefex agar. The same single-band thresholding algorithm was also evaluated in this 24-h culture study as a reference algorithm.

The parameters for the single-band thresholding algorithm applied to the 24-h culture images were slightly changed as follows: for the BA case, $T_1 = 2$ and $T_2 = 5$ ($T_1 = 3$ and $T_2 = 7$ for 48-h cultures) and for the Cefex case, $T_1 = 3$ and $T_2 = 7$ ($T_1 = 3$ and $T_2 = 10$ for 48-h cultures). In the 24-h BA case, the overall classification accuracy at the pixel level was 94.54 %, about 4 % down from 98.07 % of the 48-h BA case. In the 24-h Cefex case, the overall classification accuracy at the pixel level was 85.95 %, about 11 % down from 96.81 % of the 48-h Cefex case. At the spot level, 158 out of all 169 spots (93 %) on BA plates were correctly (at least by majority) classified. Similarly, at the spot level, 107 out of all 124 spots (86 %) on

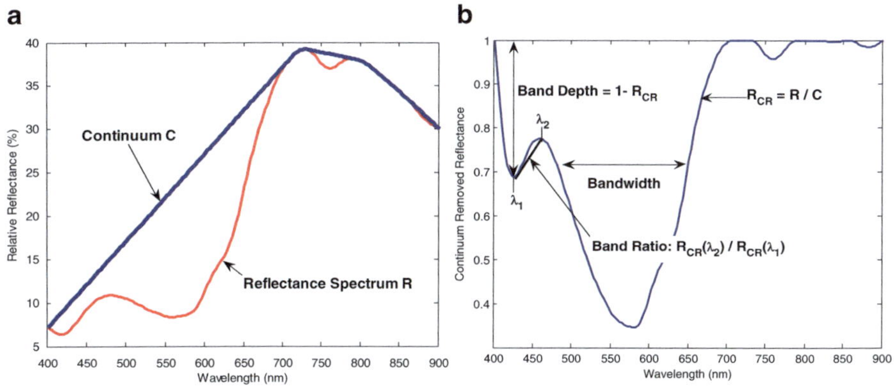

Fig. 7.8 Example of continuum removal and absorption features. (**a**) Continuum example and (**b**) Continuum-removed spectrum

Cefex plates were correctly (at least by majority) classified. Compared to the Cefex agar (85.95 % and 86 % at the pixel and spot level, respectively), the blood agar (94.54 % and 93 % at the pixel and spot level, respectively) was a better agar medium in differentiating *Campylobacter* colony spots from non-*Campylobacter* contaminants when the incubation time was reduced to 24 h. Although the detection accuracy for BA cultures was over 90 %, there was a need to improve the detection accuracy. Thus, a new algorithm using band ratio features obtained from continuum removed spectra was developed.

The continuum-removal analysis is aimed to quantify absorption bands departing from a common baseline (Clark and Roush 1984). The common baseline (i.e. continuum) is defined as the convex hull surrounding the data points of a reflectance spectrum. In other words, the continuum consists of piecewise continuous lines connecting local maximum points of the reflectance spectrum (Fig. 7.8a). Continuum removal is a procedure to isolate a particular absorption feature for analysis by dividing the reflectance spectrum at each wavelength with the continuum at the corresponding wavelength: $R_{CR} = R/C$ where R_{CR} is the continuum-removed spectrum, R is the reflectance spectrum and C is the continuum (Fig. 7.8). The continuum-removed spectral values range from 0 to 1. The first and last points in the reflectance spectrum are always local maxima for the continuum. Hence, the first and last points become 1 in the continuum-removed spectrum (Fig. 7.8b). After continuum removal was applied, the parametric quantification of the absorption band features can be done with the calculation of band depth, bandwidth and wavelength position (Kruse 1988; Kokaly and Clark 1999). As shown in Fig. 7.8b, the band depth D of each band can be calculated by subtracting the continuum-removed reflectance from 1: $D = 1 - R_{CR}$ (Kokaly and Clark 1999). Therefore, the slope information of the continuum-removed spectrum was utilized to enhance differences in the absorption features of *Campylobacter* and non-*Campylobacter* organisms. This slope information was represented by a band ratio (Fig. 7.8b).

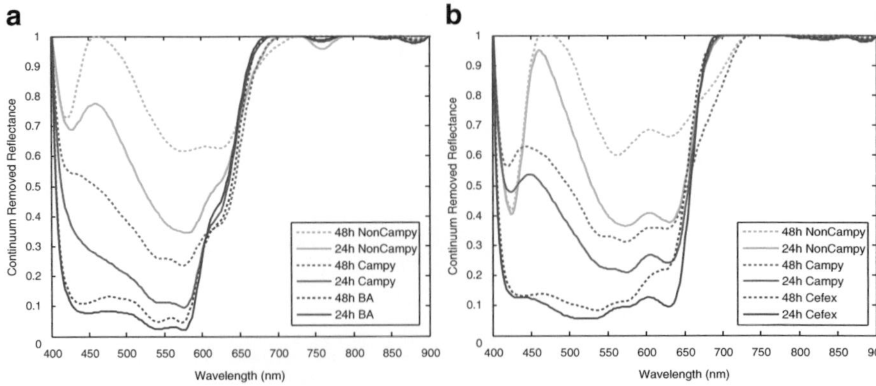

Fig. 7.9 Continuum-removed mean reflectance spectra of both 24 h and 48 h cultures on (**a**) Blood agar and (**b**) Cefex agar plates

For band ratio computation of continuum-removed spectra, it is not always necessary to apply the continuum removal to all wavelengths. For instance as in Fig. 7.8, the continuum for the band ratio computation at λ_1 nm and λ_2 nm may only need reflectance values at four wavelengths: 401 nm (approximation), λ_1 nm, λ_2 nm and 701 nm (approximation), where λ_1 nm, λ_2 nm will be determined later. The continuum removal at these four points does not affect the continuum-removed spectra as long as the value at 700 nm is a local maximum for the continuum curve and the values at λ_1 nm, λ_2 nm are not local maxima. Reduction of input wavelengths to compute the continuum may be useful for multispectral imaging. Both the hyper-spectrum (all of 193 bands) and the multi-spectrum (4 bands at 401 nm, λ_1 nm, λ_2 nm, and 701 nm) were compared for the continuum removal.

Figure 7.9 shows the continuum-removed mean reflectance spectra of *Campylobacter* and non-*Campylobacter* cultures (24-h and 48-h) in two different agar media (blood agar and Cefex), which were obtained from the mean reflectance spectra shown in Fig. 7.7. The most prominent characteristic observed from the continuum-removed spectra was the difference in slopes between *Campylobacter* and non-*Campylobacter* spectra of blood agar cultures. Specifically, in the range between approximately 420 nm and 465 nm of the spectra of the blood agar cultures, the slope of non-*Campylobacter* was positive whereas that of *Campylobacter* was negative (Fig. 7.9a). From this observation, it was straightforward to adopt a band ratio technique using two bands. The bands for the band ratio were chosen from wavelengths corresponding to local extrema of the non-*Campylobacter* spectra. The extrema were at 426 nm (λ_1, minimum) and 458 nm (λ_2, maximum). The same effect, however, was not observed from the Cefex-based cultures. Nonetheless, the same band selection strategy was applied to the case of Cefex 24-h cultures where the local extrema of the non-*Campylobacter* continuum-removed spectra were observed at 423 nm (minimum) and 461 nm (maximum).

To find the optimal threshold value for the band ratio algorithm, a histogram analysis was used. Figure 7.10 shows the histograms of the band ratio data obtained

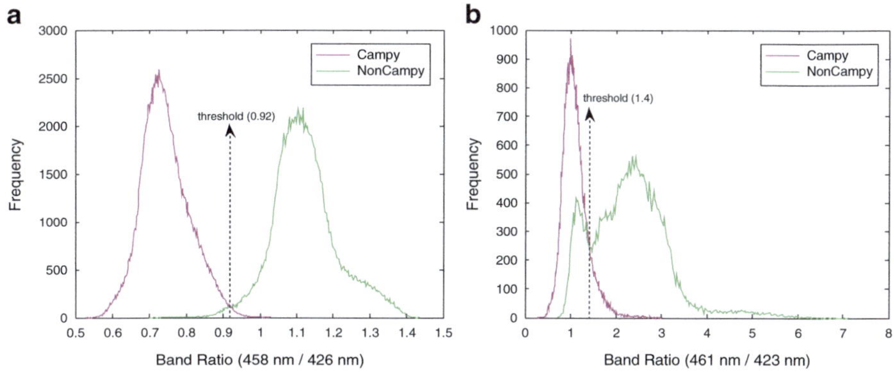

Fig. 7.10 Histogram of band ratio data. (**a**) Blood agar and (**b**) Cefex agar

by the continuum removal process. The histograms of *Campylobacter* and non-*Campylobacter* band-ratio data in 24-h blood agar showed unimodal distributions without much overlap. Thus, from Fig. 7.10a, a threshold value (0.92) dividing two distributions optimally was selected. The selected threshold value (0.92) was very close to the mean (0.93) of the measured band ratio values. The case of Cefex cultures was not simple because of the bimodal distribution of the non-*Campylobacter* cultures as shown in Fig. 7.10b. Graphically, this bimodality indicates the inefficiency of the use of band ratio in the case of Cefex cultures. Thus, it was not desirable to use this band-ratio feature for classification of 24-h Cefex culture data. The selection of the threshold at 1.4 for the case of Cefex cultures was based on the constraint to get the better detection rate of Campylobacter. One can mathematically obtain the optimal threshold value minimizing the classification errors or theoretically use a receiver operating characteristic (ROC) curve which is a graphical plot of the true positive rate versus the false positive rate for a binary classifier system as a function of its discrimination threshold. However, it was found that after experiments with different thresholds around heuristically chosen values, the selection of the true optimal threshold using a theoretic optimization framework such as the ROC curve and the Bayesian classifier was not necessary. The best threshold values were determined after these preliminary experiments.

The overall classification accuracy of the developed band ratio algorithm on the 24-h blood agar-based cultures was 99.38 % (the hyperspectral continuum removal case) and 97.21 % (the multispectral continuum removal case). In the hyperspectral continuum removal (CR) case, two bands for band ratio were directly obtained from the continuum removed spectral bands. However, in the multispectral continuum removal (CR) case, note that four bands were first obtained from reflectance hyperspectral bands at 401 nm, 426 nm (λ_1), 458 nm (λ_2), and 701 nm without continuum removal, and then the continuum-removal was applied to these four reflectance spectral bands. The band image at 503 nm was used for segmentation of locations of blood agar, grown colonies and plates. Then, the band ratio data were

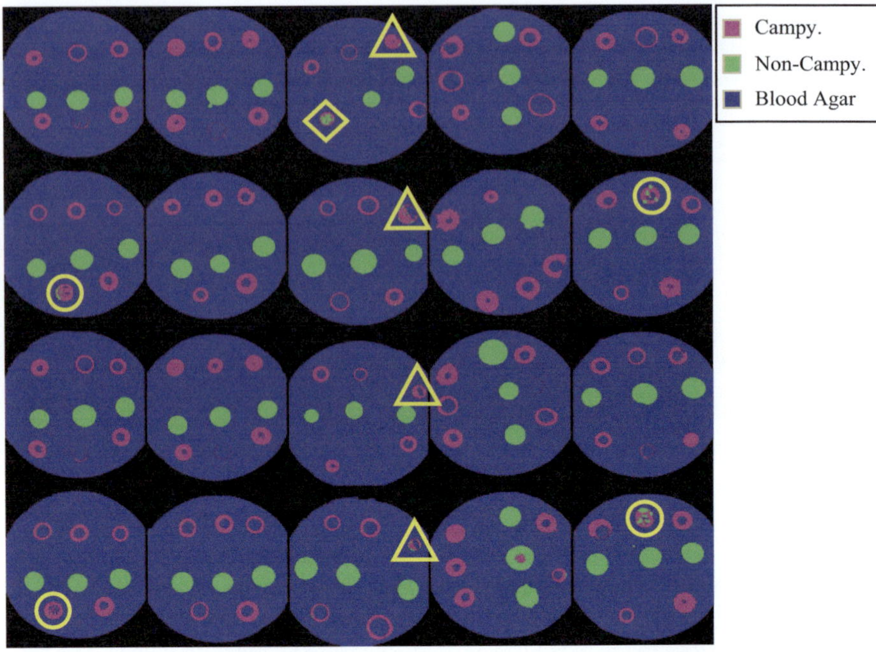

Fig. 7.11 Blood agar 24-h cultures: Classification results of the developed band ratio algorithm using two bands obtained from continuum removal of hyperspectral bands (193 bands). The markers indicate the spots that were classified incorrectly (*box*), partially incorrectly (Campy: *triangle* and non-Campy: *diamond*). A *circle* encloses a cross-contaminated spot

classified by a single threshold value. The results suggested that the developed band ratio technique could detect *Campylobacter* species and non-*Campylobacter* contaminants with 97–99 % accuracy in 24 h of incubation.

Figure 7.11 shows the classification result images of the developed band ratio algorithm in the hyperspectral CR mode. All 169 spots were correctly classified either entirely or at least by vast majority (100 %). Four *Campylobacter* spots (enclosed with triangles in the third column) showed a small amount of non-*Campylobacter* pixels. All four cross-contaminated *Campylobacter* spots (circled) were correctly classified.

In conclusion, a hyperspectral image processing algorithm was developed for detecting spot colonies of *Campylobacter* species and non-*Campylobacter* contaminants incubated for 24 h. The continuum-removed mean reflectance spectra of blood agar-based cultures showed prominent difference in slopes between 426 and 458 nm. The band ratio algorithm using two continuum-removed spectral bands at 426 and 458 nm was a key detection technique. The continuum-removed band ratio method showed 97–99 % classification accuracy. Blood agar was the better culture medium than Campy-Cefex agar in terms of *Campylobacter* detection accuracy. The experimental results suggest the developed band ratio algorithm can detect *Campylobacter* species and non-*Campylobacter* contaminants with up to 99 % accuracy in 24 h of incubation. The developed imaging protocol is applicable to spread plating techniques using chicken carcass rinses or other pathogen detection studies.

7.4 Detection of Non-O157 STEC

This section describes the development of hyperspectral imaging techniques for differentiating the "Big Six" non-O157 STEC serogroups grown on spread plates with Rainbow agar. Following the similar protocol developed for the *Campylobacter* detection study mentioned previously, a study (Windham et al. 2012) using spot plating was conducted for non-O157 STEC detection and a prediction model was developed to classify the big six STEC serogroups on pure spot plates, where 5 μL individual spots of each serogroup were used. However, when applied to an independent test set obtained from spread plates, the model calibrated with spot plates was sensitive to changes and variations in STEC populations and growth conditions of spread plates. This problem was, in part, due to the fact that the model suffered from insufficient spectral and spatial sampling of data over many different colony populations. In this regard, spread plates provide more diverse and realistic colony populations. Therefore, there was a need to study hyperspectral imaging with spread plates. In this section, two studies using spread plating are described. The first study was concerned with hyperspectral imaging for pure cultures separately spread on each plate. The second study was hyperspectral imaging for mixed cultures.

In spot plating, shadows cast over agar were not a big problem because the selection of ROIs was not greatly affected by the shadows. However, it was not the case with spread plating because the colonies on semi-transparent agar cast a lot of shadows, as shown in Fig. 7.12. Thus, a transparent sample holder was designed to minimize shadows cast by colonies on spread plates. Other than the new sample holder, hyperspectral image acquisition was performed with the same push-broom line-scan VNIR hyperspectral imaging system (Fig. 7.1) that was used for the *Campylobacter* study. The working distance from the objective lens to the Petri dish was about 40 cm. On-camera binning was set to 2 (spatial) × 2 (spectral) with a 30-ms integration time.

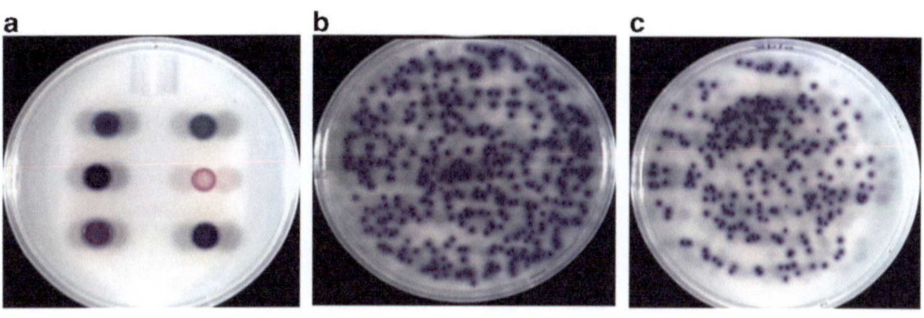

Fig. 7.12 RGB color-composite images of (**a**) spot plate of the "Big Six" non-O157 STEC serogroups and (**b**) spread plates of pure O103 and (**c**) O145 STEC cultures. The images have shadows from earlier system configuration. Colonies in (**b**, **c**) show an example of similarities within the STEC population

Acquired hyperspectral images were calibrated with the 75 % reflectance Spectralon target, and the size of each image was reduced. Then, the relative reflectance value R was transformed to absorbance ($\log_{10}(1/R)$) in order to reduce non-linearity in reflectance measurements, and absorbance was used for model development. The spectral dimension of each image was reduced to 473 spectral bands ranging from 400 to 1,000 nm by removing extreme wavelength bands. Thus, the resulting image size became 688 (W) × 500 (H) × 473 (λ). Finally, spectral noise was reduced by a Savitzky-Golay smoothing filter (window size: 25; order of moment: 4) at each pixel position (Savitzky and Golay 1964). After the aforementioned operations, all hyperspectral images were stitched together into a single image mosaic.

To obtain absorbance spectra from colonies, regions-of-interest (ROIs) were semi-automatically obtained with an interactive thresholding tool available in Fiji (http://fiji.sc, an open source image processing package based on ImageJ). The 428-nm image was used for colony ROI segmentation because 428 nm had good contrast between colony and background agar pixels. Glint pixels with specular reflectance were not included in the ROIs. Touching segmentation objects were separated with the ROI tool in ENVI software (Exelis Visual Information Solutions, Boulder, CO, USA).

In supervised hyperspectral-image classification, sampling is concerned with the selection of a subset of pixels. A total of six additional ROI sets were derived by sub-sampling at each colony ROI in order to characterize spatial and spectral variations in the pixel population of each colony. Figure 7.13 shows an example

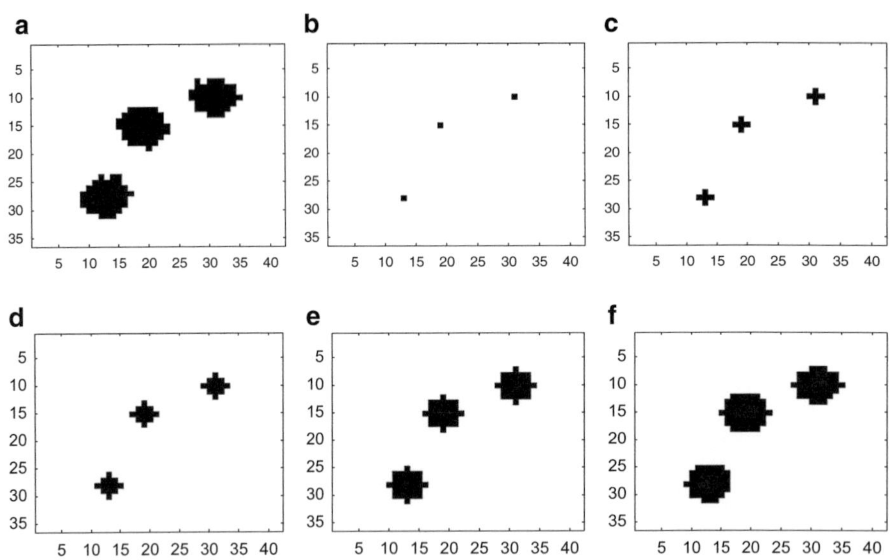

Fig. 7.13 Example of colony ROIs and their subsets. Dx ($x=1, 2, 3, 4$) include all pixels within the Euclidean distance x from the geometric center of each colony ROI in (**a**). (**a**) colony ROIs, (**b**) Center, (**c**) D1, (**d**) D2, (**e**) D3, and (**f**) D4

of colony ROIs and their corresponding subsets of ROI sets including the geometric center (i.e. centroid), four ROIs defined by Dx ($x = 1, 2, 3, 4$) that was an ROI set of pixels within the Euclidean distance x (a radius in pixels) from the geometric center of each colony ROI. When extracting spectral data from pixel locations defined by the ROIs, the data were reshaped into an $M \times N$ data matrix $\boldsymbol{X}$ whose values were associated with M observations (samples) in row and N variables (wavelengths) in column. A response vector y of 6 class labels was also created.

For model development, a data matrix $\boldsymbol{X}$ and a response vector y were partitioned into a training set and a validation set with a holdout approach without random sampling because conventional cross-validation with random sampling tends to partition homogeneous samples within each colony, thus leading to over-optimistic results. In this scheme, all combinations of mosaic columns were used to split the data into training and validation sets.

The data pre-treatment methods were applied as part of pre-processing to predictors $\boldsymbol{X}$. The pre-treatment methods used were none (absorbance only), multiplicative scatter correction (MSC), standard normal variate and detrending (SNVD), first derivative with a gap width of 11 points, moving average smoothing with a gap width of 11 points before differentiation, and MSC-corrected first derivative, and SNVD-corrected first derivative. The application order of the pre-treatment methods was MSC (or SNVD), moving average, and differentiation when all methods were used. In addition, preprocessed spectral data were transformed by principal component analysis (PCA) to reduce the dimensionality of the feature space and to perform supervised classification using either the Mahalanobis distance (MD) classifier or the k-nearest neighbor (kNN) classifier that was PC scores. Thus, the number of principal components (PCs) used for classification was also considered an important operating parameter. The optimal number of PCs was studied in a previous study (Yoon et al. 2013a), where the minimum requirement was 6 PCs and then the prediction performance was maxed out from 12 PCs. Figure 7.14 summarizes the schematic block diagram of the imaging protocol, preprocessing and classification algorithm for non-O157 STEC detection.

7.4.1 Pure Cultures on Spread Plates

The bacterial cultures were obtained from a culture collection at the Eastern Laboratory of USDA Food Safety Inspection Service (FSIS). A total of six non-O157 STEC strains were chosen with one strain from each representative O-serogroup (O26, O45, O103, O111, O121, and O145). The specific STEC strains were O26:H2 strain 4, O45:H2 strain 8, O103:H2 strain D, O111:H1 strain 16, O121:H19 strain A, and O145:H- strain K. The pathogenicity of all test strains was confirmed by the presence of two genetic targets: one of two *stx* genes (stx_1 and stx_2) and the intimin (*eae*) genes (USDA 2012). Working stocks of each culture were stored on nutrient agar slants (Becton Dickinson, Sparks, MD, USA) at 4 °C. Cell suspensions were prepared from cultures grown overnight on Blood agar

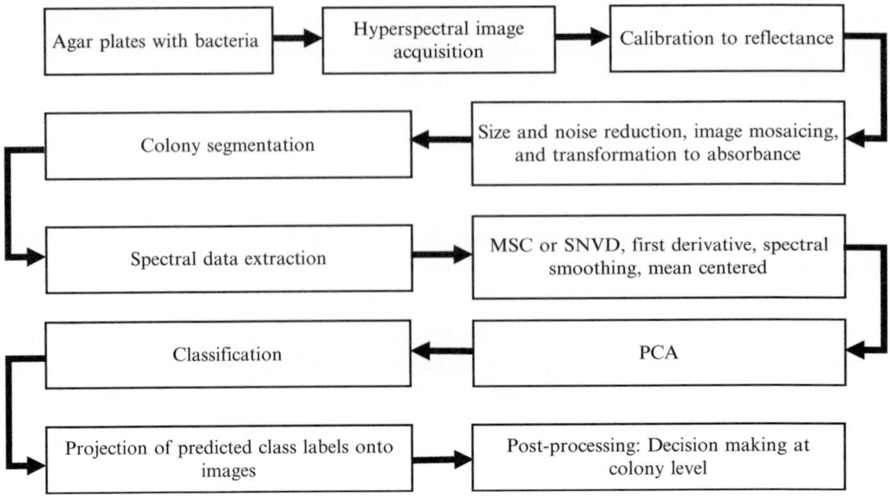

Fig. 7.14 Flowchart of multivariate hyperspectral image analysis for classification and prediction of colonies

(BA, Trypticase Soy Agar with 5 % sheep blood, Remel, Lenexa, KS, USA) at 37 °C. Cells were suspended in sterile saline (0.85 %) at an initial concentration of approximately 10^9 colony forming unit (CFU) mL^{-1} (0.50 turbidity), with a Dade Behring MicroScan Turbidity Meter (Dade Behring, West Sacramento, CA, USA). Serial dilutions of each cell suspension were prepared in sterile saline. Then approximately 50 and 100 CFU (50 and 100 μL aliquots of 10^3 $CFUmL^{-1}$ dilutions) were inoculated onto Rainbow agar plates (100-mm diameter) containing 10 mgL^{-1} novobiocin plus 0.8 mgL^{-1} potassium tellurite (RBA, Biolog, Inc., Hayward, CA, USA) by a spread plating technique. All plates were incubated at 37 °C for 24 h.

Following the above protocol, two replicate experiments were carried out over a period of 2 months. Thus, a total of 24 plates (2 dilutions × 2 replicates × 6 serogroups) were prepared for this study. All plates except O121 and O145 of replicate 1 were pure cultures. Four spread plates mixed with O121 and O145 cultures were prepared for the replicate 1 experiment only. All calibrated 24 hyperspectral images were stitched together into a single image mosaic, as shown in Fig. 7.15. The hyperspectral data cubes were arranged according to serogroups in rows, two serial dilutions (duplicates) and data collection dates (replicates) in columns. The mosaic file size was about 15 GB.

The total number of colony ROIs was 1,421. The total number of pixels (i.e. observations or samples) in all colony ROIs was 51,173. Figure 7.16 shows example RGB-color images that highlight the improved image quality, as well as the ambiguity in the colony color and morphology within the six non-O157 STEC serogroups. The first row shows color-composite reflectance images of each serogroup, the second row shows absorbance and the third row shows colony

Fig. 7.15 Non-O157 STEC image mosaics (color-composites) of (**a**) reflectance, R and (**b**) absorbance, $\log_{10}(1/R)$ images

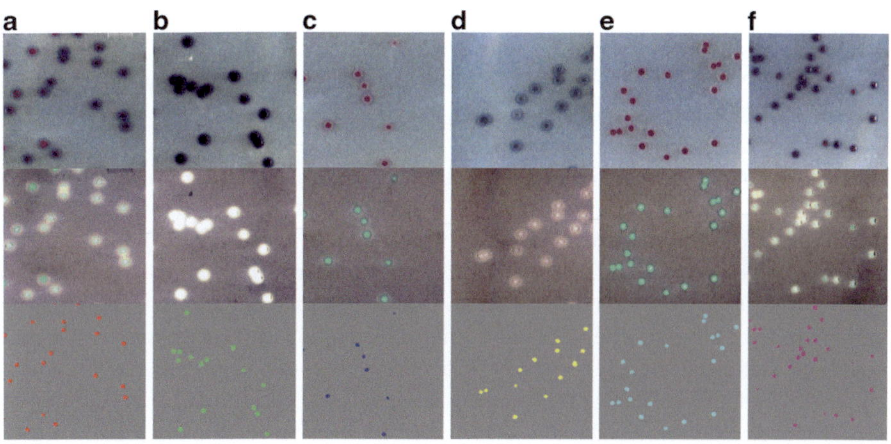

Fig. 7.16 Example RGB-color images of the "big six" non-O157 STEC colonies: *Top*: reflectance, *middle*: absorbance, *bottom*: colony ROIs. (**a**) O26, (**b**) O45, (**c**) O103, (**d**) O111, (**e**) O121, and (**f**) O145

ROIs with color codes. Compared to Fig. 7.12, most shadows were removed although boundaries of each colony were still fuzzy due to the limitation of optics in the hyperspectral imager. A quick visual observation of reflectance images revealed that O26 and O145 colonies grew into two color tones (light magenta and dark magenta), and O103 and O121 colonies showed similar pink colors

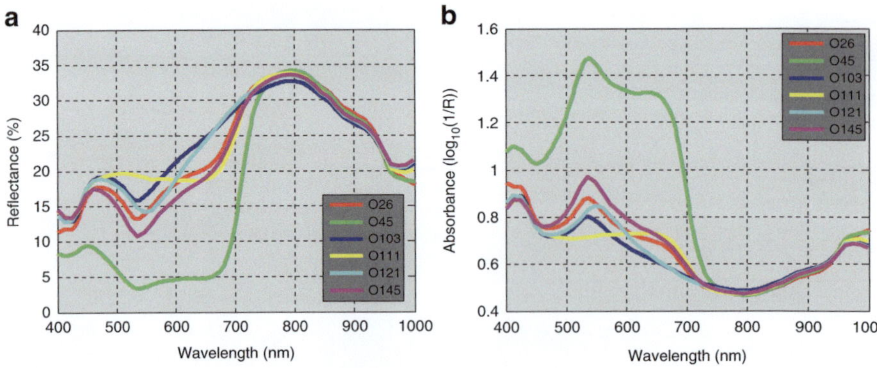

Fig. 7.17 Mean (**a**) reflectance and (**b**) absorbance spectra of six non-O157 STEC serogroups from ground-truth colony regions (i.e., ROIs)

whereas O45 (close to black) and O111 (gray) were somewhat different from the others in both color and morphology.

The mean ROI spectra of each serogroup are shown in Fig. 7.17. The serogroups O26 (red) and O145 (magenta) had generally similar spectral characteristics to each other with some differences in absorption (or reflectivity) from 470 to 650 nm. This similarity also was consistent with the aforementioned visual assessment (Fig. 7.16a, f). The serogroups O103 (blue) and O121 (cyan) had almost identical spectral curves between 400 and 540 nm, but displayed subtle differences between 540 and 650 nm. The spectral similarity of O103 (blue) and O121 (cyan) at color bands (e.g. 450, 550, 650 nm) produced similar color tones to human eyes as shown in Fig. 7.16c, e. It was also interesting to observe that the absorption peak of O121 was near 550 nm (Fig. 7.7b), which was shifted to the right about 10 nm from the absorption peaks of the other serogroups except the spectrally flat O111.

The classification results obtained by Mahalanobis distance and kNN classifiers were mapped onto the agar plate images (not shown). Serogroups O111 and O121 showed consistently over 99 % accuracy regardless of the classification algorithm and preprocessing treatments. However, the classification accuracies of serogroups O26, O45, O103, and O145 showed varying results from 84 % up to 100 %, depending on the preprocessing treatment adopted. Classification accuracy improvement due to scatter correction was about 10 % on average. The preprocessing treatment with SNVD, first derivative with 11-point gaps and moving average with 11-point gaps provided the best mean classification performance (95.06 %) when applied to the 5-pixel ROI set defined around the centroid of each colony ROI.

In conclusion,, a VNIR hyperspectral imaging technique for detection of non-O157 STEC serogroups on agar plates was developed by building spectral libraries and developing classification models. The detection accuracy of the classification model was about 95 %.

7.4.2 Mixed Cultures on Spread Plates

A mixed culture is a laboratory culture that contains two or more identified species or strains of microorganisms. Spread plates of mixed cultures may produce diverse and realistic colony populations mimicking actual microbial populations of contaminated food samples although mixed cultures are still laboratory control samples. However, in hyperspectral imaging of colonies from mixed cultures, performance of a classification model is much more difficult to validate than pure cultures because it is unknown where specific bacteria grow on an agar plate due to spreading of liquid cultures and bacterial competition for growth and survival (Fredrickson 1977; Hibbing et al. 2010), and thus it is almost impractical to confirm the identity of every colony with a genetic and/or a biochemical confirmation method simply for validating classification models. This difficulty is in part because there are too many (typically about 50–300) colonies per plate.

Cell suspension mixtures containing equal portions (500 μL aliquots of 10^3 CFU/mL) of serogroups O45, O111, and O121 were prepared from the individual STEC serogroup serial dilutions. An equivalent concentration of a fourth serogroup (O26, O103 or O145) was inoculated into the three strain mixture. The reason why the mixed cultures were prepared with the mixture formula of three easy serogroups plus each one of the three difficult serogroups, O45, O111, and O121, was due to its simplicity in performance validation of the developed classifiers. The aforementioned mixture formula was designed to build ground-truth maps only from the measured images. For each mixture, 50 and 100 μL aliquots were spread onto individual Rainbow agar plates. All plates were incubated at 37 °C for 24 h.

Following the above protocol, one experiment was carried out. Thus, a total of 6 plates (2 cell concentrations × 3 mixtures) with mixed cultures and 12 plates (2 cell concentrations × 6 serogroups) with pure cultures were used to evaluate the developed classification models. The acquired hyperspectral images were processed in the same way as the model creation process (see Fig. 7.15). All data collected were used as independent test (interchangeably with validation) data for the prediction models that were trained with the pure culture data mentioned in the previous section. Figures 7.18 and 7.19 show the mosaics of the mixed and pure cultures, i.e., the test (validation) set.

The circular forms (i.e. colony shapes) were observed from all colonies. Outer boundaries of O121 colonies were more distinctive and less fuzzy than the others. The color of O45 colonies was almost black and visually very different from the other serogroups. Figure 7.20 shows the examples of colony appearance typically observed from the measured reflectance and absorbance (transformed from reflectance) images. The color of all colonies except O45 (dark green to black) and some of O111 colonies (grayish blue tone similar to the agar background) was purple varying from bright to dark. The center area of each colony was darker than the perimeter. O111 colonies were grayish color on the agar plates with less cell concentration (left column images of the mosaic) and light purple color on the

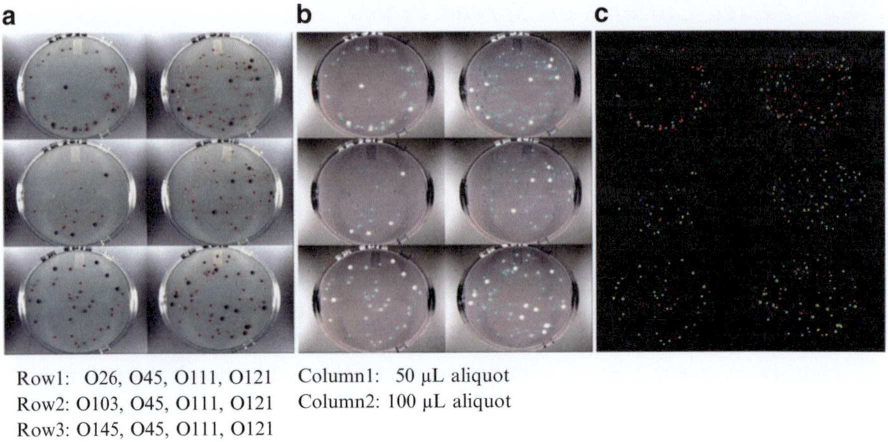

Fig. 7.18 Mixed cultures: Image mosaics (color-composites) of (**a**) reflectance, R, (**b**) absorbance, $\log_{10}(1/R)$ images, and (**c**) ROIs (*red*: O26, *green*: O45, *blue*: O103, *yellow*: O111, *cyan*: O121, and *magenta*: O145)

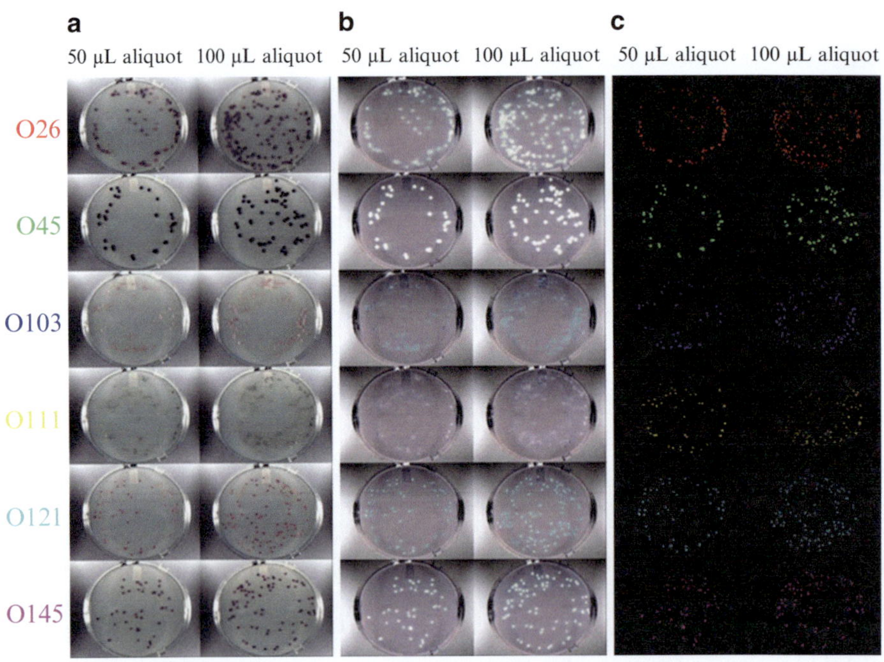

Fig. 7.19 Pure cultures: Image mosaics (color-composites). (**a**) Reflectance, R, (**b**) Absorbance, $\log_{10}(1/R)$, and (**c**) ROIs

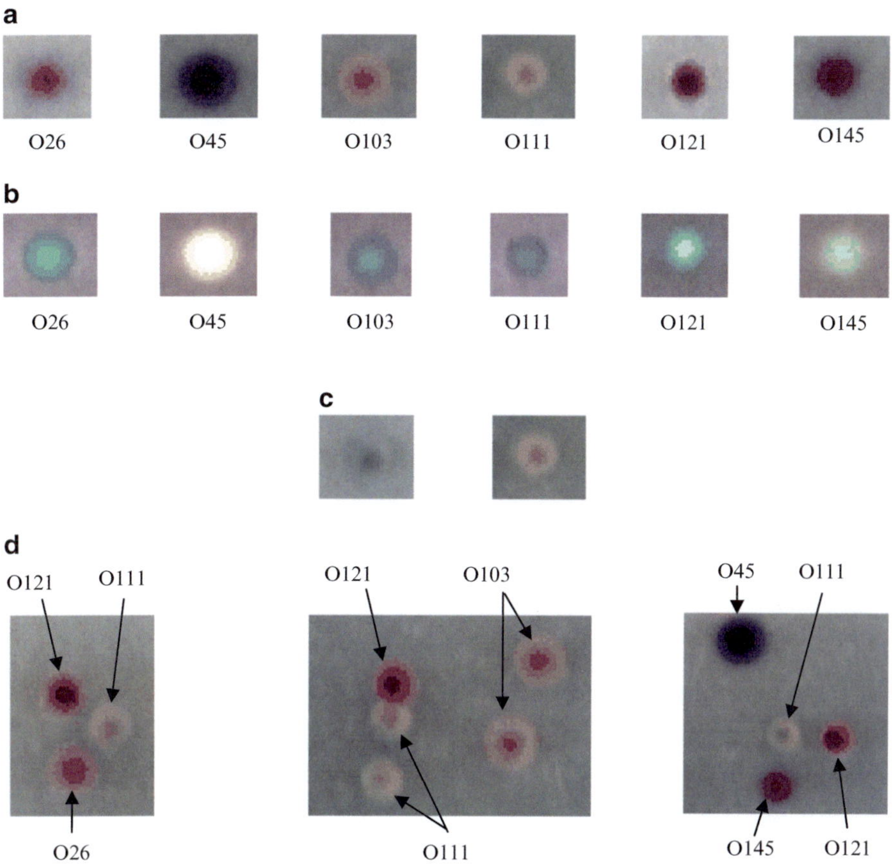

Fig. 7.20 Color composite examples of colonies. (**a**) Reflectance (color-composite): Example colonies, (**b**) Absorbance (color-composite): Example colonies, (**c**) Variability: Color difference of O111 colonies, and (**d**) Appearance differences of colonies

agar plates with more cell concentrations (right column images of the mosaic). A further study is necessary to find the importance factors such as texture, surface causing the differences in colony appearance and to incorporate them into the multivariate classification models.

Figure 7.21 shows the mean spectra of each serogroup obtained from the ROIs of the training and validation sets including both pure and mixed cultures. Overall except O26, the spectral responses of the two validation sets were more similar than the training set, which confirmed the previous study finding that replication of experiments was the largest uncertainty to the predictive performance of the classification models. The 600–700 nm shoulders of the pure O26 cultures disappeared when O26 was mixed. One possible explanation for why O26 in mixed cultures showed the spectral difference between 600 and 700 nm was the bacterial competition for survival and growth (Fredrickson 1977; Hibbing

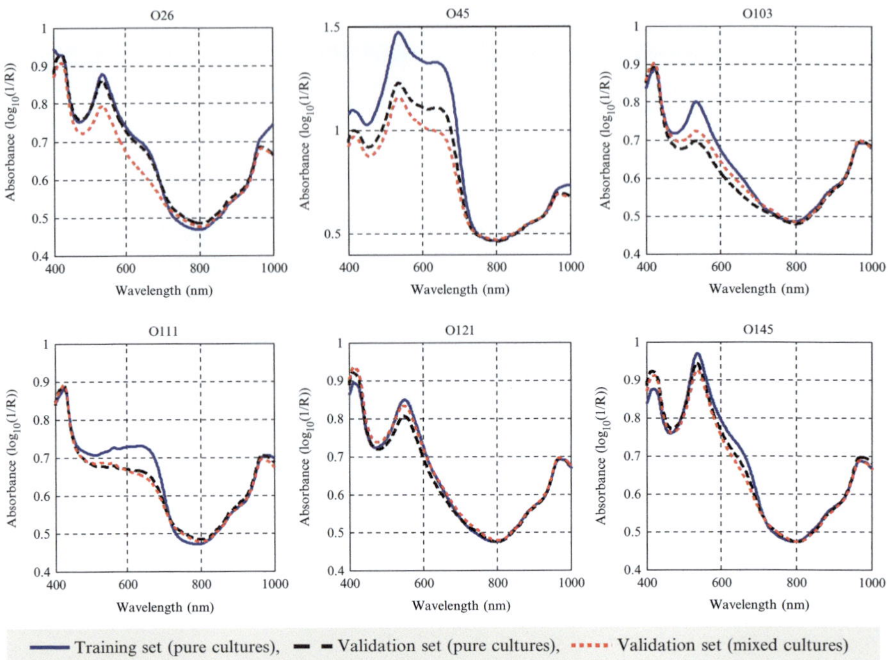

Fig. 7.21 Mean absorbance spectra of non-O157 STEC in pure and mixed cultures

et al. 2010). Although no quantitative analysis was made to measure the differences or variability between the training and validation sets, it was assumed, from Fig. 7.21, that the differences in mean-spectral responses between the training and validation sets were not large enough to re-train the models.

A total of 16 prediction models (interchangeable with classification models) from four preprocessing methods (MSC1, MSC2, SNV1 and SNV2), two classifiers (Mahalanobis distance and kNN), and two detection levels (pixel and colony) were evaluated with classification accuracy against the two validation sets of pure and mixed cultures, respectively. In the case of the positive control set of pure cultures, all 16 prediction models produced over 95 % prediction performance, ranging from 96 to 99.88 %. The average classification accuracy was 98.31 %. The colony-level decision making algorithm showing 99.42 % accuracy was approximately 2 % better than the pixel-level decision making. The performance variability among the four preprocessing models was less than 1 %. The performance difference between two classifiers was trivial (0.23 %). The best model was SNV1 or SNV2 with kNN and colony-level decision making. In the case of the mixed cultures, the best performance was obtained from the model adopting SNV2 (SNVD-corrected first derivative with a gap width of 11 points, moving average with a gap width of 11 points) and kNN ($k = 3$). The overall classification accuracy at pixel level was about 95.6 % with Kappa coefficient 0.9457. Only eight out of 311 colonies in the mixed culture plates were misclassified (thus, 97.58 % accuracy at the spot level decision making).

In conclusion, the multivariate classification models developed using a training set of pure spread plates were validated against two independent test sets with pure and mixed cultures. The prediction power of the models was over 97 % in terms of classification accuracy measured at the colony level.

7.5 Conclusions

In this chapter, the use of VNIR hyperspectral imaging was demonstrated for detection of foodborne pathogens, such as *Campylobacter* and non-O157 STEC, on agar plates. A thresholding algorithm using a spectral band at 501 nm for Campy-Cefex agar or at 503 nm for blood agar was effective to differentiate *Campylobacter* and common background microflora incubated for 48 h with over 99 % accuracy. A band ratio method using continuum removed reflectance spectra was the most effective way for early detection of *Campylobacter* colonies incubated for 24 h in blood agar. The limitation of these studies was to use spot plates. Thus, future research will need to explore hyperspectral imaging to detect and differentiate *Campylobacter* and background microflora colonies on spread plates. Prediction models using hyperspectral imaging were developed from 1,421 non-O157 STEC pure cultures spread on 24 individual Rainbow agar plates. The best overall mean classification accuracy of 95.06 % was achieved by a prediction model that adopted a k-nearest neighbor classifier of principal component scores. The developed models were validated with mixed and pure cultures spread on 6 and 12 Rainbow agar plates, respectively. The results showed 95 % overall detection accuracy at pixel level and 97 % at colony level. Future research needs to apply the models to detect and identify non-O157 STEC spiked into food samples. The developed hyperspectral imaging technology showed potential to increase the speed and accuracy of presumptive-positive screening of foodborne pathogens with direct plating. The fully developed imaging system is expected to automatically locate and identify foodborne pathogens grown on Petri dishes and can be expanded to detect other pathogens like Salmonella grown on agar media.

References

Banada PP, Huff K, Bae E, Rajwa B, Aroonnual A, Bayraktar B, Adil A, Robinson JP, Hirleman ED, Bhunia AK (2009) Label-free detection of multiple bacterial pathogens using light-scattering sensor. Biosens Bioelectron 24:1685

Bayraktar B, Banada PP, Hirleman ED, Bhunia AK, Robinson JP, Rajwa B (2006) Feature extraction from light-scatter patterns of Listeria colonies for identification and classification. J Biomed Opt 11:34006

Beauchamp CS, Sofos JN (2009) Diarrheagenic Escherichia coli. In: Juneja VK, Sofos JK (eds) Pathogens and toxins in foods: challenges and interventions. ASM, Washington, DC, p 82

Bosilevac JM, Koohmaraie M (2011) Prevalence and characterization of non-O157 Shiga toxin producing Escherichia coli isolates from commercial ground beef in the United States. Appl Environ Microbiol 77(6):2103–2112

Brooks JT, Sowers EG, Wells JG, Greene KD, Griffin PM, Hoekstra RM, Strockbine NA (2005) Non-O157 Shiga toxin-producing Escherichia coli infections in the United States, 1983–2002. J Infect Dis 192(8):1422–1429

Centers for Disease Control and Prevention http://www.cdc.gov/foodborneburden/trends-in-foodborne-illness.html

Clark RN, Roush TL (1984) Reflectance spectroscopy: quantitative analysis techniques for remote sensing applications. J Geophys Res 89:6329

Dwivedi HP, Jaykus LA (2011) Detection of pathogens in foods: the current state-of-the-art and future directions. Crit Rev Microbiol 37:40

Food Safety and Inspection Service (FSIS) (2006) Detection and enumeration method for Campylobacter jejuni/coli from poultry rinses and sponge samples. http://www.fsis.usda.gov/PDF/Baseline_Campylobacter_Method.pdf

Fredrickson AG (1977) Behavior of mixed cultures of microorganisms. Annu Rev Microbiol 31:63–87

Gracias KS, McKillip JL (2004) A review of conventional detection and enumeration methods for pathogenic bacteria in food. Can J Microbiol 50:883

Griffin PM (1998) Epidemiology of Shiga toxin-producing Escherichia coli infections in humans in the United States. In: Kaper JB, O'Brien AD (eds) Escherichia coli 0157:H7 and other Shiga toxin-producing E. coli strains. ASM, Washington, DC

Hibbing ME, Fuqua C, Parsek MR, Peterson SB (2010) Bacterial competition: surviving and thriving in the microbial jungle. Nat Rev Microbiol 8:15–25

Hirleman ED, Guo S, Bae E, Bhunia AK (2008) System and method for rapid detection and characterization of bacteria colonies using forward light scattering. U.S. Patent No. 7,465,560

International Organization for Standardization (ISO) (2006) Microbiology of food and animal feeding stuffs—horizontal method for detection and enumeration of Campylobacter spp.—part 1: detection method. ISO 10272-1

Kokaly RF, Clark RN (1999) Spectroscopic determination of leaf biochemistry using band-depth analysis of absorption features and stepwise multiple linear regression. Remote Sens Environ 67:267

Kruse FA (1988) Use of airborne imaging spectrometer data to map minerals associated with hydrothermally altered rocks in the northern Grapevine mountains, Nevada and California. Remote Sens Environ 24:31

Lazcka O, Campo F, Muñoz FX (2007) Pathogen detection: a perspective of traditional methods and biosensors. Biosens Bioelectron 22:1205

Line JE (2001) Development of a selective differential agar for isolation and enumeration of Campylobacter spp. J Food Protect 64(11):1711–1715

Mandal PK, Biswas AK, Choi K, Pal UK (2011) Methods for rapid detection of foodborne pathogens: an overview. Am J Food Technol 6:87

Meng J, Doyle MP (1998) Microbiology of Shiga toxin-producing Escherichia coli in foods. In: Kaper JB, O'Brien AD (eds) Escherichia coli 0157:H7 and other Shiga toxin-producing E. coli strains. ASM, Washington, DC, p 97

National advisory committee on microbiological criteria for foods (NACMCF) (2007) Analytical utility of Campylobacter methodologies. J Food Protect 70(1):241–250

Savitzky A, Golay MJE (1964) Smoothing and differentiation of data by simplified least squares procedures. Anal Chem 36:1627

Scallan E, Hoekstra RM, Angulo FJ, Tauxe RV, Widdowson MA, Roy SL, Jones JL, Griffin PM (2011) Foodborne illness acquired in the United States-major pathogens. Emerg Infect Dis 17(1):7–15

Siragusa GR, Line JE, Brooks LL, Hutchinson T, Laster JD, Apple RO (2004) Serological methods and selective agars to enumerate Campylobacter from broiler carcasses: data from inter- and intralaboratory analyses. J Food Protect 67(5):901–907

Stern NJ, Fedorka-Cray P, Bailey JS, Cox NA, Craven SE, Hiett KL, Musgrove MT, Ladely S, Cosby D, Mead GC (2001) Distribution of Campylobacter spp. in selected U.S. poultry production and processing operations. J Food Protect 64(11):1705–1710

USDA, FSIS (2012) Detection and isolation of non-O157 Shiga-toxin producing Escherichia coli (STEC) from meat products, Microbiology Laboratory Guidebook, MGL 5B.02 http://www.fsis.usda.gov/PDF/Mlg_5B_02.pdf

Velusamy V, Arshak KI, Korostynska O, Oliwa K, Adley C (2010) An overview of foodborne pathogen detection: in the perspective of biosensors. Biotechnol Adv 28:232

Windham WR, Yoon SC, Ladley SR, Heitschmidt GW, Lawrence KC, Park B, Narang N, Cray WC (2012) The effect of regions of interest and spectral pre-processing on the detection of non-O157 Shiga-toxin producing Escherichia coli serogroups on agar media by hyperspectral imaging. J Near Infrared Spectrosc 20:10

Yoon SC, Lawrence KC, Siragusa GR, Line JE, Park B, Feldner PW (2009) Hyperspectral reflectance imaging for detecting a foodborne pathogen: campylobacter. Trans ASABE 52:651

Yoon SC, Lawrence KC, Line JE, Siragusa GR, Feldner PW, Park B, Windham WR (2010) Detection of Campylobacter colonies using hyperspectral imaging. J Sens Instrum Food Qual Saf 4:35

Yoon SC, Windham WR, Ladely SR, Heitschmidt GW, Lawrence KC, Park B, Narang N, Cray WC Jr (2013a) Hyperspectral imaging for differentiating colonies of non-O157 Shiga-toxin producing Echerichia coli (STEC) serogroups on spread plates of pure cultures. J Near Infrared Spectrosc 21:81–95

Yoon SC, Windham WR, Ladely SR, Heitschmidt GW, Lawrence KC, Park B, Narang N, Cray WC Jr (2013b) Differentiation of big-six non-O157 Shiga-toxin producing Escherichia coli (STEC) on spread plates of mixed cultures using hyperspectral imaging. J Food Meas Charact 7(2):47–59. doi:10.1007/s11694-013-9137-4

Chapter 8
Measurement of Food Optical Properties

Renfu Lu and Haiyan Cen

8.1 Introduction

The term *optical properties* has many different connotations or interpretations within the scientific community. To many researchers in food and agriculture, optical properties are often referred to as reflectance or transmittance measurements. Reflectance and transmittance, like force or pressure, are extrinsic measurements because they, although related to the intrinsic properties, depend on type of instrument, sensing mode and setup, and sample size and shape. The optical techniques based on this school of measurement principles are, herein, referred to as *empirical* in order to differentiate from the other class of techniques that are based on the fundamental radiation transfer theory. Accordingly, near-infrared spectroscopy (NIRS), which is widely used for composition analysis, quality inspection, and process monitoring and control of food and agricultural products, is an empirical technique because it provides extrinsic measurements on the light reflected back from or transmitted through the sample.

Mention of commercial products or trade names in the chapter is solely for providing factual information and does not imply recommendation or endorsement by the U.S. Department of Agriculture.

R. Lu (✉)
U.S. Department of Agriculture, Agricultural Research Service, East Lansing, MI, USA
e-mail: renfu.lu@ars.usda.gov

H. Cen
Department of Biosystems and Agricultural Engineering, Michigan State University, East Lansing, MI, USA

College of Biosystems Engineering and Food Science, Zhejiang University, Hangzhou, China
e-mail: hycen@zju.edu.cn

B. Park, R. Lu (eds.), *Hyperspectral Imaging Technology in Food and Agriculture*, Food Engineering Series, DOI 10.1007/978-1-4939-2836-1_8

The second, more rigorous definition of optical properties is based on the fundamental principle of light interaction with turbid or diffusive media and the radiation transfer theory [broadly speaking, we may also include those phenomenological models like the famous model developed by Kubelka and Munk (1931)]. With this definition, light transfer is primarily dependent on two fundamental or intrinsic optical parameters: *absorption* and *scattering coefficients* (when anisotropy factor can be incorporated into the scattering coefficient, which is discussed in more detail in the following sections). This fundamental definition of optical properties is widely adopted in the biomedical research community and has led to completely different measurement principles and techniques (Wang and Wu 2007). In this chapter, we are mainly focused on the fundamental approach of measuring the spectral absorption and scattering properties of food and agricultural products.

With the fundamental approach, we attempt to measure and separate the absorption and scattering properties. In principle, this approach would give us more complete characterization of the optical properties of biological materials, compared to empirical techniques like NIRS. However, because the fundamental approach generally requires more sophisticated instrumentation and computational algorithms, there still exist considerable technical challenges in achieving accurate, consistent measurements from intact biological tissues or food products. Intensive research has been carried out over the past three decades in the development of noninvasive techniques to measure the optical properties of biological tissues in the biomedical field (Bykov et al. 2006; Welzel et al. 2004; Wilson and Patterson 2008). The methods that have been developed in the past include: time-resolved (Patterson et al. 1989), frequency-domain (Patterson et al. 1991), and spatially-resolved (Groenhuis et al. 1983b). Time-resolved technique is based on measuring the attenuation, broadening and delay of a short light pulse, caused by the absorption and scattering events during photon propagation in highly scattering media. It seeks to use the pathlength information implied by the time of escape to estimate or determine the optical properties of tissue traveled by the photons. Frequency-domain technique, on the other hand, provides information equivalent to that obtained in the time domain. In the frequency domain, the propagation and measurement of light are accomplished through sinusoidally modulated sources. Spatially-resolved technique requires measuring the spatial distribution profiles of diffuse reflectance generated by a continuous-wave (or steady-state) light source, from which scattering and absorption coefficients are extracted. All three methods have been extensively studied and each has its merits and shortcomings (Tuchin 2000; Wang and Wu 2007). Overall, time-resolved and frequency-domain methods require more sophisticated instrumentation and have limited wavelength range selections, but they are more suitable for measuring the optical properties of tissues at a greater depth. Spatially-resolved method, on the other hand, is simpler in instrumentation with broader wavelength range selections, but its measurements are likely to be more influenced by the superficial layer of the medium.

So far, only limited research has been reported on measuring the optical absorption and scattering properties of food and agricultural products. In early studies, researchers used the phenomenological models, like the Kubelka–Munk model

(Kubelka and Munk 1931), to determine the optical absorption and scattering properties of food and agricultural products (Birth 1978; Birth et al. 1976; Law and Norris 1973). These early studies showed promising results in using the absorption and scattering coefficients to predict quality of foods. But little further research was reported in the late 1980s and 1990s. Two factors might have contributed to the subdued interest in measuring the fundamental optical properties of food and agricultural products during that time period. First, although the phenomenological models, like the Kubelka–Munk model, are widely accepted, they lack generality and can only be used under restricted experimental conditions with special sample preparation procedures. Second and perhaps more importantly, it is much more challenging in instrumentation and algorithm implementation for accurate measurement of optical absorption and scattering properties for food and agricultural products.

In the past decade, we have seen renewed interest in characterizing and measuring the optical properties of food and agricultural products, due to the latest advances in optical and computer technologies. Cubeddu et al. in Italy (2001) applied time-resolved reflectance spectroscopy that was originally developed for non-food applications to measure the optical properties of fruits. Over the years, the Italy group has done extensive studies, in collaboration with horticulturists, to promote the time-resolved technique as a new means for quality evaluation of horticultural products (Rizzolo et al. 2010). Nicolai et al. (2008) used the time-resolved technique developed by that group to predict soluble solids content (SSC) and firmness of pears. Although a highly nonlinear relationship between the reduced scattering coefficients at 900 nm and firmness was observed, no satisfactory calibration model could be established between the absorption coefficient spectrum and SSC. Another research group from the University of Missouri at Columbia, Missouri, USA did a series of original research in using a spatially-resolved method with a NIRS and a fiber optical probe to measure the optical absorption and scattering properties of beef muscles for tenderness prediction (Xia et al. 2007). The group also developed an imaging-based technique to quantify light scattering patterns in meat analogs for assessing the structural or textural characteristics (Yao et al. 2004). Other researchers used destructive techniques (i.e., total reflectance and transmittance) to measure the spectral properties of apple fruit for the spectral region of 350–2,200 nm (Saeys et al. 2008).

In the past few years, our research team has been using a different approach to measure the optical properties of food and agricultural products (Cen and Lu 2010; Qin and Lu 2006, 2008). This approach is based on the spatially-resolved principle, coupled with hyperspectral imaging technology, to achieve fast and effective measurement of the absorption and reduced scattering spectra for the region of 400–1,000 nm. Hyperspectral imaging, which combines conventional imaging and spectroscopy techniques to acquire spatial and spectral information simultaneously, is ideally suited for measuring spatially-resolved diffuse reflectance profiles for a broad spectral region. Through extensive research (Cen and Lu 2010; Cen et al. 2009; Qin and Lu 2006, 2007, 2008), we demonstrated that the hyperspectral imaging-based spatially-resolved technique can provide accurate measurement of the spectral properties of food and agricultural products, and it is useful for nondestructive quality evaluation of horticultural and food products (Cen and Lu 2012; Qin and Lu 2008).

In this chapter, we provide an overview of our recent research in the development and application of hyperspectral imaging-based spatially-resolved technique for measuring the optical absorption and scattering properties of food products. We first introduce the principle and theory of spatially-resolved spectroscopic technique. We then present the development of a new optical property measuring instrument and application examples of using the optical properties to assess the maturity/quality of fruit products. Finally, recommendations are given for further research in the measurement of optical properties for food and agricultural products.

8.2 Principle and Theory of Spatially-Resolved Spectroscopic Technique

8.2.1 Light Absorption and Scattering

Light consists of a packet of particles, called photons, which carry electromagnetic energy and momentum but have no rest mass. Light transfer in turbid or diffusive biological tissues or food products is a complicated phenomenon, which involves both absorption and scattering (Tuchin 2000). The fundamental optical parameters for turbid biological materials include absorption coefficient, scattering coefficient, and anisotropy factor. *Absorption coefficient* (μ_a) quantifies the conversion of light energy into other forms of energy such as heat, electricity, or chemical energy. Absorption or the decrease in the amount of electromagnetic radiation is proportional to the incident light intensity and the distance over which the absorption takes place in an absorbing-only medium with the unit in mm^{-1} or cm^{-1} (Fig. 8.1a). A relationship between the absorption and the chemical composition may be established, which could be used to evaluate quality, ripeness, and defects of agricultural and food products. Scattering is a physical process that takes place

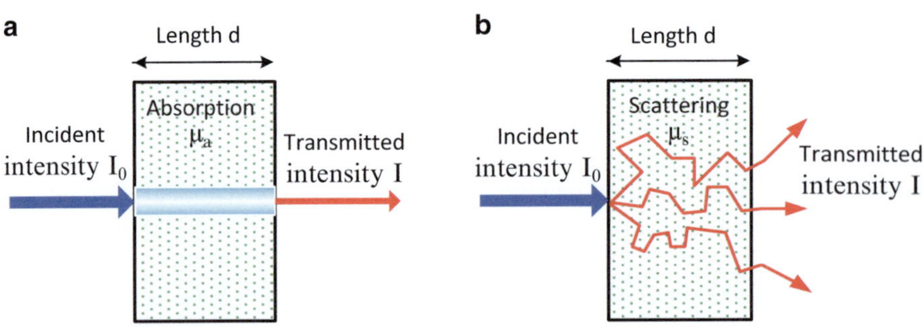

Fig. 8.1 Light interaction with matter: (**a**) absorption and (**b**) scattering

when light interacts with scattering media, and the travelling path of the photons is no longer direct as shown in Fig. 8.1b. *Scattering coefficient* (μ_s) quantifies the probability of photons scattering unit path length, and is the inverse of the average distance that light travels among scattering events. μ_s represents a probability per unit length of a photon being scattered, which has the same units as μ_a in mm^{-1} or cm^{-1}. Light scattering in the tissue depends on many variables including the size of scattering particles, the wavelength of the light, and the variation of the refractive indices of the various tissue components. In agricultural and food products, scattering is closely related to the cellular structures and characteristics, and therefore it could provide useful information about their condition and quality. *Anisotropy factor* (g) is defined as a measure of the amount of photons retained in the forward direction after a single scattering event. In many biological materials, scattering is dominant during a light transport process, which is known as the diffusion regime. Because the photons encounter many scattering events in small steps before an absorption event takes place, the total scattering could be considered as isotropic. Hence the exact value of the anisotropy factor is no longer needed for the description of light propagation in the tissues, and the *reduced scattering coefficient* $\mu_s' = (1 - g)\mu_s$ is commonly used. As a result, μ_a and μ_s' are the only optical parameters in the diffusion regime.

8.2.2 *Diffusion Theory*

Light propagation in biological materials is governed by the radiation transport equation. Exact solutions to the equation are found only under a few very restricted conditions. For most biological materials in which scattering is dominant (i.e., $\mu_s' \geq \mu_a$), diffusion approximation is valid (Durduran et al. 1997; Ishimaru 1978). If we further assume that the source-detector distance is greater than the transport mean free path [mfp′, which is the inverse of total attenuation coefficient, or $1/(\mu_s' + \mu_a)$, representing the mean free path between interactions] and the source term $Q(\mathbf{r}, \hat{s}, t)$ is isotropic with an equal probability of scattering in all solid angles combined with a net flux, then the radiation transport equation can be simplified to the following diffusion equation (Haskell et al. 1994)

$$\frac{1}{c}\frac{\partial \Phi(\mathbf{r}, t)}{\partial t} - D\nabla^2\Phi(\mathbf{r}, t) + \mu_a\Phi(\mathbf{r}, t) = S(\mathbf{r}, t) \tag{8.1}$$

where $\Phi(\mathbf{r}, t) = \int_{4\pi} L(\mathbf{r}, \hat{s}, t)d\Omega$ is the fluence rate, $D = [3(\mu_a + \mu_s')]^{-1}$ is the diffusion coefficient, and $S(\mathbf{r}, t) = \int_{4\pi} Q(\mathbf{r}, \hat{s}, t)d\Omega$ represents an isotropic source.

Under steady-state conditions (i.e., the light source remains constant or continuous-wave over time), the first term on the left side of Eq. 8.1 is zero, and the term on the right side of Eq. 8.1 is also zero when no light source exists in the medium. This leads

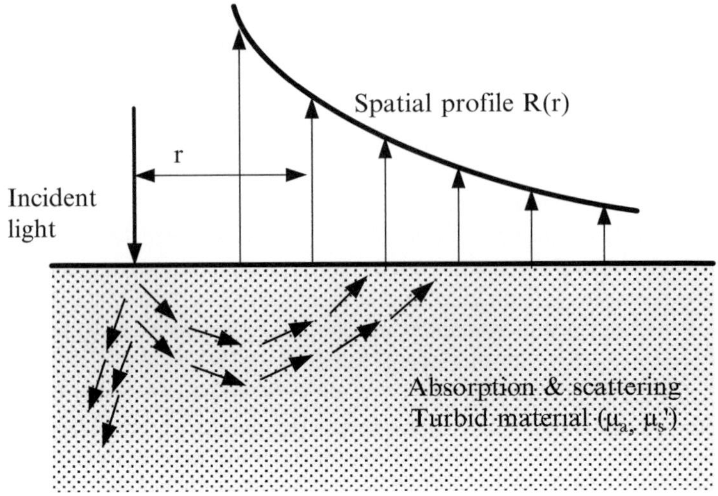

Fig. 8.2 Measurement principle for spatially-resolved technique

to the steady-state diffusion equation, which is the theoretical basis for spatially-resolved technique. Reynolds et al. (1976) first used the spatially-resolved theory to quantify light propagation in turbid media. Later, Langerholc (1982) and Marquet et al. (1995) suggested that spatially-resolved measurement could be used to determine the optical properties of biological tissues. Figure 8.2 shows the principle of spatially-resolved technique. As a small continuous-wave light beam perpendicularly illuminates the sample's surface, photons will scatter in different directions or be absorbed in the medium. Some of the photons will remit from the area close to the light incident point. By measuring the remitted light at different distances from the light source, we can extract the optical coefficients using an appropriate analytical solution of the diffusion equation coupled with an inverse algorithm.

For the case of steady-state spatially-resolved reflectance for a homogeneous semi-infinite turbid medium, Farrell et al. (1992) derived an analytical solution from the diffusion equation (Eq. 8.1) using the extrapolated boundary conditions, at which the fluence is forced to zero by introducing a negative 'image source'. The diffuse reflectance from the medium is calculated as the current across the boundary, and it is originated from a single isotropic point source located at a depth of one transport mean free path in the medium. The final expression of the reflectance R at the surface of the semi-infinite turbid medium is

$$R(r) = \frac{a'}{4\pi}\left[\frac{1}{\mu_t'}\left(\mu_{eff} + \frac{1}{r_1}\right)\frac{\exp(-\mu_{eff} r_1)}{r_1^2} + \left(\frac{1}{\mu_t'} + \frac{4A}{3\mu_t'}\right)\left(\mu_{eff} + \frac{1}{r_2}\right)\frac{\exp(-\mu_{eff} r_2)}{r_2^2}\right] \tag{8.2}$$

where r is the source-detector distance, $a' = \mu_s'/(\mu_a + \mu_s')$ is the transport albedo, $\mu_{eff} = [3\mu_a(\mu_a + \mu_s')]^{1/2}$ is the effective attenuation coefficient, and $\mu_t' = \mu_a + \mu_s'$ is the total attenuation coefficient, $r_1 = \left(z_0^2 + r^2\right)^{1/2}$ and $r_2 = \left[(z_0 + 2z_b)^2 + r^2\right]^{1/2}$ are the distances from the observation point at the interface to the isotropic source and the image source, and $z_0 = \left(\mu_a + \mu_s^{'}\right)^{-1}$, and $z_b = 2AD$, in which D is the diffusion coefficient and $A = 0.2190$ for $n = 1.35$, a typical value for most biological materials, is the internal reflection coefficient related to the relative index of the tissue-air interface n, which can be calculated from an empirical equation developed by Groenhuis et al. (1983a).

Later, Kienle and Patterson (1997) proposed an improved analytical solution by expressing the reflectance as the integral of the radiance over the backward hemisphere based on the study of Haskell et al. (1994). In this case, the radiance can be expressed as the sum of isotropic fluence rate and the flux, which is given below

$$R(r) = \frac{C_1}{4\pi D}\left[\frac{\exp(-\mu_{eff}r_1)}{r_1} - \frac{\exp(-\mu_{eff}r_2)}{r_2}\right] + \frac{C_2}{4\pi}\left[\frac{1}{\mu_t^{'}}\left(\mu_{eff} + \frac{1}{r_1}\right)\frac{\exp(-\mu_{eff}r_1)}{r_1^2} + \left(\frac{1}{\mu_t^{'}} + 2z_b\right)\left(\mu_{eff} + \frac{1}{r_2}\right)\frac{\exp(-\mu_{eff}r_2)}{r_2^2}\right] \tag{8.3}$$

where $z_b = 2D\left(1 + R_{eff}\right)\left(1 - R_{eff}\right)^{-1}$ for the extrapolated boundary condition, R_{eff} is the effective reflection coefficient, which is equal to 0.4498 for the refractive index n = 1.35. A thorough discussion of calculating R_{eff} can be found in Haskell et al. (1994). $C_1 = \frac{1}{4\pi}\int_{2\pi}\left[1 - R_{fres}(\theta)\right]\cos\theta \, d\Omega$ and $C_2 = \frac{3}{4\pi}\int_{2\pi}\left[1 - R_{fres}(\theta)\right]\cos^2\theta \, d\Omega$ are constants determined by the relative refractive index mismatch at the tissue-air interface, in which $R_{fres}(\theta)$ is the Fresnel reflection coefficient for a photon with an incident angle θ relative to the normal to the boundary, and Ω is the solid angle. For $n = 1.35$, $C_1 = 0.1277$ and $C_2 = 0.3269$ in Eq. 8.3.

In practice, spatially-resolved measurement employs a point light source or a narrow collimated beam of the constant intensity and multiple detectors at different source-detector distances. Optical fiber arrays and non-contact reflectance imagery are two commonly used sensing configurations in spatially-resolved measurement systems (Doornbos et al. 1999; Fabbri et al. 2003; Jones and Yamada 1998; Pilz et al. 2008). The former requires multiple spectrometers or a single imaging spectrometer to measure diffuse reflectance at different distances from the light incident point. Optical properties at multiple wavelengths or over a specific spectral region can be obtained using this method. Yet the measurements need good contact between the detecting probes and the sample, which may not be suitable for agricultural and food products. The second method usually uses a CCD (charge-coupled device) camera to acquire diffuse reflectance from the scattering medium generated by a point light beam. The measurement can be achieved without contacting the investigated medium, which is particularly advantageous for food and agricultural products because of the safety and sanitation requirements. However, most research on non-contact reflectance imagery mode can only provide optical property information at single or several wavelengths.

8.3 Instrumentation for Measuring the Optical Properties

Accurate measurement of scattering and absorption parameters depends on a number of factors in the algorithm implementation and instrument design. A proper inverse algorithm to the diffusion equation (Eq. 8.2 or Eq. 8.3) needs to be developed and optimized. The system should be able to measure the optical properties of turbid foods and biological materials for the visible and short-wave near-infrared region of 500–1,000 nm, a spectral region that is rich in information about the chemical composition and quality of food and agricultural products. Since the shape and size of the light beam can directly affect measurement accuracies, it is important to examine and optimize the light beam. Moreover, an appropriate source-detector distance, including minimum source-detector distance and maximum source-detector distance, is critical for determining the range of the spatially-resolved reflectance profile.

8.3.1 Optimization of the Inverse Algorithm and Optical Designs

Extensive studies were carried out to evaluate and optimize the inverse algorithm and optical designs for the development of a hyperspectral imaging-based spatially-resolved system (Cen et al. 2009; Cen and Lu 2010).

Direct solutions to the diffusion equation (Eq. 8.2 or Eq. 8.3), called forward problems, provide a quantitative description of light interaction with biological materials, whereas estimating optical absorption and scattering properties from the diffusion equation is an inverse problem (or parameter estimation). Inverse light transport problems are much more complicated than forward problems, and sometimes they are ill-posed. Therefore, it is necessary to investigate the intrinsic properties of the diffusion equation, develop an appropriate inverse algorithm and evaluate the feasibility of estimating all parameters and the uniqueness of the solution. Determination of optical properties based on the diffusion equation is formulated as a nonlinear least squares optimization problem, which is implemented based on several statistical assumptions that errors are constant, uncorrelated and Gaussian distribution. The measured spatially-resolved reflectance profiles of many biological materials exhibit a dramatic decrease along the source-detector distance, which may violate the assumptions of Gaussian distribution errors and constant variance errors when the nonlinear least squares inverse algorithm is applied. Our optimization study showed that proper data transformation and weighting methods can overcome the problem of assumption violations since those methods can change the data pattern without changing the nature of the best-fit curve (Cen et al. 2009). Several methods, including logarithm and integral transformation and relative weighting, were tested for 29 combinations of absorption and scattering coefficients that are typical for food and horticultural products, to preprocess the raw reflectance profiles prior to the curve fitting. The results showed that the errors of estimating μ_a and μ_s' dramatically decreased when the data transformation and weighting methods were applied to the 29 groups of μ_a and μ_s'. The logarithm transformation gives the best results with the average errors of estimating μ_a and μ_s' being 10.4 % and 6.6 %, respectively. In addition, sensitivity analysis was performed by calculating the sensitivity coefficients (Cen et al. 2009) to determine the feasibility of estimating the two parameters and/or their uniqueness. It was found that the reduced scattering coefficient can be estimated more accurately than the absorption coefficient.

In a separate study (Cen and Lu 2010), optimizations of light beam and the source-detector distance were carried out for the hyperspectral imaging-based spatially-resolved system. In the real system design, a finite size beam is usually used to illuminate the sample, which deviates from the solution of the diffusion equation that is derived for an infinitely small beam. Therefore, it is important to investigate the effect of the incident beam on the determination of the optical properties. The light beam optimization study was carried out using Monte Carlo simulations, and the results showed that the error produced by the finite beam relative to the infinitely small beam was less than 1 % for the beam size of less than 0.5 mm (Cen and Lu 2010). However, the error increased linearly with the beam size larger than 0.5 mm. In general, a 1-mm beam would introduce around 5 % error in μ_a and μ_s' compared to the infinitely small beam. Hence the light beam in the system should be less than 1 mm in size in order to control the error to within 5 %. Laboratory tests were performed to characterize the actual beam size and profile (Cen and Lu 2010). Figure 8.3 shows the measured 3-D beam profiles and 2-D intensity contours at the wavelengths of 650 and 950 nm for a laboratory optical property measuring system. The beam at the visible and short near-infrared region has a good Gaussian distribution and its shape is circular with the roundness $Rd = 0.986$ (≈ 1). Based on the commonly accepted method for defining the size of Gaussian beam, the beam size in this system is 0.83 mm, which would have caused less than 4 % error in estimating μ_a and μ_s'. A further optimization study was performed on the source-detector distance including maximum source-detector distance and minimum source-detector distance, using 12 model liquid samples made up of Intralipid scattering material and two types of dye (Cen and Lu 2010). It was found that the optimal minimum source-detector distance should be about 1.5 mm, and the optimal maximum source-detector distance should be equivalent to 10–20 transport mean free paths or determined by the minimum signal-to-noise ratio of 20.

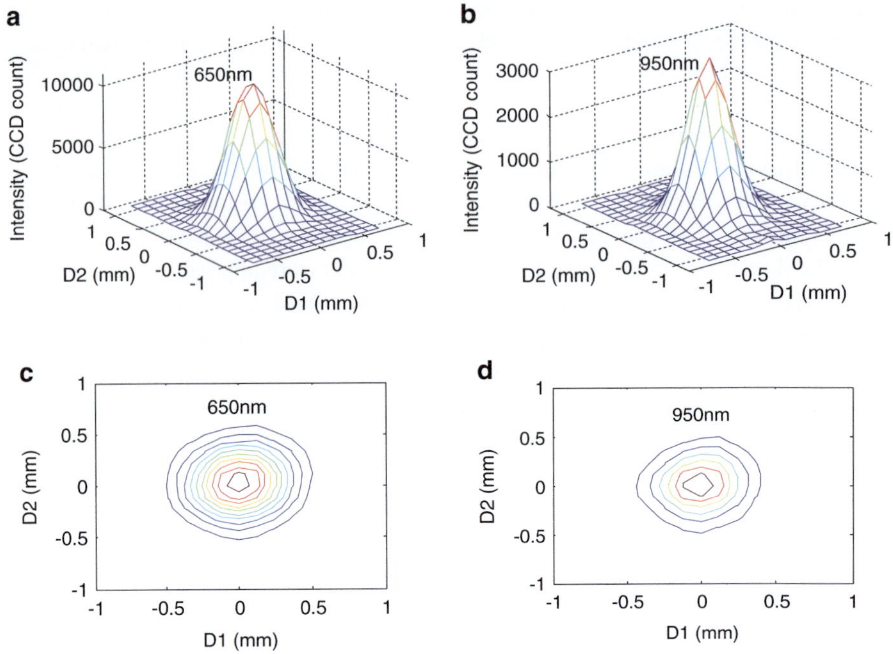

Fig. 8.3 Three-dimensional profiles (**a**, **b**) and two-dimensional (**c**, **d**) contours of the incident light beam at wavelengths of 650 and 950 nm, where D1 is the direction along the scan line of the optical system and D2 is perpendicular to the scan line (Cen and Lu 2010)

8.3.2 Instrument Development

An optical property measuring instrument, named 'Optical Property Analyzer' (OPA) (Fig. 8.4), was assembled by incorporating the optimal algorithm and optical design parameters.

The OPA is a multi-purpose optical instrument; it can measure the optical properties of food and biological materials or function as a regular hyperspectral imaging system for acquiring hyperspectral images. This is achieved through the use of two separate light sources: one line light is used for hyperspectral image acquisitions and the other point light is specially designed for optical properties measurements. Since this chapter is focused on optical properties measurement, we will skip the hyperspectral imaging acquisition design features/functions here.

The hardware of the OPA includes three main parts: imaging, illumination, and sample positioning unit (Fig. 8.4b). The imaging unit is operated in line-scanning mode, and is composed of a high performance electron-multiplying CCD (i.e., EMCCD) camera (Luca^{EM} R604, Andor Technology plc., South Windsor, CT, USA), an imaging spectrograph (ImSpector V10E, Spectral Imaging Ltd., Oulu,

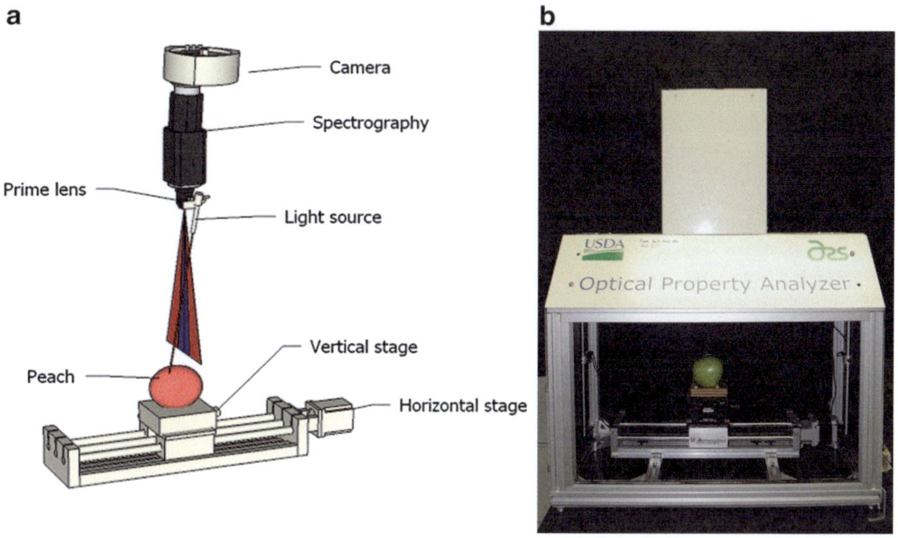

Fig. 8.4 The multi-purpose Optical Property Analyzer (or OPA): (**a**) schematic and (**b**) the actual instrument

Finland), and a prime lens. A tungsten halogen light bulb with the output of 20 W (HL-2000-HP, Ocean Optics, Dunedin, FL, USA) is connected to a DC regulated controller chip to provide point light, which covers a broad wavelength range of 369–2,000 nm. An optical fiber coupled with a specially-designed focusing lens is used for delivering a circular beam with the diameter of 1 mm at the focal point. The incident beam is arranged 1.5 mm away from the scanning line with the angle of 15° with respect to the vertical axis, and it is parallel to the scanning line. This lighting arrangement allows maintaining a constant offset distance between the light incident point and the scanning line position, even when the height of the sample changes slightly during scanning or when the sample is not placed at the exact predetermined height. The sample positioning unit includes a motorized linear horizontal stage, a manually adjustable vertical stage, and a sample holder for positioning samples to the predetermined position. During the measurement, each sample is first moved to the pre-determined height via the vertical stage, and it then starts to move horizontally in synchronization with image acquisitions by the imaging system. To improve the repeatability of measurements, the system is set in default to take 19 scans from each sample at an increment of 0.5 mm horizontal displacement for a total distance of 9 mm. The instrument provides automatic dark subtractions and corrections for fruit geometry and non-uniform instrument response.

The OPA software was developed using Microsoft Visual C#. It provides functions for system control (light source, camera and stage), image acquisition, real-time data analysis and on-screen display of 2-D images, 1-D scattering profiles,

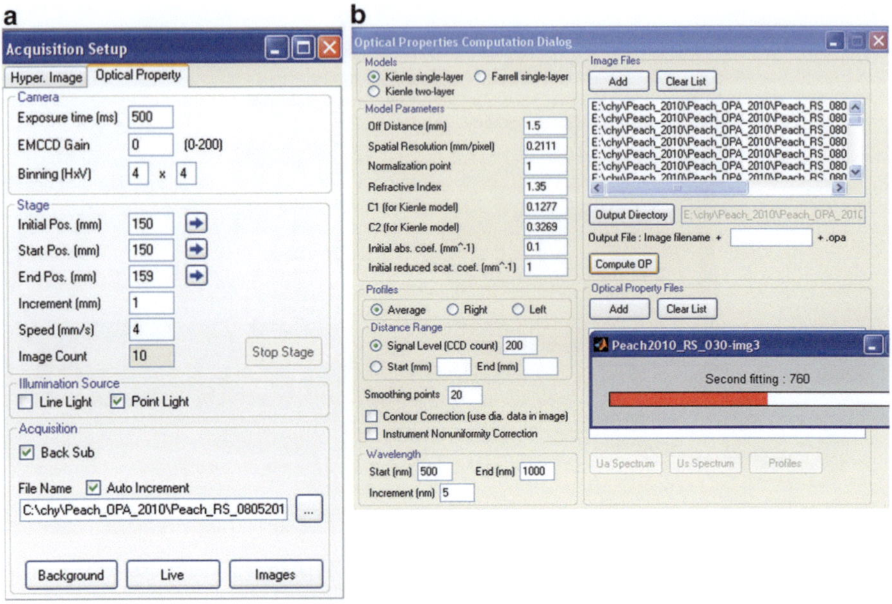

Fig. 8.5 Display windows of the Optical Property Analyzer (OPA) for (**a**) imaging acquisition setup and (**b**) optical properties computation

and the measured spectra of absorption and reduced scattering coefficients. Figure 8.5a shows the window for setting parameters related to the camera, stage, illumination and image saving functions for the optical properties measurement. A user-friendly interface in the software allows the user to choose one of the two diffusion models, i.e., Farrell model (Eq. 8.2) and the modified Kienle model (Eq. 8.3), to calculate the spectra of absorption and reduced scattering coefficients. The user first sets up all parameters in the left column of 'Optical Properties Computation Dialog' and then selects the output directory and file name in the right column of the dialog. Spectral properties computations are triggered with a processing message window by clicking 'Compute OP'.

Figure 8.6a shows a typical hyperspectral reflectance image for peach fruit acquired by the OPA. A raw reflectance spectrum representing a vertical line at the center spatial position is presented in Fig. 8.6b. Each horizontal line taken from the image represents one spatially-resolved reflectance profile at a specific wavelength (Fig. 8.6c), and hence the entire image, in effect, consists of 101 spatially-resolved reflectance profiles for the wavelengths of 500–1,000 nm at an interval of 5 nm.

Since the spatially-resolved reflectance profiles are symmetric to the light incident point, the two sides are first averaged in the extraction of optical properties (Fig. 8.6d). Smoothing and normalization using the peak value for each profile are also applied to the averaged profiles to reduce the noise and avoid the need for

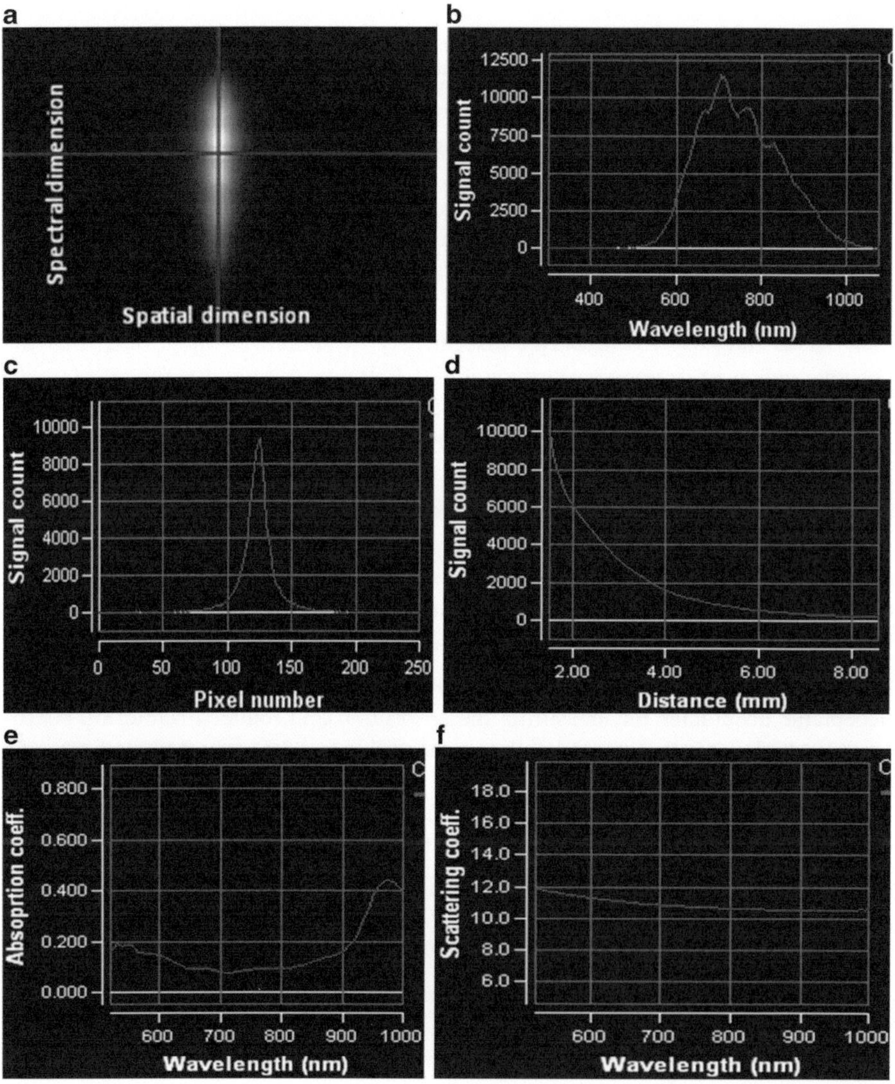

Fig. 8.6 Hyperspectral reflectance image and optical absorption and scattering spectra of a peach sample: (**a**) 2-D display of the original reflectance image, (**b**) a raw spectrum extracted for a specific scattering distance, (**c**) a spatially-resolved reflectance profile extracted for a selected wavelength, (**d**) pre-processed or averaged spatially-resolved reflectance profile at the selected wavelength, (**e**) the spectrum of absorption coefficient (μ_a), and (**f**) the spectrum of reduced scattering coefficient (μ_s')

absolute reflectance measurement. Each pre-processed spatially-resolved reflectance profile for every wavelength is then fitted by the chosen diffusion model using the least squares inverse algorithm, from which the spectra of absorption and reduced scattering coefficients are obtained. The final μ_a and μ_s' spectra are the average over 19 scan measurements for each sample (Fig. 8.6e, f).

The OPA was evaluated for accuracy, stability, precision/reproducibility, and sensitivity by using liquid model samples and following the procedures described in Cen and Lu (2010). The average estimated errors for all model samples at 530–850 nm were 24 % for μ_a and 7 % for μ_s'. The system reproducibility (or precision) and the coefficient of variation (or CV) in the absorption peak at 555 nm were less than 10 % and 4 % for μ_a and μ_s', respectively. It should be mentioned that the absolute values of μ_a were very small at 700–850 nm for blue dye model samples, and at 530–600 nm for green dye model samples, thus causing the relatively large error of estimating μ_a compared with that of μ_s'. The main error sources for estimating μ_a and μ_s' could have come from the light beam, source-detector distance, and inverse algorithm, according to our optimization studies of the inverse algorithm and optical designs (Cen and Lu 2010; Cen et al. 2009). The result of sensitivity measurements on μ_a is presented in Fig. 8.7. The minimum detectable value of μ_a was 0.0117 cm^{-1}. The sensitivity of μ_s', as determined by the CV values, was always less than 3 % because μ_s' is much larger than μ_a for the investigated range of $7.0 \le \mu_s' \le 39.9\,cm^{-1}$. These results showed that the OPA has achieved acceptable accuracies for measuring the absorption and reduced scattering coefficients, which are either comparable to or superior to other reported studies using time-resolved, frequency-domain, or other types of spatially-resolved instruments (Spichtig et al. 2009; Svensson et al. 2005).

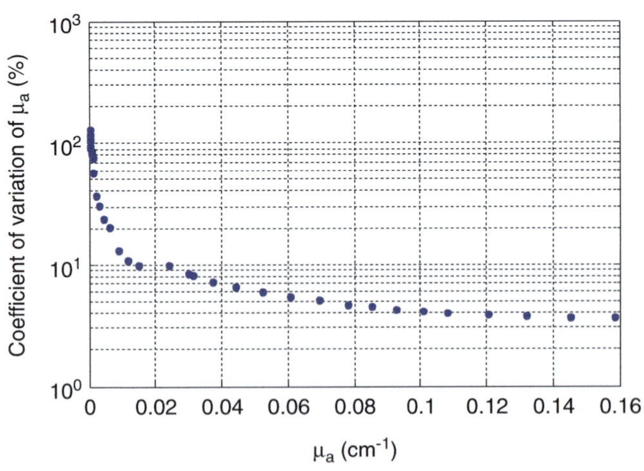

Fig. 8.7 Coefficient of variation versus reordered ascending absorption coefficients of a model sample at different wavelengths (Cen 2011)

8.4 Optical Assessment of Fruit Maturity/Quality

In this section, we first present some typical absorption and scattering spectra for a variety of food products (fruits, vegetables, meats and liquid food products) and then show two application examples, in which the hyperspectral imaging-based spatially-resolved system or the OPA was used to assess the maturity and quality of peaches and apples.

Figure 8.8 shows the spectral absorption and scattering coefficient spectra for five fruit and vegetable samples, three meat samples and three liquid food samples. The μ_a spectra of 'Golden Delicious', 'Delicious', and 'Granny Smith' apples and a 'Redstar' peach fruit (Fig. 8.8a) had absorption peaks at 675 nm, which corresponds to the chlorophyll absorption waveband, and the μ_a values ranged from 0.10 to 0.48 cm^{-1} at this wavelength. The 'Granny Smith' apple had the highest chlorophyll absorption due to its greenish skin and flesh. The ripened tomato did not show an absorption peak at 675 nm, because chlorophyll in fully ripened tomatoes decreases greatly or even disappears completely. In fully ripened tomatoes, anthocyanin becomes the dominant pigment, which absorbs light at 535 nm, as shown in Fig. 8.8a. The absorption values of the fruit and vegetable samples for 720–900 nm were relatively small and consistent, but they increased dramatically above 900 nm and peaked at 970 nm due to the water absorption. For the beef, chicken and pork samples, we also observed two prominent absorption peaks at 560 and 970 nm (Fig. 8.8b). The absorption peak at 560 nm could be attributed to the combined effect of myoglobin, oxymyoglobin, and metmyoglobin (Xia et al. 2007), while absorption at 970 nm was due to the water in the meat samples. For two orange juice samples, the absorption spectra over 550–900 nm were relatively flat, and only one prominent absorption peak was found at 970 nm due to absorption by the water; while for the milk sample, there are several small absorption peaks and one dominant peak at 970 nm caused by the water absorption.

Compared to μ_a spectra, the reduced scattering coefficient spectra of these samples were relatively flat and had fewer features. For most of the tested samples, their $\mu_s^{'}$ values decreased steadily with the increasing wavelength. This pattern of changes is consistent with Mie scattering theory and other reported studies that scattering is wavelength-dependent (Keener et al. 2007; Michels et al. 2008). The apple samples had higher $\mu_s^{'}$ values (9.0–17.0 cm^{-1}) over the entire spectral region from 500 to 1,000 nm, while the tomato had the lowest $\mu_s^{'}$ values (4.5–6.0 cm^{-1}) (Fig. 8.8a). The three meat samples also had different $\mu_s^{'}$ values over the entire wavelengths; the beef had the highest $\mu_s^{'}$ values, while the chicken sample had the lowest (Fig. 8.8b). In addition, the milk sample had distinctly higher $\mu_s^{'}$ values than the orange juice samples over the spectral region of 500–1,000 nm. Fat globules and casein micelles are the main, excellent scattering particles in milk. The ability of scattering particles to scatter light depends on their density and size, according to the empirical equation $\mu_s^{'} = a\lambda^{-b}$, which describes the relationship between $\mu_s^{'}$ and wavelength λ, where a is proportional to the density of the scattering particles, and

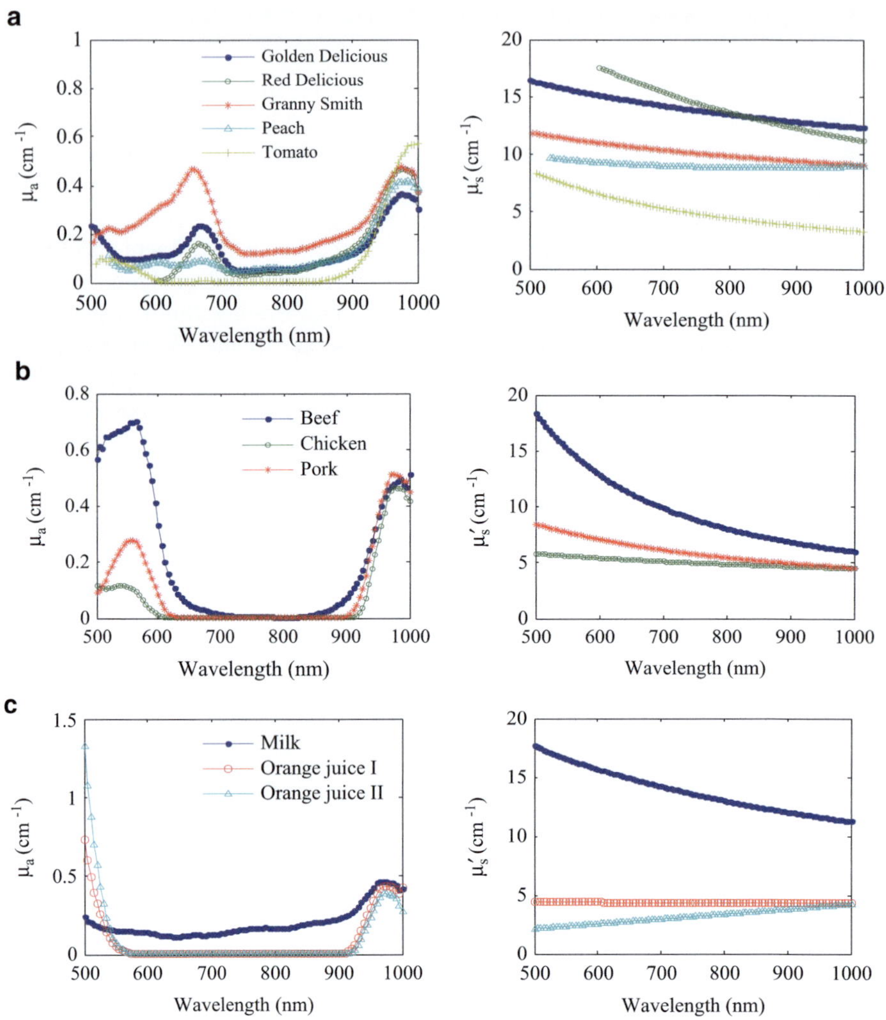

Fig. 8.8 Absorption (*left pane*) and reduced scattering coefficient (*right pane*) spectra for (**a**) three apples of different varieties, a peach fruit, and a tomato fruit; (**b**) beef, chicken, and pork; and (**c**) a full fat milk sample and two orange juice samples from two different sources

b depends on the particle size (Mourant et al. 1997). Hence the value of μ_s' and its pattern of change with wavelength can provide useful information about the structural and physical characteristics of these samples, such as firmness in fruit, tenderness in meat, and fat content in milk. In the following, we show two application examples about using the optical properties to assess the maturity/quality of peaches and apples.

Experiments were carried out in the summer and fall harvest season of 2010. Peach and apple samples were hand picked during the normal harvest time from orchards of Michigan State University's Clarksville Horticultural Experiment Station in Clarksville, Michigan. For the maturity study, 500 'Redstar' peaches were harvested three times within a 1-week period, and optical and reference maturity measurements were performed within the same day. For the apple quality study, two cultivars, 'Golden Delicious' (GD) and 'Delicious' (DL), were harvested once a week during six consecutive weeks. Eighty apples for each cultivar were tested next day after each harvest, and the remaining apples were kept in refrigerated air storage at 0 °C. Tests for the stored apples were begun 1 week after the last harvest for up to 6 weeks. A total of 1,039 GD and 1,040 DL apples were used in the experiment. Optical measurements were first performed on the peach or apple samples using the OPA with the procedures described in the previous section. Thereafter, standard destructive measurements, including Magness–Taylor firmness test, Brix refractometry for soluble solids content (SSC), and color measurement for peach skin and flesh using a digital chroma meter, were performed to provide reference fruit maturity/quality measures. A detailed description of the experimental procedures and destructive quality measurements is given in Cen and Lu (2012a and 2012b).

To use the optical parameters to predict the maturity parameters of peach samples, calibration models were developed using partial least squares (PLS) regression for each optical parameter (μ_a and $\mu_s^{'}$) and each of the three combinations, i.e., μ_a & $\mu_s^{'}$, $\mu_a \times \mu_s^{'}$ and μ_{eff}, where μ_a & $\mu_s^{'}$ refers to the simple cascading of the two parameter spectra into one spectrum for each sample, $\mu_a \times \mu_s^{'}$ is the multiplication of the two parameters wavelength by wavelength, and $\mu_{eff} = \left[3\mu_a\left(\mu_a + \mu_s^{'}\right)\right]^{1/2}$ is the effective attenuation coefficient which reflects the light penetration ability. The combinations of μ_a and $\mu_s^{'}$ were used for the calibration model development, in view of the fact that the physiological process and hence the maturation/ripening of peaches are normally accompanied with simultaneous changes in the absorption and scattering properties. Finally, the calibration models were used to predict the remaining samples (1/4 of the total samples) that were not used in the calibration.

Table 8.1 shows PLS prediction results for the firmness, SSC, and skin and flesh color of 'Redstar' peaches using μ_a, $\mu_s^{'}$ and their combinations (μ_a & $\mu_s^{'}$, $\mu_a \times \mu_s^{'}$ and μ_{eff}). μ_a and $\mu_s^{'}$ showed various levels of correlation with the maturity parameters, with values of the correlation coefficient (r) varying from 0.420 to 0.855 for μ_a and from 0.204 to 0.840 for $\mu_s^{'}$. In most cases, prediction results obtained from μ_a spectra were better than those from $\mu_s^{'}$ spectra. However, the PLS prediction models developed for $\mu_s^{'}$ are simpler with fewer factors, compared to those for μ_a. As shown in Table 8.1, using the combined data of μ_a and $\mu_s^{'}$ generally improved prediction results with r = 0.724 (standard error of prediction or SEP = 18.13 N) for firmness, r = 0.458 (SEP = 0.96 °Brix) for SSC, r = 0.893 (SEP = 3.54) for the skin color parameter L^*, and r = 0.722 (SEP = 3.32) for the flesh color L^*. Most of

Table 8.1 Partial least squares (PLS) prediction results for the maturity of 'Redstar' peaches using absorption (μ_a) and reduced scattering (μ_s') coefficients and their combinations[a] [Cen and Lu (2012a)]

	μ_a			μ_s'			μ_a & μ_s'			$\mu_a \times \mu_s'$			μ_{eff}		
Maturity index	Fact.	r	SEP	Fact.	r	SEP	Fact.	r	SEP	Fact.	r	SEP	Fact.	r	SEP
Firmness	10	0.713	18.47	6	0.494	10	12	0.720	18.38	11	0.722	18.43	10	**0.724**	18.13
SSC	15	0.420	0.99	5	0.315	13	12	**0.458**	0.96	14	0.451	0.96	13	0.419	0.99
Skin color															
L*	13	0.855	4.07	6	0.837	13	15	0.884	3.65	14	0.881	3.70	13	**0.893**	3.54
h°	16	0.717	2.82	6	0.795	13	16	0.765	2.58	16	0.738	2.70	13	**0.778**	2.50
C*	14	0.847	4.19	6	0.840	13	13	0.883	3.66	13	0.882	3.69	13	**0.886**	3.63
Flesh color															
L*	16	0.660	3.62	6	0.630	21	15	0.674	3.63	18	0.711	3.40	21	**0.722**	3.32
h°	8	0.457	1.99	4	0.204	6	10	0.433	2.03	7	0.447	1.99	6	**0.462**	1.98
C*	25	0.580	5.33	6	0.575	20	10	0.640	4.87	19	0.640	4.86	20	**0.645**	4.84

[a] μ_a & μ_s' = the cascading of the two optical parameter spectra into one spectrum for each sample

$\mu_a \times \mu_s'$ = multiplication of the two optical parameters wavelength by wavelength

$\mu_{eff} = \left[3\mu_a\left(\mu_a + \mu_s'\right)\right]^{1/2}$ is the effective attenuation coefficient measuring the light penetration ability

SSC soluble solids content

L*, h°, and C* = lightness, hue angle, and chroma, respectively

Fact. = number of factors or latent variables used in the calibration model

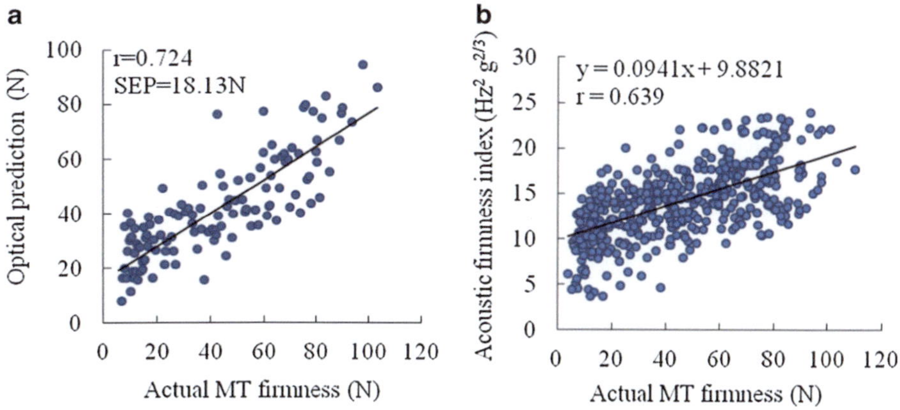

Fig. 8.9 (**a**) Prediction of Magness–Taylor (MT) firmness of 'Redstar' peaches using the effective attenuation coefficient (μ_{eff}) and (**b**) correlation between the acoustic and MT firmness measurements [Cen and Lu (2012a)]

the improved results were obtained using the effective attenuation coefficient (μ_{eff}), which is reciprocal of light penetration depth, and describes the extent to which the light can penetrate a medium. Hence, it is not surprising to have better predictions for the peach maturity parameters using μ_{eff} (except for SSC).

In this experiment, we also compared the OPA with a commercial desk-top acoustic firmness sensor for measuring the firmness of peaches. The acoustic sensor (AWETA, Nootdorp, the Netherlands) assesses the firmness of fruit by measuring its resonant frequency and mass. The results in Fig. 8.9 showed that the OPA gave better firmness predictions ($r = 0.724$ and SEP = 18.13 N) than the acoustic sensor ($r = 0.639$). While results for peach maturity prediction still need improvement, they have clearly demonstrated that the hyperspectral imaging-based spatially-resolved technique is potentially useful for assessing the maturity of peaches.

For the apple study, the same calibration and prediction procedures as that for peaches were used. Prediction results for the firmness of GD and DL apples using μ_a and μ_s', and their combinations (μ_a & μ_s', $\mu_a \times \mu_s'$, and μ_{eff}) for the freshly harvested, after-storage and combined groups are presented in Table 8.2. Likewise, SSC prediction results using the optical properties for each group are summarized in Table 8.3. Both μ_a and μ_s' were correlated with the apple firmness and SSC for each cultivar. Firmness prediction results obtained from μ_a spectra ($r = 0.687$–0.885 for GD, and $r = 0.744$–0.844 for DL) were better than those from μ_s' spectra ($r = 0.630$–0.793 for GD, and $r = 0.702$–0.768 for DL) for each of the three sample groups. This was also true for the SSC evaluation. The combinations of μ_a and μ_s' again improved the prediction of apple firmness and SSC in most cases. With the best combination of μ_a and μ_s', the correlations for the firmness of DL and GD apples for the three test groups were 0.692–0.892 and 0.788–0.863, respectively, and they were 0.741–0.791 and 0.536–0.842 for SSC, respectively.

Table 8.2 Firmness prediction results for 'Golden Delicious' and 'Delicious' apples for the freshly harvested, after-storage and combined groups[a] [Cen and Lu (2012b)]

Optical parameter		Golden Delicious			Delicious		
		Factors	r	SEP	Factors	r	SEP
Freshly harvested	μ_a	16	0.687	6.10	20	0.785	8.43
	μ_s'	6	0.630	6.49	7	0.764	8.68
	μ_a & μ_s'	12	0.651	6.37	18	0.809	7.91
	$\mu_a \times \mu_s'$	17	**0.692**	**6.06**	19	0.813	7.83
	μ_{eff}	16	0.687	6.10	22	**0.822**	**7.67**
After-storage	μ_a	17	0.726	6.11	25	0.744	9.36
	μ_s'	7	0.689	6.44	7	0.702	9.90
	μ_a & μ_s'	14	0.712	6.24	17	**0.788**	**8.57**
	$\mu_a \times \mu_s'$	17	**0.734**	**6.02**	20	0.765	9.00
	μ_{eff}	18	0.730	6.07	29	0.762	9.10
Combined	μ_a	34	0.885	8.14	34	0.844	9.56
	μ_s'	9	0.793	10.60	9	0.768	11.35
	μ_a & μ_s'	38	0.881	8.29	19	0.852	9.31
	$\mu_a \times \mu_s'$	39	**0.892**	**7.89**	35	0.857	9.12
	μ_{eff}	33	0.884	8.16	38	**0.863**	**8.94**

[a]*SEP* standard error of prediction. See the footnotes of Table 8.1 for explanation of the optical parameters

Table 8.3 Prediction results for the soluble solids content of 'Golden Delicious' and 'Delicious' apples for the freshly harvested, after-storage and combined groups[a] [Cen and Lu (2012b)]

Optical parameter		Golden Delicious			Delicious		
		Factors	r	SEP	Factors	r	SEP
Freshly harvest	μ_a	16	0.787	0.70	17	0.823	0.76
	μ_s'	6	0.489	0.99	7	0.784	0.84
	μ_a & μ_s'	12	0.781	0.70	17	**0.842**	**0.73**
	$\mu_a \times \mu_s'$	17	**0.791**	**0.69**	20	0.816	0.78
	μ_{eff}	16	0.777	0.72	22	0.821	0.77
After-storage	μ_a	18	0.713	0.95	14	0.533	0.88
	μ_s'	7	0.561	1.12	6	0.460	0.92
	μ_a & μ_s'	14	**0.741**	**0.92**	13	0.518	0.89
	$\mu_a \times \mu_s'$	18	0.732	0.93	11	0.502	0.90
	μ_{eff}	19	0.726	0.94	18	**0.536**	**0.87**
Combined	μ_a	20	0.760	0.84	23	0.812	0.88
	μ_s'	8	0.544	1.09	8	0.750	0.99
	μ_a & μ_s'	16	0.768	0.83	19	**0.825**	**0.85**
	$\mu_a \times \mu_s'$	23	**0.778**	**0.82**	21	0.804	0.89
	μ_{eff}	22	0.773	0.83	27	0.805	0.89

[a]*SEP* standard error of prediction
See the footnote of Table 8.1 for explanation of the optical parameters

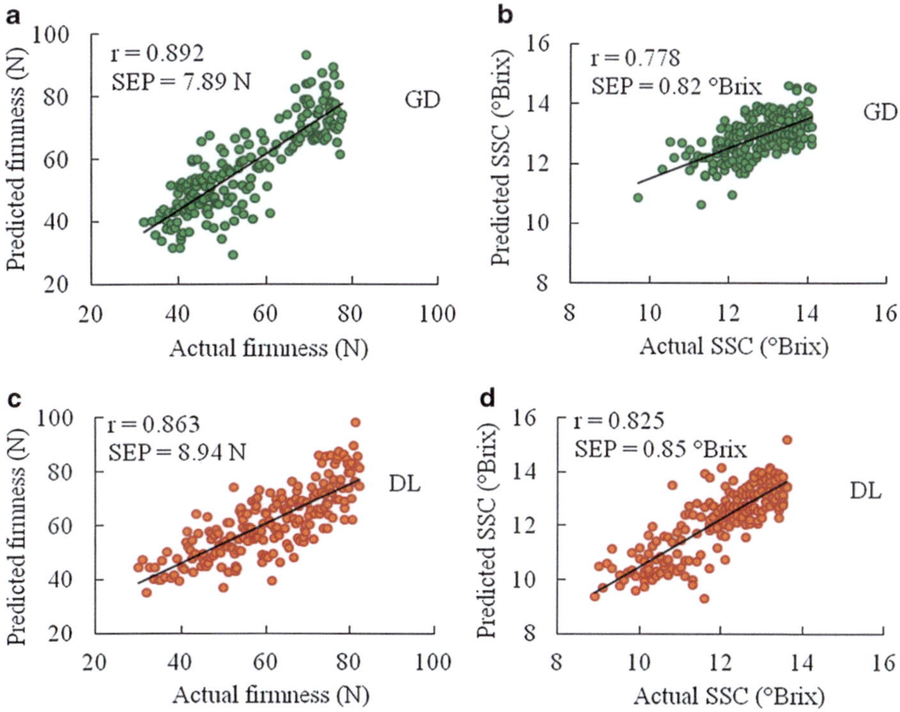

Fig. 8.10 Prediction of fruit firmness (**a**, **c**) and soluble solids content or SSC (**b**, **d**) using the best combinations of μ_a and μ_s' for the combined group of 'Golden Delicious' (GD) and 'Delicious' (DL) apples [Cen and Lu (2012b)]

For the freshly harvested and after-storage groups, the correlations were relatively low because of smaller firmness variations for each group. However, when the data from these two groups were pooled, improved correlations for the firmness prediction of both cultivars were obtained, although the SEP values increased slightly due to larger firmness variations in the combined group. For SSC, the best predictions were achieved for the freshly harvested group with r = 0.791 (SEP = 0.69 °Brix) for GD and r = 0.842 (SEP = 0.73 °Brix) for DL. Comparable results were obtained for the combined group because the SSC in the apple fruit did not change significantly during storage. Figure 8.10 shows the firmness and SSC predictions for GD and DL apples obtained with the best combination of μ_a and μ_s' for the combined group. Better predictions of firmness for GD and DL apples with the correlation of 0.892 and 0.863, respectively, were obtained than for SSC predictions with r = 0.778 and 0.825. These results are comparable with other reported studies using hyperspectral scattering technique (Mendoza et al. 2011; Qin et al. 2009).

8.5 Concluding Remarks

In the chapter, we have described hyperspectral imaging-based spatially-resolved technique for measuring the optical properties of food products and other biological materials. Compared to other techniques like time-resolved and frequency-domain, this technique is faster and simpler in instrumentation and also covers a broader spectral region. It also demonstrated the ability to assess the maturity and quality of peaches and apples. Further research is, however, still needed in order to make this new technique available for property and quality evaluation of a wide range of food and agricultural products.

Optical measurements using the current system still show relatively large variability for the same sample. This variability is related to several factors, including the roughness and geometric irregularities on the surface of fruit and the inherent shortcomings (e.g., use of small light beam) with the method itself. Hence further study should be carried out to evaluate and improve the measurement accuracy and repeatability of the technique and minimize the effect of sample surface condition (i.e., roughness, unevenness, or irregularity) on optical measurements. Moreover, the performance of the diffusion model is also dependent on the scattering and absorption characteristics of samples because the model is based on certain assumptions (i.e., scattering dominant, isotropic source, and source-detector distance being greater than one transport mean free path, etc.). Hence we need to further evaluate the limitation and application scope of the diffusion model for food and agricultural products.

Many biological materials, such as fruit, are composed of two layers of distinctive tissues (e.g., skin and flesh). The current system is only suitable for evaluating homogenous media. Further research is thus needed to develop an effective method for measuring the optical properties of heterogeneous or multi-layer biological materials and to reduce the computational complexity and time for estimating the absorption and reduced scattering coefficients.

References

Birth GS (1978) Light-scattering properties of foods. J Food Sci 43(3):916–925

Birth GS, Davis CE, Townsend WE (1976) Scattering coefficient as a measure of pork quality. J Anim Sci 43(1):237

Bykov AV, Kirillin MY, Priezzhev AV, Myllyla R (2006) Simulations of a spatially resolved reflectometry signal from a highly scattering three-layer medium applied to the problem of glucose sensing in human skin. Quantum Electron 36(12):1125–1130

Cen H (2011) Hyperspectral imaging-based spatially-resolved technique for accurate measurement of the optical properties of horticultural products. Michigan State University, Biosystems and Agricultural Engineering, East Lansing Unpublished Ph.D. dissertation, 204pp

Cen H, Lu R (2010) Optimization of the hyperspectral imaging-based spatially-resolved system for measuring the optical properties of biological materials. Opt Express 18(16):17412–17432

Cen H, Lu R (2012a) Hyperspectral imaging-based spatially-resolved technique for peach maturity and quality assessment. Trans ASABE. Transactions of the ASABE 55(2):647–657
Cen H, Lu R (2012b) Analysis of absorption and scattering spectra for assessing the internal quality of apple fruit. Acta Horticulturae 945:181–188
Cen H, Lu R, Dolan K (2009) Optimization of inverse algorithm for estimating optical properties of biological materials using spatially-resolved diffuse reflectance. Inverse Probl Sci Eng 18 (6):853–872
Cubeddu R, D'Andrea C, Pifferi A, Taroni P, Torricelli A, Valentini G, Dover C, Johnson D, Ruiz-Altisent M, Valero C (2001) Nondestructive quantification of chemical and physical properties of fruits by time-resolved reflectance spectroscopy in the wavelength range 650–1000 nm. Appl Optics 40(4):538–543
Doornbos RMP, Lang R, Aalders MC, Cross FW, Sterenborg H (1999) The determination of in vivo human tissue optical properties and absolute chromophore concentrations using spatially resolved steady-state diffuse reflectance spectroscopy. Phys Med Biol 44(4):967–981
Durduran T, Yodh AG, Chance B, Boas DA (1997) Does the photon-diffusion coefficient depend on absorption? J Opt Soc Am A Opt Image Sci Vis 14(12):3358–3365
Fabbri F, Franceschini MA, Fantini S (2003) Characterization of spatial and temporal variations in the optical properties of tissuelike media with diffuse reflectance imaging. Appl Optics 42 (16):3063–3072
Farrell TJ, Patterson MS, Wilson B (1992) A diffusion-theory model of spatially resolved, steady-state diffuse reflectance for the noninvasive determination of tissue optical-properties in vivo. Med Phys 19(4):879–888
Groenhuis RAJ, Ferwerda HA, Tenbosch JJ (1983a) Scattering and absorption of turbid materials determined from reflection measurements. 1. Theory. Appl Optics 22(16):2456–2462
Groenhuis RAJ, Tenbosch JJ, Ferwerda HA (1983b) Scattering and absorption of turbid materials determined from reflection measurements. 2. Measuring method and calibration. Appl Optics 22(16):2463–2467
Haskell RC, Svaasand LO, Tsay TT, Feng TC, McAdams MS (1994) Boundary-conditions for the diffusion equation in radiative-transfer. J Opt Soc Am A Opt Image Sci Vis 11(10):2727–2741
Ishimaru A (1978) Wave propagation and scattering in random media, vol 1, Single scattering and transport theory. Academic, New York
Jones MR, Yamada Y (1998) Determination of the asymmetry parameter and scattering coefficient of turbid media from spatially resolved reflectance measurements. Opt Rev 5(2):72–76
Keener JD, Chalut KJ, Pyhtila JW, Wax A (2007) Application of Mie theory to determine the structure of spheroidal scatterers in biological materials. Opt Lett 32(10):1326–1328
Kienle A, Patterson MS (1997) Improved solutions of the steady-state and the time-resolved diffusion equations for reflectance from a semi-infinite turbid medium. J Opt Soc Am A Opt Image Sci Vis 14(1):246–254
Kubelka P, Munk F (1931) Ein beitrag zur optik der farbanstriche. Z Tech Phys 12:593–601
Langerholc J (1982) Beam broadening in dense scattering media. Appl Optics 21(9):1593–1598
Law SE, Norris KH (1973) Kubelka-Munk light-scattering coefficients of model particulate systems. Trans ASAE 16(5):914–917, 921
Marquet P, Bevilacqua F, Depeursinge C, Dehaller EB (1995) Determination of reduced scattering and absorption-coefficients by a single charge-coupled-device array measurement. 1. Comparison between experiments and simulations. Opt Eng 34(7):2055–2063
Mendoza F, Lu R, Ariana DP, Cen H, Bailey BB (2011) Integrated spectral and image analysis of hyperspectral scattering data for prediction of apple fruit firmness and soluble solids content. Postharvest Biol Technol 62(2):149–160
Michels R, Foschum F, Kienle A (2008) Optical properties of fat emulsions. Opt Express 16 (8):5907–5925
Mourant JR, Fuselier T, Boyer J, Johnson TM, Bigio IJ (1997) Predictions and measurements of scattering and absorption over broad wavelength ranges in tissue phantoms. Appl Optics 36 (4):949–957

Nicolai BM, Verlinden BE, Desmet M, Saevels S, Saeys W, Theron K, Cubeddu R, Pifferi A, Torricelli A (2008) Time-resolved and continuous wave NIR reflectance spectroscopy to predict soluble solids content and firmness of pear. Postharvest Biol Technol 47(1):68–74

Patterson MS, Chance B, Wilson BC (1989) Time resolved reflectance and transmittance for the noninvasive measurement of tissue optical-properties. Appl Optics 28(12):2331–2336

Patterson MS, Moulton JD, Wilson BC, Berndt KW, Lakowicz JR (1991) Frequency-domain reflectance for the determination of the scattering and absorption properties of tissue. Appl Optics 30(31):4474–4476

Pilz M, Honold S, Kienle A (2008) Determination of the optical properties of turbid media by measurements of the spatially resolved reflectance considering the point-spread function of the camera system. J Biomed Opt 13(5):054047

Qin J, Lu R (2006) Hyperspectral diffuse reflectance imaging for rapid, noncontact measurement of the optical properties of turbid materials. Appl Optics 45(32):8366–8373

Qin J, Lu R (2007) Measurement of the absorption and scattering properties of turbid liquid foods using hyperspectral imaging. Appl Spectrosc 61(4):388–396

Qin J, Lu R (2008) Measurement of the optical properties of fruits and vegetables using spatially resolved hyperspectral diffuse reflectance imaging technique. Postharvest Biol Technol 49 (3):355–365

Qin J, Lu R, Peng Y (2009) Prediction of apple internal quality using spectral absorption and scattering properties. Trans ASABE 52(2):499–507

Reynolds L, Johnson C, Ishimaru A (1976) Diffuse reflectance from a finite blood medium—applications to modeling of fiber optic catheters. Appl Optics 15(9):2059–2067

Rizzolo A, Vanoli M, Spinelli L, Torricelli A (2010) Sensory characteristics, quality and optical properties measured by time-resolved reflectance spectroscopy in stored apples. Postharvest Biol Technol 58(1):1–12

Saeys W, Velazco-Roa MA, Thennadil SN, Ramon H, Nicolai BM (2008) Optical properties of apple skin and flesh in the wavelength range from 350 to 2200 nm. Appl Optics 47(7):908–919

Spichtig S, Hornung R, Brown DW, Haensse D, Wolf M (2009) Multifrequency frequency-domain spectrometer for tissue analysis. Rev Sci Instrum 80(2):024301

Svensson T, Swartling J, Taroni P, Torricelli A, Lindblom P, Ingvar C, Andersson-Engels S (2005) Characterization of normal breast tissue heterogeneity using time-resolved near-infrared spectroscopy. Phys Med Biol 50(11):2559–2571

Tuchin V (2000) Tissue optics: light scattering methods and instruments for medical diagnosis. SPIE, Bellingham

Wang LV, Wu H-I (2007) Biomedical optics: principles and imaging. Wiley, Hoboken

Welzel J, Reinhardt C, Lankenau E, Winter C, Wolff HH (2004) Changes in function and morphology of normal human skin: evaluation using optical coherence tomography. Br J Dermatol 150(2):220–225

Wilson BC, Patterson MS (2008) The physics, biophysics and technology of photodynamic therapy. Phys Med Biol 53(9):R61–R109

Xia JJ, Berg EP, Lee JW, Yao G (2007) Characterizing beef muscles with optical scattering and absorption coefficients in VIS-NIR region. Meat Sci 75(1):78–83

Yao G, Liu KS, Hsieh F (2004) A new method for characterizing fiber formation in meat analogs during high-moisture extrusion. J Food Sci 69(7):E303–E307

Chapter 9
Quality Evaluation of Plant Products

Jasper G. Tallada, Pepito M. Bato, Bim P. Shrestha, Taichi Kobayashi, and Masateru Nagata

9.1 Introduction

Hyperspectral imaging or imaging spectrometry combines the strengths of computer vision technology with optical spectroscopy. It is primarily suited for measurement of parameters that vary spatially both at the external surface of samples and internally within the samples. The parameters may be physical features such as incipient bruises or surface contamination, or chemical constituents such as sugar and acidity. While the acquisition of images generally follows the procedures of machine vision, adding a spectral dimension would require the rigor of multivariate statistics, also known as chemometrics, to find functional relationships between the measured values and target parameters. Its application to agriculture, particularly to post-harvest processing, has recently been explored by university research laboratories in order to develop new techniques for non-destructive measurement of quality.

Hyperspectral imaging has opened up new opportunities and challenges for measurement of internal and external quality parameters for fruits and vegetables.

J.G. Tallada (✉)
Cavite State University, Indang, Philippines

U.S. Department of Agriculture, Agricultural Research Service, Manhattan, KS, USA
e-mail: jtallada@gmail.com

P.M. Bato
University of the Philippines Los Baños, Los Baños, Philippines

B.P. Shrestha
Kathmandu University, Dhulikhel, Nepal

T. Kobayashi
University of Miyazaki, Miyazaki, Japan

M. Nagata
Faculty of Agriculture, University of Miyazaki, Miyazaki, Japan
e-mail: mnagata@sky.miyazaki-mic.ac.jp

B. Park, R. Lu (eds.), *Hyperspectral Imaging Technology in Food and Agriculture*,
Food Engineering Series, DOI 10.1007/978-1-4939-2836-1_9

For instance, accurately measuring the spatially varying sweetness and sourness as primary components of taste of fruits would enable more efficient grading, thus increasing the confidence in marketing the produce. The increasing preferences of the Japanese consumers for the best quality fruits and vegetables are not only related to their changing personal interests but are also rooted in their cultural tradition of giving the best gift as possible. Thus guaranteeing the premium quality of fruits is of utmost importance. There are many dimensions of quality for different kinds of commodities, which often require new and better non-destructive measurement techniques. For example, the increasing awareness for the health benefits of anthocyanins by certain segments of the market had prompted research work on accurately quantifying the internal pigment contents in strawberries, carrots and certain varieties of sweet potatoes. Increasing concerns for food safety have also ushered development of more powerful technologies that can be placed in-line at a packing house operation. Incipient bruises in fruits can lead to early rot, which can potentially lead to the rejection of not only the suspicious fruits but of the entire batch. Since bruises are not easily recognized at the fruit surface, hyperspectral imaging and its complementary technology, multi-spectral imaging, offer a unique way of detecting these types of fruit damage.

The main goal of this chapter is to present several specific techniques of hyperspectral imaging for non-destructive quality measurement of fruits and vegetables. Discussions are given on the hardware design, components selection, preparatory protocols and analytical techniques.

9.2 Hyperspectral Imaging Setup

The design and construction of an imaging system is crucial for the success of any hyperspectral imaging study. There are two distinct types of hyperspectral imaging systems, namely the pushbroom and the area scan. The pushbroom system is exemplified by the ImSpector device (Spectral Imaging Ltd., Oulu, Finland) having a prism-grating-prism optical assembly that permits collection of spectra for a line of pixels from an object on a moving stage. The speed of the moving stage defines the spatial resolution of the system in the moving direction. One-time spatial corrections are needed in order to view the entire object for each wavelength. The area scan system, on the other hand, places an electronic variable filter unit in front of a monochrome camera; each time the system takes two-dimensional image for a narrow waveband selected by the filter. This is similar to the more common special effects technique that is used in popular photography, for example by placing an interference green or long-pass infrared filter in front of the camera lens. The electronic filter may either consist of a liquid crystal tunable filter (LCTF) or an acousto-optical tunable filter (AOTF). The Cambridge Research and Instrumentation Inc. (Woburn, MA, USA) produces LCTFs under its Varispec line of filters as shown in Fig. 9.1. There are two models available: one model operates in the visible range from 400 to 700 nm with a 10 nm FWHM (full width at half maximum)

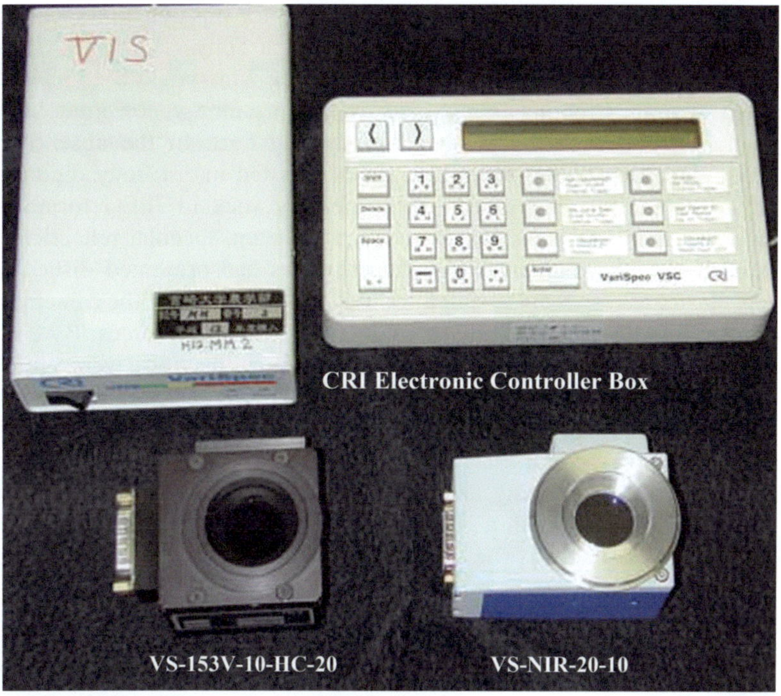

Fig. 9.1 A Varispec liquid crystal tunable filter

bandwidth (VS-V153-10-HC-20) and another operates in the visible-near infrared range from 650 to 1,100 nm also with a 10 nm FWHM bandwidth (VS-NIR-20-10). The center wavelengths of these filters are set by keying in the values in their accompanying hand-held CRI Electronic Controller Box. Alternately, the controller may be connected to a serial port (RS-232 protocol) of a personal computer with a program that remotely sets the wavelength.

There are some positive points in using an area scan system. Firstly, the number and interval resolution of the acquisition can be easily configured. For example, one may decide to get images, say, from 400 to 700 nm inclusive at 1 nm intervals to produce 301 sets of spectral images for a particular sample, or in another application from 500 to 650 nm at 10 nm intervals (16 images total). This would have dramatic effect on the speed and time in getting the hyperspectral image cube sets. Secondly, no complicated geometric correction has to be made since the object remains stationary throughout the acquisition process (Lawrence et al. 2003). The overall system design is not overly complicated, compared to moving stages in a pushbroom system. Control of components would be much easier and failure of parts can be placed at a minimum cost. Finally, since the center wavelength is randomly selectable, the hyperspectral imaging area scan system can be easily re-configured into a multi-spectral imaging system especially if the research aims to identify optimal wavelengths for discrimination or measurement.

Our laboratory assembled two imaging systems for the two Varispec LCTFs. The visible range hyperspectral system comprises a Sony AVC-D5 CCD video camera having a Cosmicar television lens with the Varispec VS-153-10-HC-20 filter fitted in front. Lighting is provided by two machine vision grade tungsten halogen lights with an infrared filter and cooling fans. In the absence of an appropriate digitizing board, the image signal was fed into a Sony digital video camera (DCR-PC 100) and saved into a memory stick in JPEG format (8-bit resolution). During the early stage of building the setup, specular reflections from shiny surfaces of strawberry and eggplant samples had presented difficulties in acquiring clean images. A polarizing filter was placed in front of the camera lens to complement with the pre-built polarizing filters in the light sources. By correctly adjusting the filters through trial and error, specular reflections were eliminated from the images. The enclosure was covered around by black poster boards, and the front by a black textile to seal off the system against ambient light.

The second hyperspectral imaging system (Fig. 9.2) was built for the visible-near-infrared range comprising an Apogee AP2E camera (Apogee Instruments, Inc., Auburn, CA, USA) with a Nikkor f/1.2 optical lens to which the Varispec VS-NIR-20-10 was attached. The camera was mounted to a standard photography

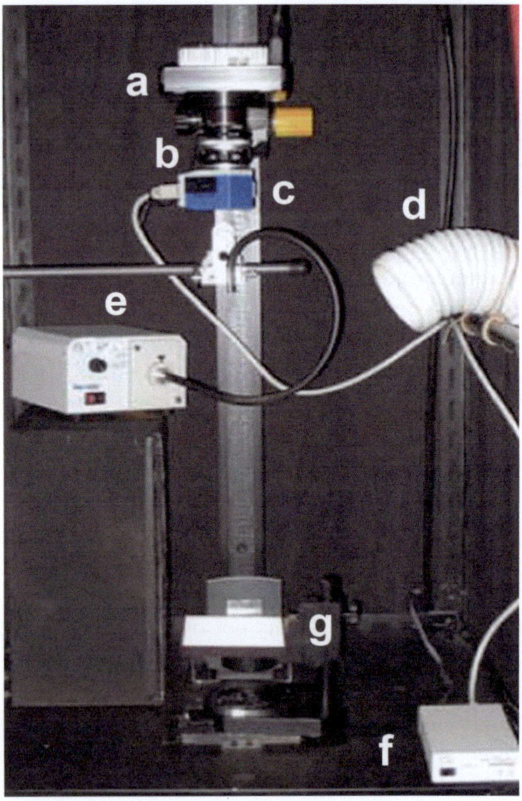

Fig. 9.2 Hyper-spectral imaging setup consisted of (**a**) Apogee AP2E camera, (**b**) Nikkor lens, (**c**) CRI NIR Varspec LCTF, (**d**) cooler exhaust, (**e**) Dolan-Jenner Fiber-Lite light source, (**f**) Varispec Controller Box, and (**g**) sample stage

frame which allows vertical and horizontal height adjustments. Lighting was provided by a Dolan-Jenner Fiberlite PL950 DC-regulated Illuminator (Dolan-Jenner Industries, St. Lawrence, MA, USA) with a 150 W tungsten-halogen bulb through its fiberoptic light guide. This illuminator allowed smooth and continuous wavelengths of light from the visible to the near-infrared. An air-conditioned cooling system maintains the chamber temperature (at about 22 °C) to minimize the effects of varying temperature on the image quality. Similarly, the Varispec was controlled by software through the serial RS-232 connection and image data are acquired using the accompanying Maxim/DL software of the Apogee. Image files were saved using the FITS format (Flexible Image Transport System) to maintain the 14-bit raw resolution of image data. The whole system was placed in a rigid metal frame surrounded by a photographic dark curtain to completely isolate it from stray lights.

9.3 Procedures for Acquiring Hyperspectral Image Cubes

The basic principles used in collecting hyperspectral imaging data are similar to the general methods employed by optical spectroscopy. Since there are two spatial dimensions (x and y) in the images in addition to the spectral dimension, the one-dimension single-point spectroscopic methods must be extended to the spatial region. Generally, the procedures follow discrete sequential steps. Firstly, the system must be adequately stabilized since the electronic components (light source, the LCTF and camera particularly) vary in performance with temperature. The components should be allowed to stabilize for 20–60 min after they are powered on.

Second, dark images are acquired to account for the electronic DC voltage offset of the sensor. This is achieved by either switching off the light source or through a simpler approach of placing a dark opaque cover onto the lens (such as the lens cap) or keeping close the shutter mechanism of the camera. Switching off the light source would require longer times to stabilize its intensity. While it is quite feasible to use a single dark image for the entire wavelength range, conventionally we get separate dark images for each of the center wavelengths designed for the study.

Thirdly, light reference images are acquired to establish a reference but not necessarily the maximum signal each pixel of sensor has at each wavelength. This will define not only the spectral pattern of the light source but also the interaction of this spectral pattern with the other components of the imaging system especially the lens sub-system. Reference images are taken from a reference panel usually made of a very stable diffusely emitting material such as a Spectralon (Labsphere Inc., North Sutton, NH, USA) or Teflon (poly-tetrafluoroethylene) material. Normally, these materials have well defined reflectance characteristics traceable to NIST or other measurement standards. There can be several gray levels of these materials which should be accounted for in the computation. For example, the Spectralon panels are prepared by LabSphere in 99.9, 80, 50, and 20 % average diffuse reflectance over the visible and near-infrared range. Typically, the 99.9 %

reflectance standard is used for imaging studies. Since the electronics will age during the acquisition of imaging data, both reference and dark image sets have to be collected regularly to account for any electronic drifts. The researcher may decide the re-collection either at fixed time intervals or after working on a certain number of samples.

Finally, images of the samples are acquired by either manually keying in the center wavelengths in the handheld controller or in an automated fashion using a controller program. The same collection procedure should be used with dark and reference images acquisition, that is, all images are obtained using the same exposure or integration time.

Prior to experimentation, optimization procedures should be done to define the geometric parameters of the image acquisition. Firstly, the camera and lens field of view must be inspected to define the camera height, focus and shutter opening. Along this, the correct background of the samples (usually black) on the stage must also be defined. Secondly, the size of the fruit samples must be considered in the optimization of heights of camera and light source, and more importantly, the Spectralon panel should be large enough to accommodate the largest of the samples or the target area of the samples. Some of the mango fruit samples that we have dealt with were too large for the available Spectralon panel that primarily justified our move then to limit analyses to an area of interest around the central equatorial portion of the fruits. Thirdly, the collection wavelength intervals must be carefully decided upon which will dictate the total collection time for the samples and the resolution of the spectral acquisition. Note that there is a very high correlation of reflectance values between adjacent wavelengths.

An important procedure prior to experimentation is the establishment of the exposure time. This is an important optimization step to avoid saturation and blooming of pixels of the sensor. To accomplish this, at a certain fixed exposure time, reference images are taken off the reference panel across at constant wavelength interval. Using the image viewing program, the wavelength image that gives the maximum sensor value is selected. In a trial-and-error fashion, more images are acquired at this wavelength at different exposure times to identify the optimal time. Typically, the optimal time is selected that will correspond to either two-thirds or three-fourths of the maximum sensor response. For example, the Apogee AP2E camera has 14 bits of resolution corresponding to a maximum of 16,383 value ($2^{14} - 1$ or $16{,}384 - 1$), so the maximum pixel values should be about 12,000–14,000 digital value. A fixed level of exposure time is usually used for all image acquisitions (dark, reference and samples) at the entire wavelength range during experimentation. A shortcoming of using a single exposure time, however, is that the dynamic response of the sensor gets only optimized at a single wavelength as explained above, which would penalize the dynamic response efficiency of the system for the other center wavelengths since the camera sensor can have wider response at longer exposure times. Thus, another method is needed to establish the appropriate exposure time at each center wavelength. This will involve greater preparation efforts since each wavelength interval or a short range of wavelengths will be inspected separately. Also the exposure time has to be plugged into the

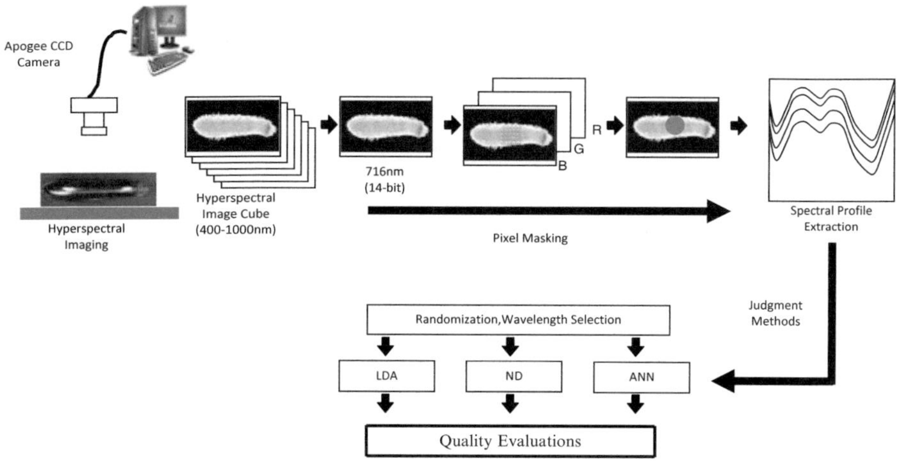

Fig. 9.3 A diagram of hyperspectral image data acquisition and analysis for identification of optimal wavelengths for fruit and vegetables

camera control software which again will take some efforts of the researcher during operation, or the controlling program must be modified to accommodate varying exposure times.

9.4 General Approaches to Analysis

A diagram showing the general processing of the hyperspectral imaging data is shown in Fig. 9.3, which depicts a combination of simple methods of image processing, spectroscopy, and mathematical or statistical procedures.

9.5 Image Processing

Matlab has an image processing toolbox that greatly facilitates the analysis of hyperspectral images. A basic initial step in the analysis is to isolate or segregate the pixels of the target object from the background by simple image grayscale thresholding technique. From this step a binary mask image (a 1 represents the target object and a 0 as the background) for the entire image cube is generated from a selected wavelength image in the hyperspectral image cube. Such mask image must define either the extents of the entire sample or an area-of-interest of the samples. The images from the Sony camera can be processed using one of the planes of the RGB image. On the other hand, the Apogee images had a resolution of 14 bits and must be converted into 8-bit bitmapped grayscale images (BMP extension name) prior to thresholding. The critical level of threshold can be decided through a logical experimentation or through an automated means (e.g., Gryathresh procedure of Matlab). Along with thresholding, it is sometimes necessary to erode or shrink the masked object especially if the entire fruit object is the target. This will remove noisy image data in the edges due to shadow effects or complicated acquisition geometry. A simple image erosion using a disc pattern at a fixed radius (imerode method) will accomplish the task.

Specific to the bruised fruits study, using the Microsoft Paint program, the pixels are selected by manually identifying and coloring the bruised pixels (in red color) and non-bruised pixels (in blue color). In a similar manner, the target area for the sugar content or firmness studies was normally confined to a circular area on the equatorial region of the fruits. The mask was defined by manually colorizing the area so that spectral extraction can be automated.

9.6 Spectral Processing

The next general step is to extract the spectral data and prepare them for the mathematical and statistical analyses. Guided by the mask image, the x and y (horizontal and vertical) coordinates of the target pixels will identify the points from which the relative reflectance data are computed upon. Either each pixel will comprise one spectral set (for example for the bruising study) or the average of the reflectance for all the masked pixels (for example in sugar contents study). Known as the flat field corrected relative reflectance spectra, the data is computed as follows:

$$I_{norm}(x, y) = \frac{I_{sample}(x, y) - I_{dark}(x, y)}{I_{reference}(x, y) - I_{dark}(x, y)} \cdot m \tag{9.1}$$

where, $I_{norm}(x, y)$ is the normalized pixel value (relative reflectance) at pixel location (x, y); $I_{sample}(x, y)$, $I_{reference}(x, y)$, $I_{dark}(x, y)$ are the pixel values for the sample, reference and dark images, respectively, at the same location (x, y); and m is a factor with a value of 1.0 based on the Spectralon reference panel that has average diffuse reflectance of 99.9 % through a wide range of wavelength of light. There are other reference panels of intermediate reflectance properties. Thus, the m factor will change proportionately. The spectral data will then be merged with the constituent values to prepare them for consequent analysis.

9.7 Mathematical and Statistical Analysis

Using some chemometric methods, predictive models for the target constituents or discrimination are derived from the spectral data. For simple constituent prediction, stepwise multiple linear regression will not only identify a limited set of wavelengths but also compute for the model coefficients to define the predicting equation. In the bruising study, stepwise linear discriminant analysis was also used to identify the important wavelengths. Prior to these analyses, spectral pre-treatment such as mean centering, smoothing, multiplicative scatter correction and derivatives may be applied on the data as suggested by earlier researches.

In some cases, the dataset may be divided into calibration and prediction or validation sets to better assess the performance of the predictive models. The validation set may be obtained from the samples not utilized in the formulation of the models. Some samples may be separately obtained to specifically serve the purpose for validation. Using this approach, the following parameters were computed:

$$\mathrm{SEC} = \sqrt{\frac{\sum_{i=1}^{N_C} \left(Y_i - \hat{Y}_i\right)^2}{N_C - p - 1}} \tag{9.2}$$

$$\mathrm{SEP} = \sqrt{\frac{\sum_{i=1}^{N_P} \left(Y_i - \hat{Y}_i - \mathrm{bias}\right)^2}{N_P - 1}} \tag{9.3}$$

$$\mathrm{bias} = \frac{\sum_{i=1}^{N_P} Y_i}{N_P} - \frac{\sum_{i=1}^{N_P} \hat{Y}_i}{N_P} \tag{9.4}$$

$$R_C = \frac{N_C \sum_{i=1}^{N_C} Y_i \hat{Y}_i - \left(\sum_{i=1}^{N_C} Y_i\right)\left(\sum_{i=1}^{N_C} \hat{Y}_i\right)}{\sqrt{N_C \sum_{i=1}^{N_C} Y_i^2 - \left(\sum_{i=1}^{N_C} Y_i\right)^2} \cdot \sqrt{N_C \sum_{i=1}^{N_C} \hat{Y}_i^2 - \left(\sum_{i=1}^{N_C} \hat{Y}_i\right)^2}} \tag{9.5}$$

$$R_P = \frac{N_P \sum_{i=1}^{N_P} Y_i \hat{Y}_i - \left(\sum_{i=1}^{N_P} Y_i\right)\left(\sum_{i=1}^{N_P} \hat{Y}_i\right)}{\sqrt{N_P \sum_{i=1}^{N_P} Y_i^2 - \left(\sum_{i=1}^{N_P} Y_i\right)^2} \cdot \sqrt{N_P \sum_{i=1}^{N_P} \hat{Y}_i^2 - \left(\sum_{i=1}^{N_P} \hat{Y}_i\right)^2}} \tag{9.6}$$

where, Y_i and $\hat{Y}_i$ are the measured and predicted experimental values, respectively for sample i; p is the number of wavelengths in the model, N_C and N_P are the number of samples in the calibration and prediction sets, respectively. SEC is the standard error of calibration with its corresponding correlation coefficient Rc, while SEP is the standard error of prediction that is corrected for bias and its corresponding correlation coefficient, Rp.

9.8 Detection of Bruises

9.8.1 Strawberry

The presence of incipient bruises may affect the economic value of fruits and vegetables directly because consumers will pick out only those items that are free of visible surface bruises and other imperfections. The effect may also be indirect because bruises can facilitate the growth of fungi and bacteria that can potentially lower the quality and hence the prices of the commodity. Detection of bruises is important in the early stages of post-harvest handling to increase the economic potential of the produce while avoiding any deleterious effects on public health and consumer confidence.

Strawberries may be harvested by hand at different ripeness levels depending on the judgment of the farmers. Riper fruits are more susceptible to mechanical bruising. But these bruises are not immediately visible and, thus, an effective optical technology must be employed.

An early work of Shrestha et al. (2002a, b) has shown the feasibility of detecting bruised areas on strawberries using near-infrared imaging rather than color imaging. The preliminary work identified a 960 nm long pass filter coupled to a near-infrared sensitive Hamamatsu camera. But machine vision techniques were severely limited for an efficient method to discriminate bruised from non-bruised surfaces on the fruits. Based on this preliminary work, further hyperspectral imaging studies were initiated, first on strawberries (Tallada et al. 2006a, b; Nagata et al. 2006b) and later on mangoes and peaches (Tallada 2006).

Controlled bruising forces (ranging from 0 to 3.0 N) were applied to strawberries of two ripeness levels, using a 25 mm spherical tip in an Orientec STA-1150 universal testing machine, as shown in Fig. 9.4. Immediately after the application of bruising, hyperspectral images were taken from 650 to 1,100 nm at 10 nm intervals. Images were taken again for the following four more days to examine the temporal dynamics of bruise development. As shown by the earlier work, NIR images (960 nm and longer) can reveal the presence of bruised tissues in fruits. Inspecting the hyperspectral images in the NIR region showed that the 980 nm (bandpass) can distinctly show the bruised tissues yet working on this wavelength alone cannot provide an efficient means for bruise detection because of possible variations of light conditions and geometry of image acquisition. To obtain the optimal wavebands for discriminating bruises, a systematic approach involved deriving a color image from the 980 nm image by usual image conversion methods of Matlab. The pixels of the bruised and non-bruised tissues were marked using the Microsoft Paint program (blue for bruised and red for non-bruised). Using this mask, reflectance spectra of pixels were extracted and then sampled for linear discriminant analysis. Expectedly, the 980 and 825 nm (spectral maximum) wavebands were found optimal for bruise detection, and from these bands, the efficiency of three bruise detection algorithms: linear discriminant analysis,

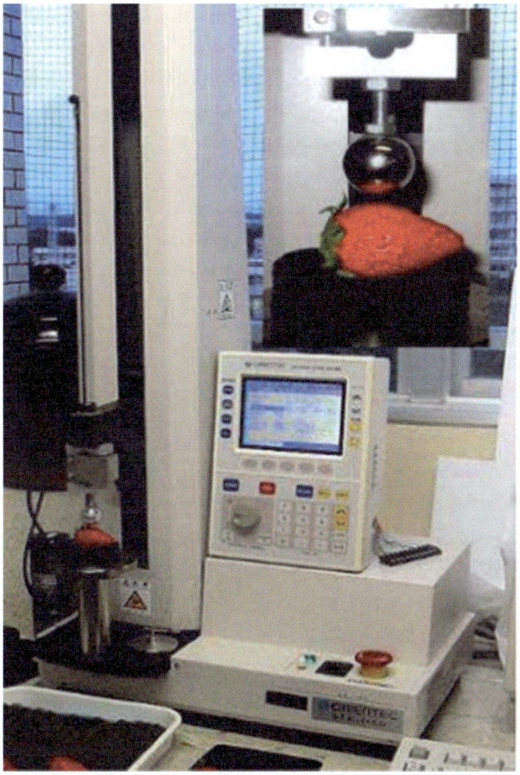

Fig. 9.4 The Orientec STA-1150 universal testing machine comprised the bruising set-up for strawberry using a 25-mm diameter ball tip (*inset*)

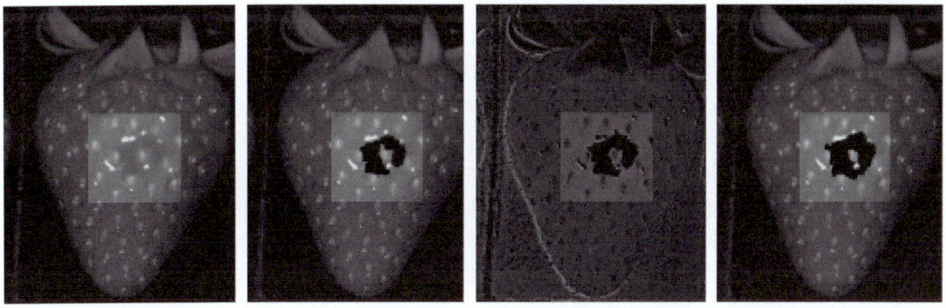

Fig. 9.5 Bruises on a strawberry fruit sample detected using three methods of judgment on a 70–80 % ripe level strawberry receiving a 2.0 N bruising force. [a]Linear discriminant analysis. [b]Normalized difference. [c]Artificial neural network

normalized difference and neural network were compared. All the three models had worked equally well as shown in Fig. 9.5, but neural network gave better detection of the extent of bruised areas, followed by normalized difference. An area-of-interest (lighted square areas in the figures) was marked because the oblate shape had resulted to false positives near the fringes of the fruits. Bruises were detected because the non-bruised tissues appeared as a uniform field in the images. When bruising happens, the expelled cellular contents spilled and concentrated into areas that attenuated the absorbance of light and hence appear darker in the NIR images. The spectral region around 980 nm has been previously identified as an absorption band for water.

The fruits were held fixed into their carrying trays to enable collection of images in the same fruit pose and position for several days. When the bruised areas were compared over time, there was a gradual reduction of detected bruised areas due to either re-absorption of the expelled water by adjacent tissues or the drying up of the bruised areas.

9.8.2 Peach and Mango

The same experiment was carried out for mangoes and peaches as shown in Fig. 9.6. The spectral images were processed using the Image Processing Toolbox of MATLAB version 6.5 (The Mathworks, Inc., Natick, MA). Through an observation, bruised areas were easily noticed from the 980 nm images such that eight-bit RGB images were derived for masking (and classifying) of pixels. Using the Microsoft Paint software, bruised tissues were manually marked with blue color while non-bruised tissues with red color as shown in Fig. 9.7.

The relative reflectance profiles for the bruised and non-bruised fruit areas for peaches and mangoes are shown in Fig. 9.8. A stronger absorption at 675 nm was observed from the peach samples than the mango samples. This is the wavelength where the absorption peak of chlorophyll occurs (McGlone et al. 2002; Merzlyak et al. 2003). Bruised tissues had consistently lower relative reflectance starting from about 700 nm and continued to the near-infrared range with the maximum difference occurring at a valley between 960 and 980 nm that represents the absorption peak of water (McGlone et al. 2002; Shrestha et al. 2002b; Zwiggelaar et al. 1996). Bruising of tissues causes an expulsion of cellular materials and water out into the extra-cellular spaces within a local area that dramatically increases the absorption of light at around 960–980 nm, thus making the region more contrastingly distinct than the surrounding areas. Greater differences were observed between bruised and non-bruised areas in peaches than in mangoes. This observation alone suggests that it would be quite challenging to discriminate bruises in mangoes based on the levels of applied forces.

Similar results were obtained for peaches probably because their thin pericarps were comparatively similar to strawberries. Detection of bruises on mangoes is difficult because of the leathery texture of the exocarp.

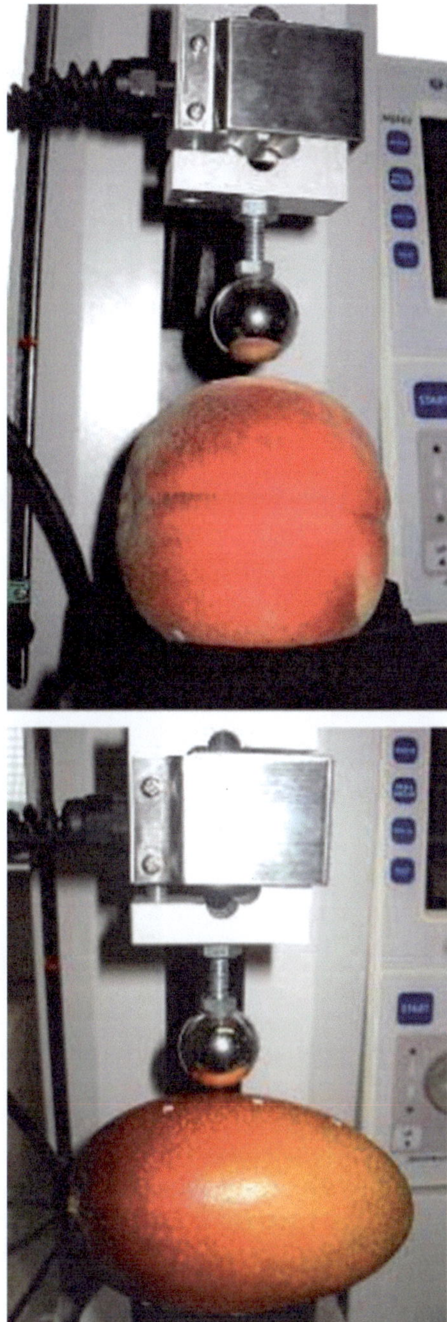

Fig. 9.6 Controlled application of bruising forces using a 25-mm ball tip on peaches (*up*) and mangoes (*down*)

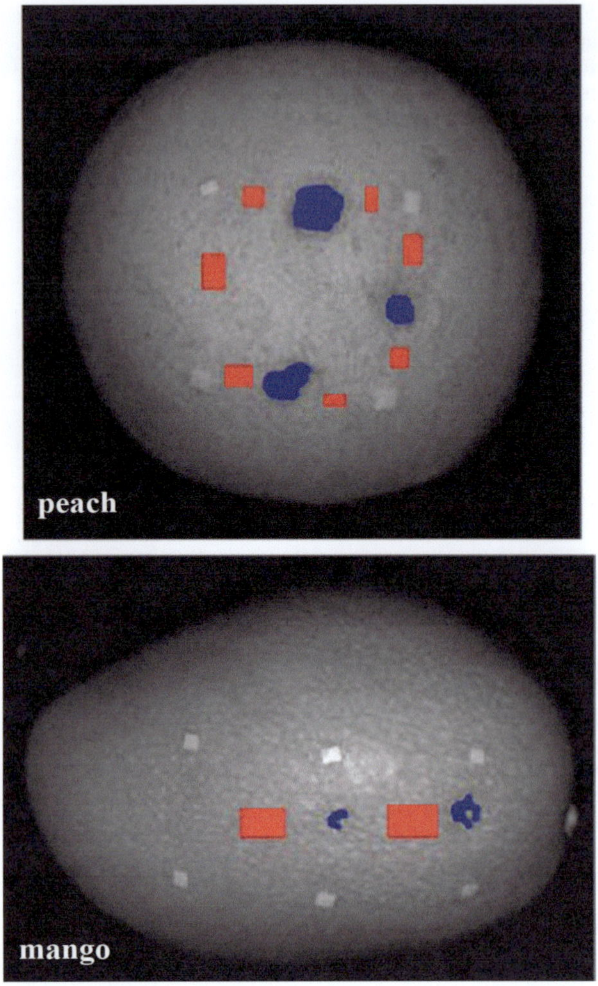

Fig. 9.7 Bruised (*blue*) and non-bruised (*red*) areas were marked using Paint software

9.9 Measurement of Firmness in Strawberry

Firmness in strawberry fruits provides an indirect measure of fruit ripeness, freshness and bruising potential. From 210 pieces of Akihime variety in three ripeness groups (50–60 % ripe, 70–80 % ripe and full ripe), hyperspectral images were taken from 650 to 1,100 nm at 5 nm intervals with varying exposure times from 0.70 to 7.25 s. Firmness measurements were taken using the same Orientec universal testing machine as shown in Fig. 9.9.

The NIR spectral profiles of three strawberries of different ripeness levels are shown in Fig. 9.10. The spectra are marked with a strong absorption by chlorophyll at around 675 nm (Chen et al. 2002; McGlone et al. 2002), and by water and starch at around 980 nm. The large reflectance data variability at 675 nm was likely caused

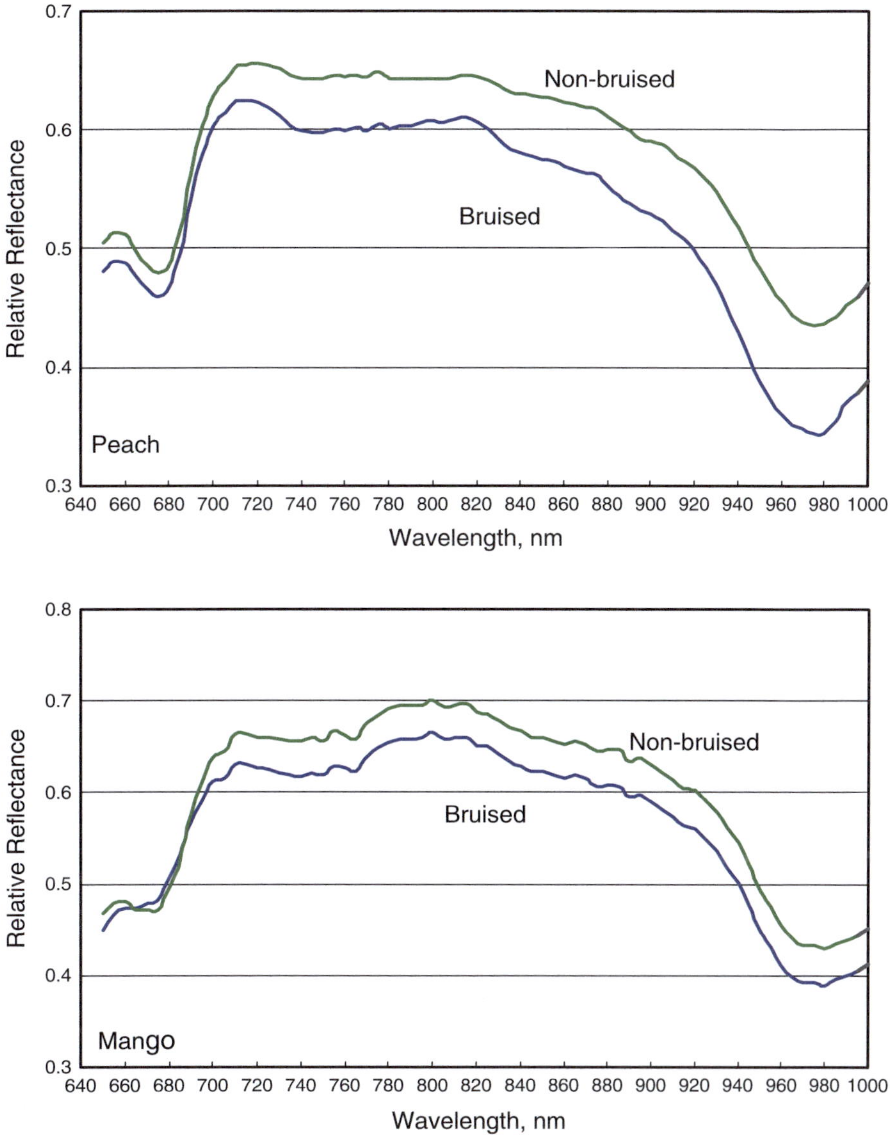

Fig. 9.8 Spectral profile of reflectance .spectra of bruised and non-bruised areas in peaches and mangoes

by different levels of ripeness of the fruits. The peak of reflectance had occurred in the near-infrared region at around 800–840 nm.

The results of the stepwise linear regression modeling are shown in Table 9.1 for different waveband combinations for the two ripeness ranges. The 50 % to full ripe range gave higher correlation but lower standard error (0.79 and 0.35 MPa

Fig. 9.9 (**a**) Firmness measurements were taken using a 3 mm diameter steel probe tip from a central point in-between fruit achenes; (**b**) relative reflectance spectral data were obtained from the same area using a circular mask

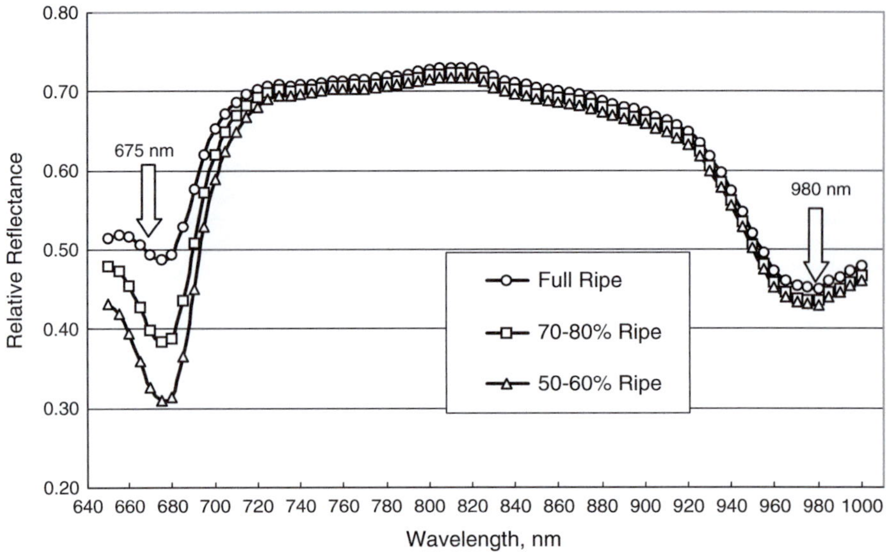

Fig. 9.10 Diffuse relative reflectance profile with respect to a white reference tile from 650 to 1,000 nm spectral images of strawberries at *open triangle* 50–60 % ripe, *open square* 70–80 % ripe and *open circle* full-ripe maturity levels

respectively) as compared to the 70 % to full ripe range (0.60 and 0.26 MPa respectively).

Validating the earlier results of the study done by Shrestha et al. (2002a), the 670–685 nm range accounted for a greater proportion of variations in firmness that again accentuated the importance of chlorophyll content as an indicator of degree of

Table 9.1 Characteristics of prediction models having different number of wavelengths for estimation of firmness

	Predictors (nm)	SEC[a]	R_C[b]	SEP[c]	R_p[d]	Bias
70 % to full-ripe maturity level fruits group						
1	680	0.252	0.702	0.241	0.645	0.025
2	680, 990	0.235	0.750	0.262	0.588	0.041
3	680, 990, 650	0.233	0.760	0.258	0.599	0.033
50 % to full-ripe maturity level fruits group						
1	685	0.356	0.783	0.344	0.796	0.046
2	685, 985	0.342	0.803	0.344	0.794	0.057
3	685, 985, 865	0.338	0.809	0.350	0.786	0.050

[a]Standard error of calibration, MPa
[b]Correlation coefficient of calibration
[c]Standard error of prediction, MPa
[d]Correlation coefficient of prediction

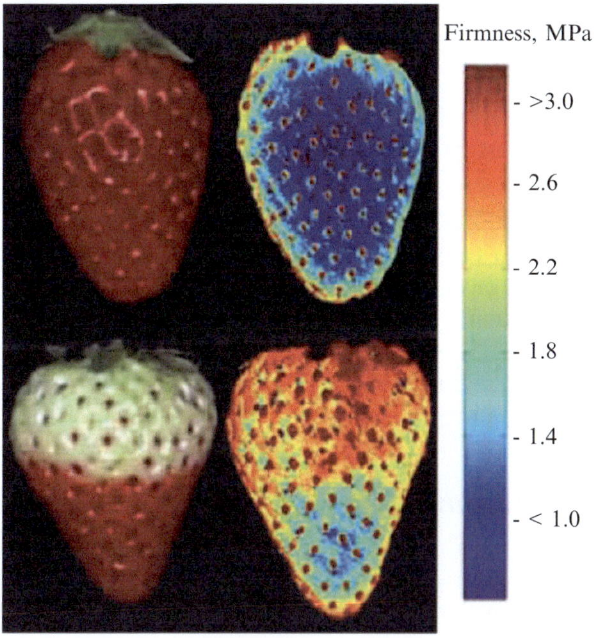

Fig. 9.11 Pseudo-color maps of firmness distribution in strawberries

ripeness, and hence the firmness of strawberries. The inclusion of the NIR range of 985–990 nm into the models further improved the prediction capability of the model. The additional model component suggested a minor but still important role of water as determinant of the turgidity of the cells that has a relationship to the firmness of fruits. The derived regression equation for the entire range of samples was:

$$Yp = -6.786\ R685 - 6.165\ R865 + 13.810\ R985 + 2.750 \quad (9.7)$$

where Yp is the predicted firmness value, and R685, R985 and R865 are the reflectance values at 685, 865 and 985 nm, respectively.

As an aid to visualize the distribution of firmness in the fruits, image maps were generated using Eq. 9.7 to show the distribution of firmness throughout the fruit body. Figure 9.11 shows some examples of these pseudo-color maps that interestingly identify hard and soft regions on the fruits.

In the final end through this firmness study, hyperspectral imaging in the visible and NIR ranges was found useful for non-destructive and non-contact estimation of the firmness of fruits. Statistical analyses showed that the three-wavelength model (685, 865, and 985 nm) could predict firmness in strawberries. More thorough regression analyses showed that combination of the three wavelength ranges of 670–685 nm, 755–870 nm, and 955–1,000 nm gave better estimate of fruit firmness.

9.10 Measurement of Soluble Solids Content in Strawberry, Peach and Mango

Soluble solids content as an estimate of fruit sweetness is an important indicator of eating quality of fruits, which can be measured non-destructively by NIR spectroscopy as an alternative to the usual chemical methods (Kobayashi et al. 2005). Experiments were carried out to measure SSC in strawberries (Nagata et al. 2004, 2005), mangoes and peaches using hyperspectral imaging within the short Vis-NIR region.

9.10.1 Strawberry

For strawberries, the spectral resolution at 5-nm intervals was explored for establishing prediction models. The results of stepwise multiple regression are shown in Table 9.2

Table 9.2 Prediction models for SSC for strawberries at 5 nm intervals

Model	Predictors	SEC	Rc	SEP	Rp	Bias
1	915	0.533	0.548	0.797	0.320	0.322
2	915, 765	0.502	0.623	0.731	0.484	0.293
3	915, 765, 870	0.457	0.711	0.645	0.637	0.356
4	915, 765, 870, 695	0.383	0.811	0.586	0.712	0.259
5	915, 765, 870, 695, 860	0.288	0.900	0.430	0.870	−0.037

for strawberries. The five-wavelength prediction models at the two resolutions are shown in Eq. 9.8.

$$\begin{aligned}\mathrm{SSCS5} = {} & 9.402 + 82.838\,\mathrm{R}_{915} + 88.848\,\mathrm{R}_{765} \\ & - 158.916\,\mathrm{R}_{870} - 122.984\,\mathrm{R}_{695} - 332.882\,\mathrm{R}_{860}\end{aligned} \tag{9.8}$$

9.10.2 *Peach*

For peaches, the resolution at 2-nm intervals was explored for forming prediction models. The results of stepwise multiple regression are shown in Table 9.3 for peaches. The five-wavelength prediction models at the two resolutions are shown in Eq. 9.9.

$$\begin{aligned}\mathrm{SSCP2} = {} & 16.450 + 54.259\,\mathrm{R}_{680} - 75.509\,\mathrm{R}_{684} \\ & - 4.931\,\mathrm{R}_{686} - 41.644\,\mathrm{R}_{666} + 65.543\,\mathrm{R}_{672}\end{aligned} \tag{9.9}$$

9.10.3 *Mango*

For mango, the resolution at 2-nm intervals was explored for forming prediction models. The results of stepwise multiple regression are shown in Table 9.4 for Mango. The five-wavelength prediction models at the two resolutions are shown in Eq. 9.10.

$$\begin{aligned}\mathrm{SSCM2} = {} & 17.053 - 46.018\,\mathrm{R}_{726} + 67.241\,\mathrm{R}_{790} \\ & + 43.261\,\mathrm{R}_{710} - 43.875\,\mathrm{R}_{718} - 22.339\,\mathrm{R}_{776}\end{aligned} \tag{9.10}$$

Table 9.3 Prediction models for SSC for peaches at 2 nm intervals

Model	Predictors	SEC	Rc	SEP	Rp	Bias
1	680	1.045	0.603	1.222	0.464	0.207
2	680, 684	0.897	0.732	1.007	0.681	0.168
3	680, 684, 686	0.874	0.750	0.982	0.699	0.142
4	680, 684, 686, 666	0.863	0.759	0.913	0.747	0.109
5	680, 684, 686, 666, 672	0.841	0.775	0.862	0.779	0.124

Table 9.4 Prediction models for SSC for mangoes at 2 nm intervals

Model	Predictors	SEC	Rc	SEP	Rp	Bias
1	726	2.274	0.397	2.114	0.504	0.222
2	726, 790	1.604	0.766	1.784	0.697	0.117
3	726, 790, 710	1.315	0.852	1.420	0.815	0.081
4	726, 790, 710, 718	1.244	0.871	1.319	0.842	0.052
5	726, 790, 710, 718, 776	1.200	0.882	1.289	0.849	−0.041

9.11 Measurement of Anthocyanins in Strawberry, Sweet Potato and Eggplant

There was good interest in the health benefits of consuming polyphenols amongst the Japanese. Anthocyanin is a class of polyphenolic pigment that is commonly found in fruits such as strawberry, vegetables such as in eggplant and root crops such as purple-fleshed sweet potato. An objective non-destructive method was developed based on hyperspectral imaging to quantify the amount of anthocyanins in strawberries (Kobayashi et al. 2006a), purple sweet potatoes (Nagata et al. 2006a), and eggplants (Kobayashi et al. 2006b). Using the visible range VS-153-10-HC-20 Varispec LCTF fitted in front of the Apogee AP2E camera, images were taken from 400 to 700 nm at 1 nm interval. Using a mask, areas were selected from the objects from which a sub-sample was taken for pigment extraction. The samples were diced and placed in 50 % acetic acid solution for at least 20 h to allow diffusion out of the pigments. Aliquot samples were placed in round cuvettes from which absorbance spectra were taken using an Avantes Avasepc-2048 spectrophotometer. Relative concentration of the pigments was computed from the peak values in the neighborhood of 525 nm of the absorbance spectra.

Following the stepwise linear regression analysis, optimal wavelengths were selected from the second derivative of the computed absorbance spectra obtained from the hyperspectral image cubes.

9.11.1 *Strawberry*

For strawberries, the spectral images were taken at 1-nm interval to form the prediction models for anthocyanin contents. The results of stepwise multiple regression are shown in Table 9.5. A five wavebands (508, 506, 507, 531 and 533 nm) had an SEC and SEP of 0.165 and 0.213 relative pigment concentration, respectively with a correlation coefficient of 0.93.

Table 9.5 Prediction models for anthocyanins for strawberries, sweet potatoes and eggplant

Model	Wavelength prediction models	SEC	Rc	SEP	Rp	Bias
Strawberries	5 (508, 506, 507, 531, 533)	0.165	0.957	0.213	0.932	0.032
Sweet potatoes	5 (523, 592, 564, 539, 516)	0.045	0.921	0.052	0.921	0.010
Eggplant	1 (716)	0.077	0.973	0.054	0.969	–

The five-wavelength prediction model is:

$$\begin{aligned}\text{Anthocyanins (Strawberries)} &= -0.313 - 33.699\,R_{508} + 22.694\,R_{506} \\ &\quad - 2.935\,R_{507} + 48.350\,R_{531} - 43.790\,R_{533}\end{aligned} \tag{9.11}$$

9.11.2 Sweet Potato

Similarly for sweet potatoes, images were acquired at 1-nm spectral resolution. The results of stepwise multiple regression are shown in Table 9.5. In the case of two varieties of purple fleshed sweet potato (Ayamurasaki and Murasakimasari), better five-wavelength prediction model (523, 592, 564, 539, and 516 nm) was obtained from Ayamurasaki with an SEC and SEP of 0.045 and 0.052 relative pigment concentration, respectively, with a correlation coefficient of 0.92. The five-wavelength prediction model is :

$$\begin{aligned}\text{Anthocyanins (Sweet Potatoes)} &= 2.111 - 5.393\,R_{523} + 15.277\,R_{592} \\ &\quad - 64.426\,R_{564} - 40.283\,R_{539} + 32.212\,R_{516}\end{aligned} \tag{9.12}$$

9.11.3 Eggplant

Finally in the case of eggplants, the single wavelength model at 716 nm (Eq. 9.13) for the anthocyanin pigment concentration had the produced the correlation coefficient of 0.969 with a standard error of prediction (SEP) of 0.054 concentration units. The results of stepwise multiple regression are shown in Table 9.5. The one-wavelength prediction models is:

$$\text{Anthocyanins (Eggplant)} = 3.471\,R_{716} + 0.596 \tag{9.13}$$

Using this model, pseudo-color distribution maps of anthocyanin pigment concentrations were derived to show the spatial variation of the pigment content throughout the eggplant. To investigate the effects of fruit sizes, the vertical distance between the camera and samples was varied by around 30 mm based on the average radius of eggplants. The results showed that varying camera heights by 30 mm did not significantly affect the measured values of anthocyanin pigment of the fruits.

9.12 Further Work

Many of the work done on hyperspectral imaging on fruits and vegetables were confined in the visible to shorter near infrared region. Near infrared spectrometric research showed that more powerful prediction and discrimination models for a wider array of constituents and non-chemical parameters can be developed in the longer regions of the NIR. With the evolution of better sensor technologies such as the use of extended indium-gallium-arsenide, lead-sulfide or lead-selenide image detectors, a wider array of applications can be opened up for hyperspectral imaging because of their increased sensitivities in the region closer to the mid-infrared part of the electromagnetic spectrum. Finer constituents can then be measured such as amino acid or fatty acid in heterogenous samples to better estimate their compositional levels. Along with improvements of imaging electronics, better approaches to model development would be essential such as the use of both linear and non-linear models. All of these will certainly make the science and technology of hyperspectral imaging more responsive to the development needs of the industry.

References

Chen Y, Chao K, Kim MS (2002) Machine vision technology for agricultural applications. Comput Electron Agric 36(2–3):173–191

Kobayashi T, Tallada J, Nagata M (2005) Basic study on measurement of sugar content for citrus unshiu fruit using NIR spectrophotometer (*In Japanese*). Bull Fac Agric Univ Miyazaki 51 (1–2):1–8

Kobayashi T, Nagata M, Tallada JG, Toyoda H, Goto Y (2006a) Study on anthocyanin pigment distribution estimation for fresh fruits and vegetables using hyperspectral imaging. Part 2. Visualization of anthocyanin pigment distribution of strawberry (*Fragaria* × *ananassa Duchesne*) (*In Japanese*). J Sci High Technol Agric 18(1):50–57

Kobayashi T, Nagata M, Tallada J (2006b) Hyperspectral imaging based assessment of color quality of fruits and vegetables: anthocyanin contents of eggplant *(In Japanese)*. Paper presented at the 2006 meeting of Agricultural Environmental Engineering Society at Hokkaido University on Sept 2006

Lawrence KC, Park B, Windham WR, Mao C (2003) Calibration of a pushbroom hyperspectral imaging system for agricultural inspection. Trans ASABE 46(2):513–521

McGlone VA, Jordan RB, Martinsen PJ (2002) Vis/NIR estimation at harvest of pre- and post-storage quality indices for 'Royal Gala' apple. Postharvest Biol Tech 25(2):135–144

Merzlyak MN, Solovchenko AE, Gitelson AA (2003) Reflectance spectral features and non-destructive estimation of chlorophyll, carotenoid and anthocyanin content in apple fruit. Postharvest Biol Tech 27(2):197–211

Nagata M, Tallada JG, Kobayashi T, Cui Y, Gejima Y (2004) Predicting maturity quality parameters of strawberries using hyperspectral imaging. ASAE Paper No. 043033. ASAE, St. Joseph

Nagata M, Tallada JG, Kobayashi T, Toyoda H (2005) NIR hyperspectral imaging for measurement of internal quality in strawberries. ASABE Paper No. 053131. ASAE, St. Joseph

Nagata M, Kobayashi T, Tallada JG, Toyoda H, Goto Y (2006a) Study on anthocyanin pigment distribution estimation for fresh fruits and vegetables using hyperspectral imaging.

Part 1. Visualization of anthocyanin pigment distribution of purple-fleshed sweet potato (*Ipomoea batatas Poir*) (*In Japanese*). J Sci High Technol Agric 18(1):42–49

Nagata M, Tallada JG, Kobayashi T (2006b) Bruise detection using NIR hyperspectral imaging for strawberry (*Fragaria* × *ananassa Duch.*). Environ Control Biol 44(2):133–142

Shrestha BP, Nagata M, Cao Q (2002a) Study on image processing for quality estimation of strawberries (part 1): detection of bruises on fruit by color image processing. J Sci High Technol Agric 13(2):115–122

Shrestha BP, Nagata M, Cao Q (2002b) Study on image processing for quality estimation of strawberries (part 2): detection of bruises on fruit by NIR image processing. J Sci High Technol Agric 13(2):115–122

Tallada JG (2006) Post-harvest quality determination in fruit and fruit-vegetables using NIR hyperspectral imaging. Ph.D. Dissertation, United Graduate School of Agricultural Science of Kagoshima University

Tallada JG, Nagata M, Kobayashi T (2006a) Detection of bruises in strawberries by hyperspectral imaging. ASABE Paper No. 063014. ASAE, St. Joseph

Tallada JG, Nagata M, Kobayashi T (2006b) Non-destructive estimation of firmness in strawberries (*Fragaria* × *ananassa Duch.*) using NIR hyperspectral imaging. Environ Control Biol 44(4):245–255

Zwiggelaar R, Yang QS, Garcis-Pardo E, Bull CR (1996) Use of spectral information and machine vision for bruise detection on peaches and apricots. J Agric Eng Res 63:323–332

Chapter 10
Quality Evaluation of Beef and Pork

Govindarajan Konda Naganathan, Kim Cluff, Ashok Samal, Chris Calkins, and Jeyamkondan Subbiah

10.1 Introduction

The meat industry is the largest food industry in the United States. There exists a need for objective, non-invasive systems for sorting meat based on quality traits to facilitate marketing. Hyperspectral imaging has a great potential to fulfill the need, as it can collect both spatial (structural) and spectral (biochemical) information on the meat surface. This section will focus on hyperspectral imaging of beef and pork.

10.2 Beef

The beef industry makes up the single largest segment of American agriculture (NCBA 2009). In 2008, the United States of America beef industry was responsible for producing $76 billion in retail sales, as well as $2.98 billion in beef exports (USDA 2008). The driving force of this industry is customer satisfaction, which in

G. Konda Naganathan • K. Cluff
Department of Biological Systems Engineering, University of Nebraska–Lincoln, Lincoln, NE, USA

A. Samal
Computer Science and Technology, University of Nebraska-Lincoln, Lincoln, NE, USA

C. Calkins
Animal Science, University of Nebraska–Lincoln, Lincoln, NE, USA

J. Subbiah (✉)
Department of Biological Systems Engineering, University of Nebraska–Lincoln, Lincoln, NE, USA

Department of Food Science and Technology, University of Nebraska–Lincoln, Lincoln, NE, USA
e-mail: jsubbiah2@unl.edu

B. Park, R. Lu (eds.), *Hyperspectral Imaging Technology in Food and Agriculture*, Food Engineering Series, DOI 10.1007/978-1-4939-2836-1_10

turn influences repurchasing. Studies have found that customers are not pleased with the inconsistencies in beef tenderness. Roughly 15–20 % of the steaks that reach consumers are tough (Miller et al. 2001) and customers can distinguish between tender and tough samples (Boleman et al. 1997). Furthermore, consumers are willing to pay more if the steak is guaranteed to be tender (Lusk et al. 2001). Consumers' willingness to pay more for guaranteed tenderness represents an opportunity to add value to tender carcasses, thereby benefiting the beef industry as well as increasing customer satisfaction. The fact that tenderness is not consistent should not be a surprise given that tenderness is not even measured during meat quality grading.

One of the difficulties in determining tenderness is that its measurement is inherently destructive. Furthermore, the only truly accurate method of measuring tenderness is to use a trained sensory panel (Shackleford et al. 2005). However, this is not a reasonable method to measure tenderness in a beef processing plant. The second best method is to measure the force required to shear the meat sample, using the Warner–Bratzler shear (WBS) force (AMSA 1995) or slice shear force (SSF) methods (Shackleford et al. 1999). The beef industry mostly follows the SSF procedure method because it is faster than the WBS procedure. In the SSF procedure, a 1-in. thick steak is excised from a beef carcass, cooked, sliced, and then sheared using a texture analyzer (Fig. 10.1). The force required to slice the sample is recorded as the SSF value. Unfortunately these methods are also not reasonable to use online in meat packing plants, because they are destructive. Hence, the industry needs a non-destructive, rapid, accurate, and online method of predicting tenderness. Light-based, non-destructive techniques may provide the solution.

Beef tenderness is an important trait related to consumer satisfaction. In fact, committees, organizations, and associations have been assembled with the primary purpose of improving customer eating experiences. According to the National Beef Quality Audit conducted in 2000, tenderness is one of the biggest challenges in meat quality (McKenna et al. 2002). During the late 1950s, the beef industry began to recognize the importance of palatability traits in meats, such as tenderness, juiciness, and flavor (Webb et al. 1964). In an attempt to control and monitor meat quality, the American Beef Cattle Performance Registry Association (ABC-PRA) was organized in 1955 (Lipsey 1999). Additional organizations were formed, such as the American Polled Hereford and Angus Associations and the Beef Improvement Federation (BIF), each with the objective to improve beef quality for the consumer. Customer satisfaction is seen as the ultimate force that drives markets to improve. In the beef industry, researchers have been trying to improve the end users experience for decades. However, inconsistency in tenderness and lack of quick, reliable methods of evaluating beef tenderness continue to be problems. This situation has caused some scientists within the U.S. beef industry to exclaim, "is our industry forever shackled to the fact that a proportion of all beef we produce will be less than satisfactory to consumers?" (Lipsey 1999). Hence it is no understatement to say that there is a thriving need to develop a system that can monitor tenderness accurately, quickly and efficiently.

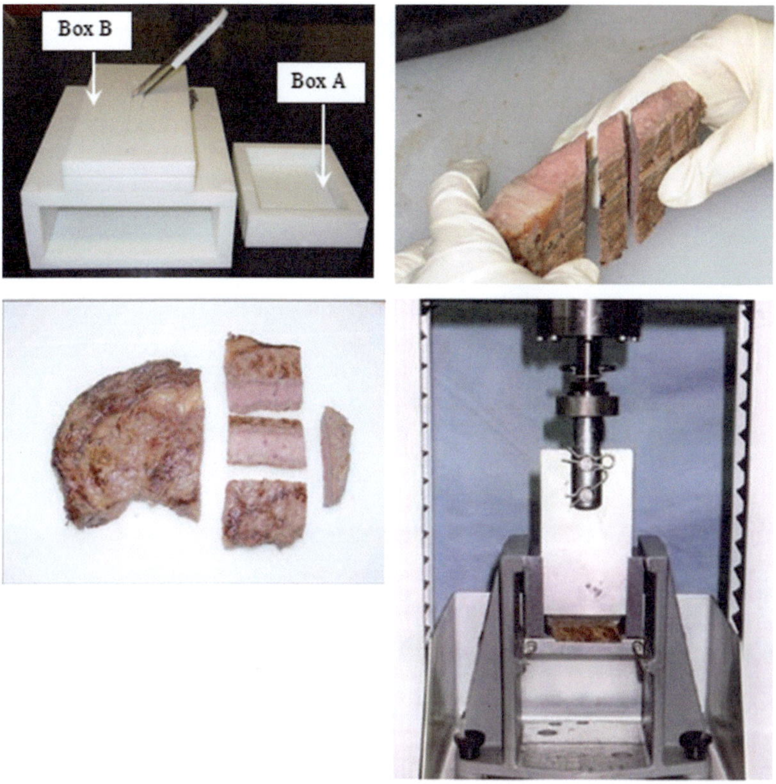

Fig. 10.1 Slice shear force (SSF) procedure to determine beef tenderness (*Source*: Subbiah 2004 and Shackelford et al. 1999)

10.2.1 Video Image Analysis

Video image analysis (VIA) has been used in an attempt to predict beef tenderness. A VIA and the RGB image generated by a VIA system are shown in Figs. 10.2 and 10.3, respectively. A system known as the BeefCam® was developed by Colorado State University to acquire RGB images of beef steak and extract L*, a*, and b* (lab color space) values from the images (Belk et al. 2000). These color features were then used to find a correlation with beef tenderness. The system was evaluated by Wyle et al. (1999, 2003) and Vote et al. (2003). They were able to predict Warner–Bratzler shear (WBS) force values with an R^2 of less than 0.21. They concluded that the BeefCam® could be useful in reducing the likelihood that a consumer would receive a tough steak, but that the system needed further refinement (Belk et al. 2000). One of the downfalls to using simple VIA analysis is that it has low spectral resolution, generally displaying only red, green, and blue bands. Therefore, it cannot be used to identify the biochemical basis for tenderness.

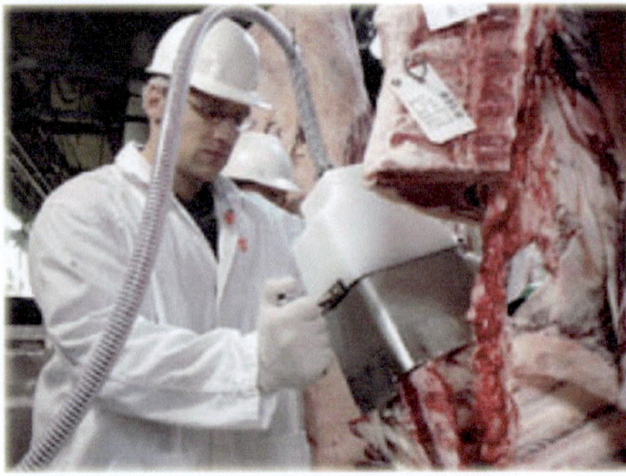

Fig. 10.2 A video image analysis for capturing RGB images of beef (*Source*: Research Management Systems, USA Inc., Fort Collins, CO)

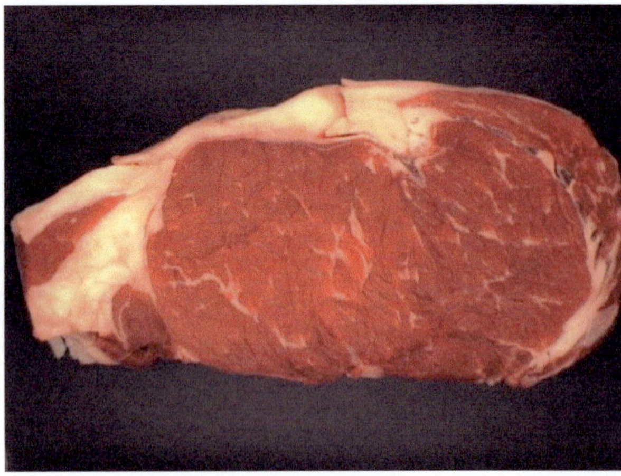

Fig. 10.3 RGB image of a beef steak (*Source*: Subbiah 2004)

10.2.2 Near Infrared (NIR) Spectroscopy

Spectroscopy is performed by measuring the reflectance or absorbance of light in a very small spatial area. The reflectance or absorbance of light is averaged over that spatial area and the spectral profile is displayed as intensity vs. wavelength. Therefore, spectroscopy results in high spectral resolution displaying how absorbance changes with respect to wavelength. Figure 10.4 shows a spectroscopic instrument collecting spectral reflectance signals of beef ribeye muscle. Tender beef reflects comparatively more light compared to the tough beef (Fig. 10.5) (Subbiah 2004).

Near infrared (NIR) spectroscopy has been tested as an instrument to predict tenderness (Byrne et al. 1998; Hildrum et al. 1994, 1995; Mitsumoto et al. 1991;

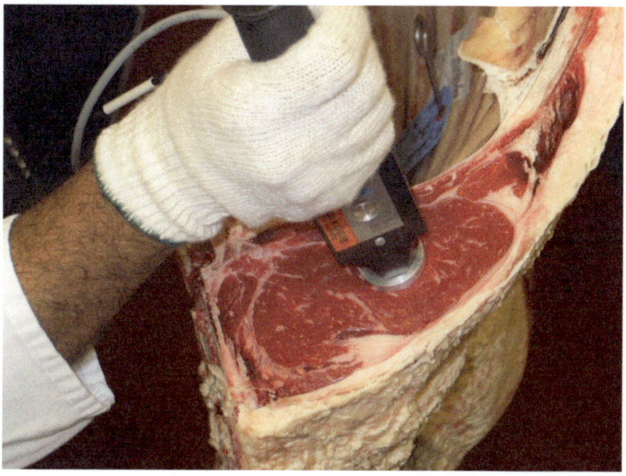

Fig. 10.4 A spectroscopic system for acquiring NIR spectra of a beef sample (*Source*: Subbiah 2004)

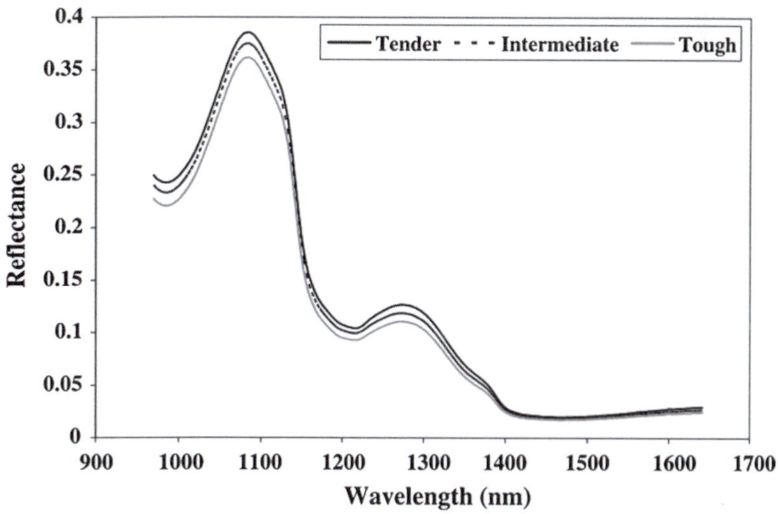

Fig. 10.5 Typical spectra of tender, intermediate, and tough beef ribeye steaks (*Source*: Konda Naganathan et al. 2008)

Naes and Hildrum 1997; Park et al. 1998). Shackleford et al. (2005) performed a study in which they collected spectra on 292 *longissimus* steaks. They validated their system with 146 steaks and were able to predict slice shear force (SSF) values with an $R^2 = 0.22$. They concluded that NIR spectroscopy might be useful in identifying tender USDA Select carcasses. Some of the downfalls of using spectroscopy are that it can collect signals from a limited sample area and the measurements can be influenced by fat flecks.

10.2.3 Hyperspectral Imaging

Hyperspectral imaging has also been used as a method of predicting beef tenderness. Hyperspectral imaging is a mode of imaging that essentially combines spectroscopy and video image analysis. Specifically, a hyperspectral image is a compilation of a set of images where each image represents reflected light in a narrow band of wavelengths. Hyperspectral imaging is unique in that it is capable of capturing both *spatial* and *spectral* information at high resolutions. With a hyperspectral image, a spectrum can be obtained for each pixel (Fig. 10.6) and a grayscale image can be obtained for each wavelength (Fig. 10.7).

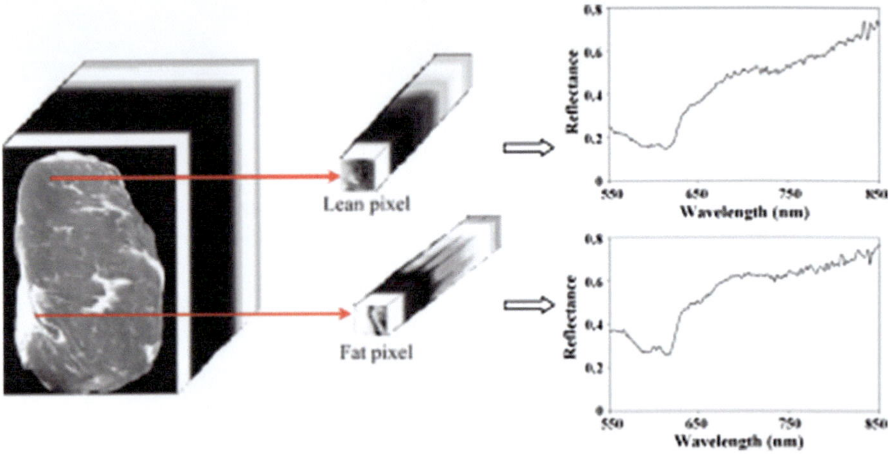

Fig. 10.6 A hyperspectral image of a beef-steak showing typical spectral signatures of lean and fat pixels (*Source*: Konda Naganathan et al. 2008)

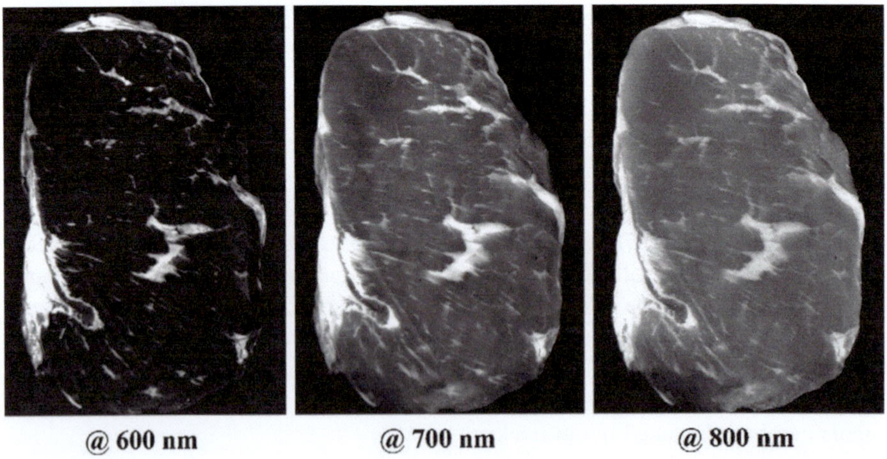

Fig. 10.7 Tonal images of a beef-steak at selected wavelengths (*Source*: Konda Naganathan et al. 2008)

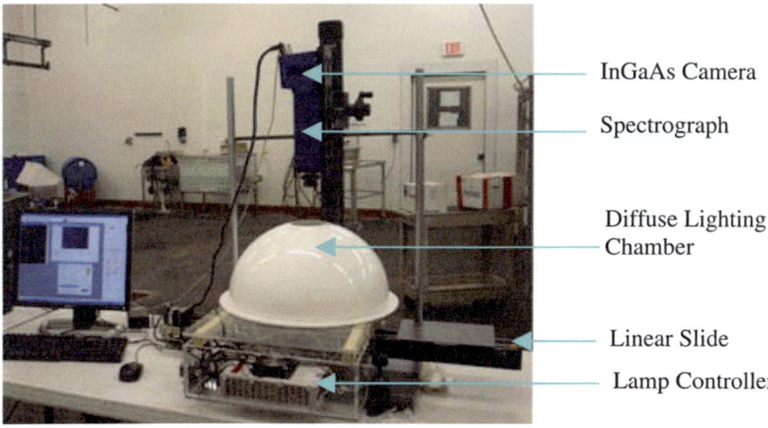

Fig. 10.8 Spectrograph-based bench-top hyperspectral imaging system (*Source*: Konda Naganathan et al. 2008)

A hyperspectral imaging system (Fig. 10.8) generally consists of a camera with a 2D sensor array and an imaging spectrograph. Spectral data are obtained by allowing only a thin line of light into a slit on the spectrograph. The spectrograph disperses the light into wavelengths and then casts the dispersed wavelengths onto the 2D sensor array of the camera. The result is an image displaying the line of light as it would appear at each different wavelength. In a line-scan hyperspectral imaging system, by moving the sample across the hyperspectral camera at a preset speed, a hyperspectral image can then be compiled. The compiled image consists of a complete 2D spatial image with a third axis containing spectra at each and every pixel. Therefore, both very high spatial and spectral resolutions can be achieved with a hyperspectral imaging system. Generally, hyperspectral imaging systems can achieve spectral resolutions ranging from macro-scale, imaging very large objects such as the Earth, down to microscopic-scales, imaging very small objects such as human tissues.

Hyperspectral imaging systems are capable of extracting both structural features and biochemical signatures from images (Konda Naganathan et al. 2006b). Structural and textural features can be extracted from a hyperspectral image because the image preserves the full spatial frame (Konda Naganathan et al. 2006a). Likewise, biochemical fingerprints can be extracted because a hyperspectral image can provide a spectrum for every pixel within the image and particular wavelengths are related to specific biochemical constituents.

Konda Naganathan et al. (2008) developed a push broom hyperspectral imaging system (Fig. 10.8) with the purpose of predicting tenderness on ($n = 111$) longissimus steaks. Their system consisted of a CCD digital video camera, linear slide, diffused flood lighting chamber, and spectrograph providing a spectral range between 400 and 1,000 nm. They used principal component analysis and gray-level co-occurrence matrix analyses to extract image textural features from the imaged hypercubes (Fig. 10.9). Finally, they used a canonical discriminant analysis to develop a model to correlate tenderness categories to the extracted textural features.

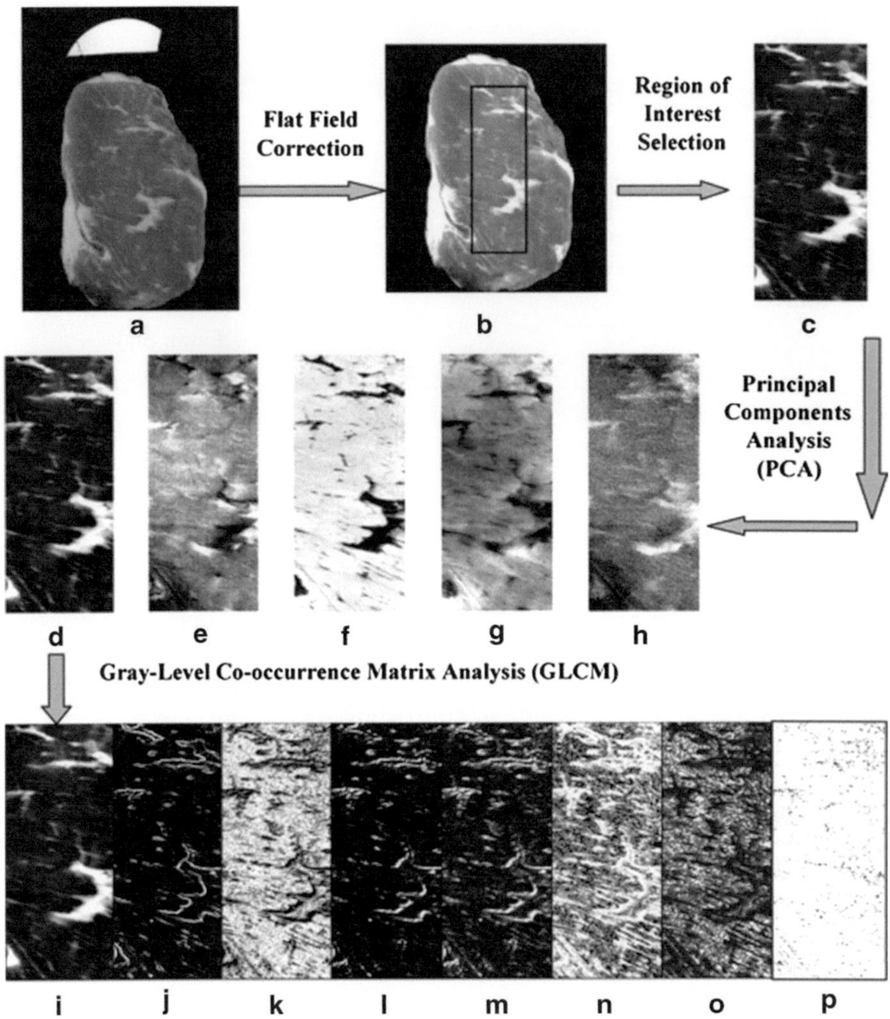

Fig. 10.9 Hyperspectral image textural feature extraction methodology. (**a**) Uncalibrated image. (**b**) Calibrated image. (**c**) Region of interest image. (**d–h**) Principal component images 1 through 5. (**i–p**) Textural images. (**i**) Mean. (**j**) Variance. (**k**) Homogeneity. (**l**) Contrast. (**m**) Dissimilarity. (**n**) Entropy. (**o**) Second moment. (**p**) Correlation (*Source*: Konda Naganathan et al. 2008)

With a leave-one-out cross-validation procedure their model was able to predict tenderness categories with 96.4 % accuracy. They concluded that this method could classify steaks into three tenderness categories based on current status of tenderness, but further study was needed to forecast aged beef tenderness.

Konda Naganathan et al. (2008) carried out a further study with a larger sample set (n = 319) in which they forecast 14-day aged cooked beef tenderness. They developed a system that consisted of an InGaAs camera, spectrograph sensitive to light in the NIR region ($\lambda = 900$–1,700 nm), a linear slide, and a diffuse flood lighting chamber. In their investigation, images of ribeye steaks were acquired at 3–5 days postmortem and then aged to 14 days postmortem. They used partial least squares regression and gray-level textural co-occurrence matrix analysis to forecast 14-day postmortem slice shear force values on the cooked beef samples. Using a leave-one-out cross-validation they were able to correctly classify tender, intermediate, and tough samples with an overall accuracy of 77 %. When they grouped the categories into tender and tough they were able to correctly identify 96.3 % of the tender samples and 62.5 % of the tough samples with an overall accuracy of 94.5 %. They concluded that hyperspectral imaging has considerable promise in beef tenderness prediction, but that further studies should be carried out with a wider range of samples to validate the model.

As hyperspectral images are acquired at a narrow wavelength interval, they tend to have redundant information. So, hyperspectral image analysis involves a dimensionality reduction method such as principal component analysis. A dimensionality reduction procedure first computes Eigen values and vectors (Fig. 10.10). The Eigen vectors are multiplied with the hyperspectral images to create transformed images such as principal component images. The peaks and valleys of the Eigen vectors may correspond to important biochemical properties. In Fig. 10.10, the peaks and valleys corresponding to protein, fat, and water are highlighted.

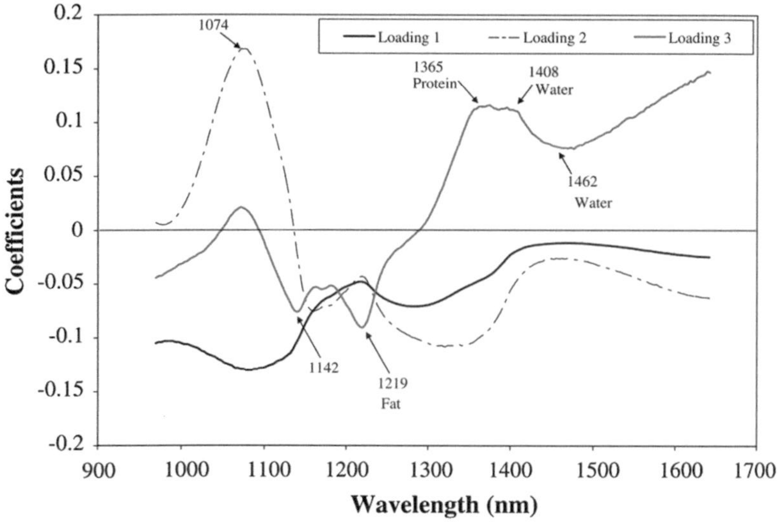

Fig. 10.10 Loading or Eigen vectors of a dimensionality reduction method such as principal component analysis (*Source*: Konda Naganathan et al. 2008)

10.2.4 Optical Scattering

Electromagnetic waves that encounter biological materials create complex interactions with charged particles (Pedrotti 2007). The complex nature of light, which has the dual ability to act as a wave of energy and a particle, presents complications in understanding the exact nature of the interactions of light with charged particles of a given medium. Despite the lack of knowledge of the exact nature of light, it is clear that interaction of electromagnetic energy radiated by luminous objects can be used to identify properties of biochemical compounds (Burns and Ciurczak 2001). Specifically, light in the ultraviolet and near-infrared regions are especially useful in identifying organic compounds. Light interaction with biochemicals at the molecular level result in changes of the energy state due to light energy being absorbed. The absorbed energy causes electrons to transition from one orbital to another.

Light scattering occurs when energy from an incident wave of light is removed by a scattering medium via absorption followed by some portion of that energy being reemitted in many directions (Pedrotti 2007). Scattering is more effective when scattering centers are small particles in comparison to the wavelength of the electromagnetic radiation. Rayleigh scattering is scattering that is predominantly due to particles whose dimensions are smaller than the radiation wavelength (Wolfson 1999). Therefore, Rayleigh scattering is not the predominate source of light scattering in beef muscle tissue. An example of Rayleigh scattering is the scattering that occurs when sunlight interacts with oxygen and nitrogen molecules in the atmosphere (Wolfson 1999). It has been found that scattering is more effective at shorter wavelengths and higher frequencies (Pedrotti 2007). This is apparent in the sky appearing as blue; higher frequency blue light is scattered more so than lower frequency red light. In fact, the scattering power of violet light ($\lambda = 400$ nm) is approximately ten times as great as that of red light ($\lambda = 700$ nm) (Pedrotti 2007).

Light scattering that occurs due to larger particles, whose size is larger relative to the wavelength of light, is known as Mie scattering. Light scattering in beef muscle tissue is predominantly due to Mie scattering. Materials that have predominately Mie scattering centers have less scattering power, however, the density of oscillators leads to considerable signal scattering that may be useful to characterize such materials (Pedrotti 2007).

10.2.5 Optical Scattering Measurements to Predict Tenderness

Thus far, limited work has been performed on investigating the interactions of optical scattering on beef muscle tissue. Electromagnetic radiation interaction with turbid biological objects produces phenomenon that includes both scattering and absorption (Xia et al. 2008). Light is able to penetrate a few millimeters into the surface of the steak. The energy from the light excites the biomolecular components

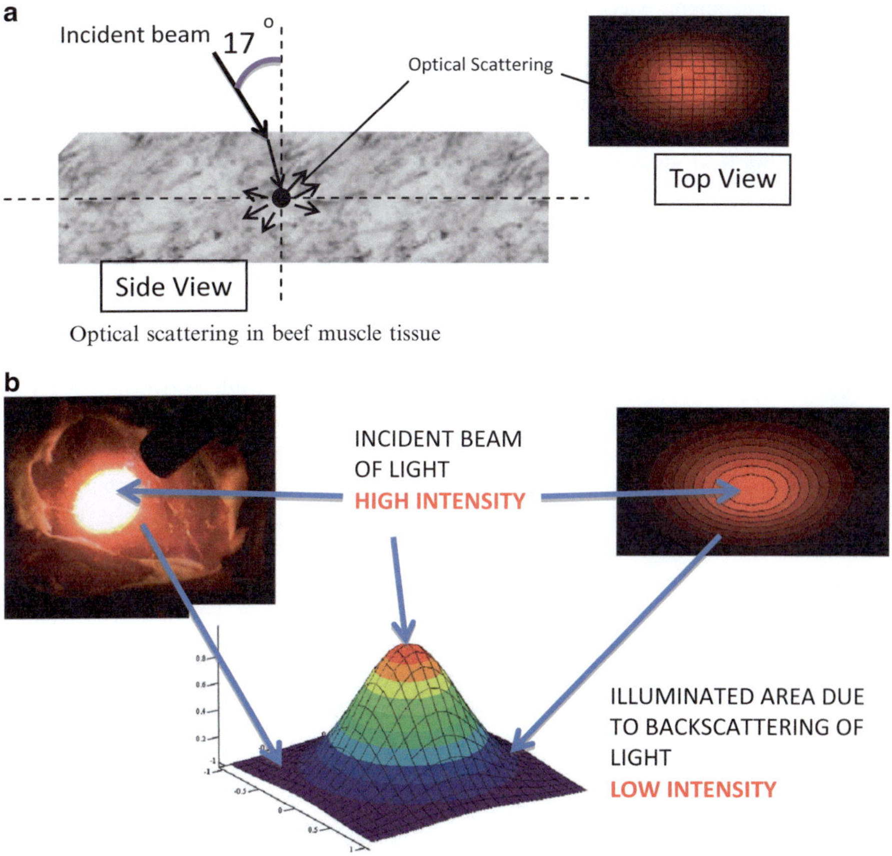

Scattering profile of light on beef muscle tissue

Fig. 10.11 Light is able to (**a**) penetrate a few millimeters into the surface of the steak. (**b**) The illuminated area around the incident beam of light is due to optical scattering (*Source*: Cluff et al. 2008)

within the muscle tissue elevating them to a higher electron state. Because chemical compounds like to exist at the lowest energy state possible they reemit the light in many directions. Hence the illuminated area around the incident beam of light is due to optical scattering, as seen in Fig. 10.11.

Xia et al. (2007) performed a study in which they used spatially-resolved optical scattering and spectroscopy to predict the Warner–Bratzler shear force of 32 cooked beef muscle samples. In their study, two fiber optic probes were used, one to cast an incident beam of light at an oblique incident angle, and the other to collect the back-scattered light. They used the oblique incident theory developed by Wang and Jacques (1995) to extract absorption and scattering coefficients from 13 points along the optical scattering on the beef muscle tissue (Wang and Jacques 1995; Lin et al. 1997). They used an optical diffuse equation where the first parameter, the absorption coefficient, was a function of the effective attenuation of scattered light in the muscle tissue. The scattering coefficient was then derived as a function of the of the absorption coefficient. The scattering and absorption coefficients essentially represented the probability that a photon was scattered or absorbed within the beef muscle tissue. Chemical compositions such as myoglobin and its derivatives influenced the beef absorption coefficient, whereas muscle structure such as sarcomere length and collagen content influenced the scattering coefficient (Xia et al. 2006, 2007). With these two parameters they were then able to predict tenderness WBS scores, in a training data set, with a coefficient of determination (R^2) of 0.59.

In another study performed by Peng and Wu (2008), hyperspectral scattering profiles were used for prediction of beef tenderness. They used a hyperspectral imaging system to measure optical scattering of beef muscle tissue. They used a laboratory hyperspectral imaging system consisting of a high performance CCD camera and imaging spectrograph with a spectral resolution of 2.8 nm. Their system was sensitive to light ranging from 400 to 1,100 nm in the electromagnetic spectrum. They used an incident beam of light of 3 mm in diameter that was cast upon the surface of a prepared sample cut (4 cm × 6 cm × 2.5 cm) of the strip loin. In total, 21 beef steak samples were used to collect hyperspectral scattering images. From the hyperspectral scattering images, they averaged all spectral data from every pixel resulting in one mean reflectance spectra for each sample steak. The mean reflectance spectra were then converted to absorbance spectra from which the first derivative was taken. Four key wavelengths were identified by plotting the correlation coefficient between the WBS scores and the first-order derivative of the averaged reflectance spectra. Multiple linear regression was then used to train a model that could predict the WBS scores. In their study, only 21 samples were used, 15 for training and 6 for validation. Their training model was able to predict WBS scores (n = 15) with a correlation coefficient of $r = 0.82$. Finally, they validated the model with n = 6 and were able to predict WBS scores with a correlation coefficient of $r = 0.94$. They concluded that their prediction results were satisfactory, but that further research was necessary with a larger set of samples and a wider range of WBS scores.

Cluff et al. (2013) developed a non-destructive method for classifying cooked-beef tenderness using hyperspectral imaging of optical scattering on fresh beef muscle tissue. Figure 10.12 presents a line scan hyperspectral imaging system ($\lambda = 922$–1,739 nm) that was used to collect hyperspectral scattering images of the *longissimus dorsi* muscle (n = 472). A modified Lorentzian function was used to fit optical scattering profiles at each wavelength (Fig. 10.13). After removing highly correlated parameters extracted from the Lorentzian function, principal component analysis was performed. Four principal component scores were used in a linear discriminant model to classify beef tenderness. In a validation data set (n = 118 samples), the model was able to successfully classify tough and tender samples with 83.3 % and 75.0 % accuracies, respectively. They also investigated the effect of fat flecks (Fig. 10.14) on the tenderness classification accuracy when various levels (0, 25, 50, and 75 %) of higher intensity fat scattering profiles were removed. They concluded that the presence of fat flecks did not have a significant effect on beef tenderness classification accuracy. Their results demonstrate that hyperspectral

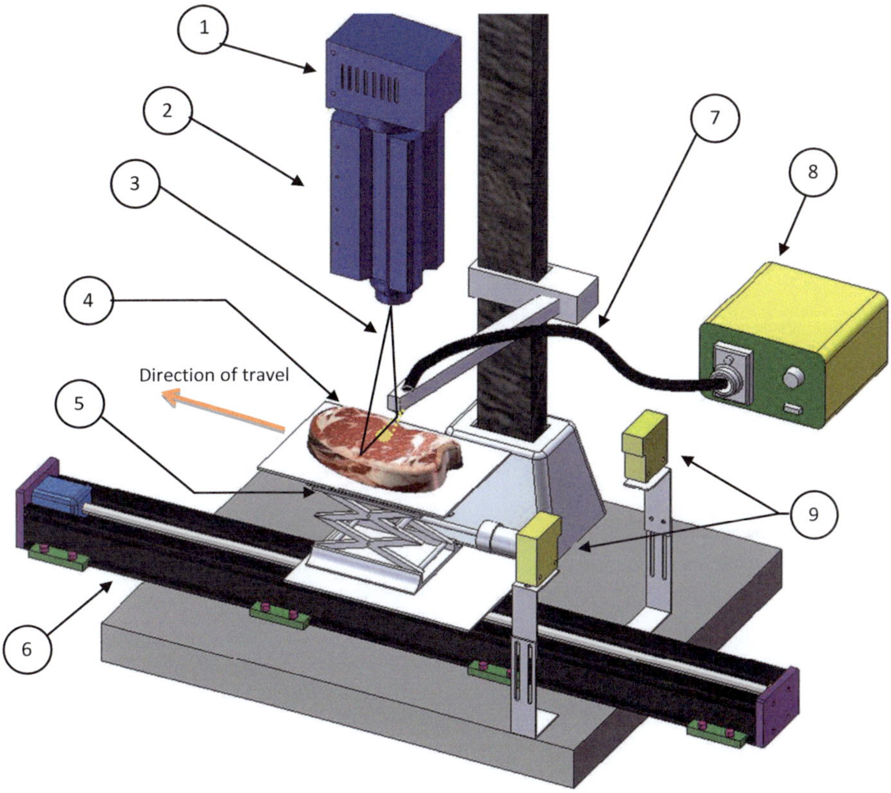

Fig. 10.12 Schematic of line scan hyperspectral imaging system used to collect hyperspectral optical scattering images of the *longissimus* steak muscle tissue. (*1*) InGaAs camera. (*2*) Spectrograph. (*3*) FOV of line scan camera offset 5 *mm* from center of incident beam of light. (*4*) Steak sample. (*5*) Automated vertical stage. (*6*) Linear slide, moved the sample under the line scan in the direction of travel. (*7*) Incident fiber optic cable. (*8*) Light source tungsten halogen lamp. (*9*) Photoelectric switch (*Source*: Cluff et al. 2013)

imaging of optical scattering is a viable technology for beef tenderness classification. The primary factors that influence optical scattering in steak are muscle fiber ultrastructure, fat content, collagen, and other organic functional groups. Similarly, the primary factors that influence beef tenderness are muscle fiber ultrastructure, fat content, and collagen content. Therefore, measurements of optical scattering with hyperspectral imaging could potentially be used as an indicator of beef tenderness.

Extensive research has been performed on evaluating video image analysis and NIR spectroscopy as non-destructive instruments to predict beef tenderness. However, these systems have not been able to attain the necessary tenderness prediction accuracies desired by the beef industry. Hyperspectral imaging is a relatively new technology and is just beginning to emerge as an instrument that can evaluate the quality of agricultural food products. Among the instruments developed for non-destructive measurement of beef tenderness, hyperspectral imaging appears to have the most promise.

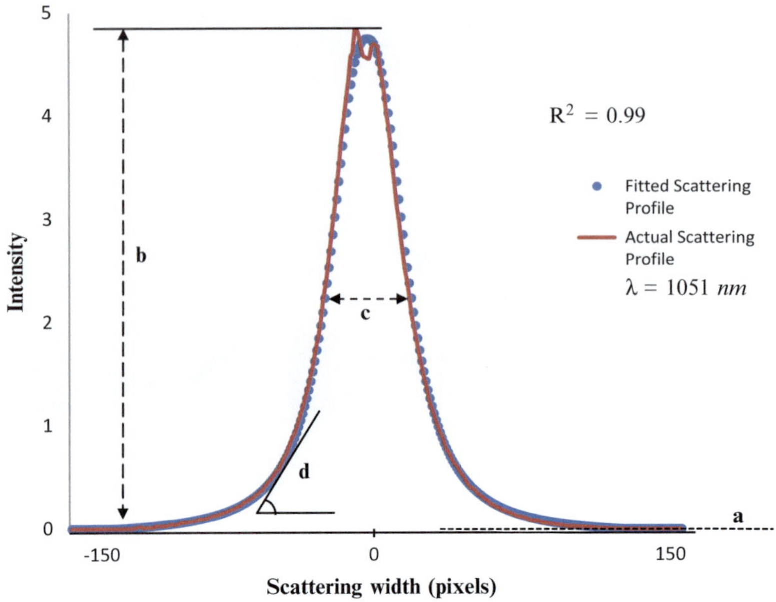

Fig. 10.13 Curve fitting of optical scattering profile to the modified Lorentzian distribution (MLD) function at λ = 1,051 nm. Fitted MLD parameters *a*, *b*, *c*, *d* were extracted from the fitted scattering profile to build the model for predicting tenderness (*Source*: Cluff et al. 2013)

10.3 Pork

Pork is the third highest consumed meat in the United States of America following beef and chicken. The economic impact that the pork industry has on the American economy has a total retail value of $51.7 billion dollars (USDS-ERS 2011). Over the last decade (2000–2008), pork consumption by Americans has averaged 50.3 lb per person (USDA-ERS 2010). Historically, pork consumption in the U.S. has only fluctuated mildly; however, a decrease in consumption is expected based on changes in America's racial/ethnic landscape, increasing elderly population, and an overall increase in health consciousness (Lin et al. 2003). Hence, there is a need within the pork industry to improve the product to minimize economic losses and maintain quality standards.

Understanding the underlying factors that influence pork consumption will enable the pork industry to provide a satisfactory product. Currently, pork is not graded with USDA quality grades and has traditionally received qualitative characteristics based on color, texture, and exudation as primary considerations in purchasing of fresh pork. As per the pork quality standards published by National Pork Board (NPB 2002), pork quality is generally characterized by four quality classes, namely RFN (reddish pink, firm, non-exudative), PFN (pale, firm, and non-exudative), RSE (reddish, soft, and exudative), PSE (pale pinkish gray, soft,

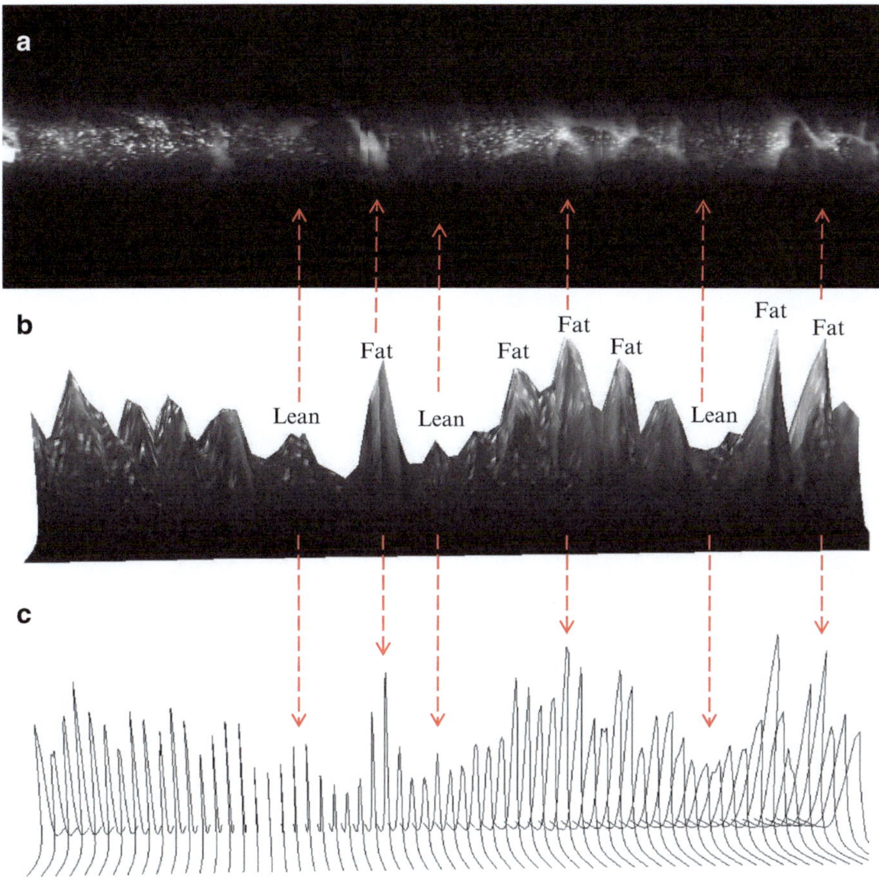

Fig. 10.14 Fat optical scattering and lean muscle optical scattering. (**a**) Incident beam of light streaked across steak surface ($\lambda = 1{,}158$ nm). (**b**) Textured surface plot of fat and lean light scattering. (**c**) Scattering profiles of fat and lean—ruled XZ wireframes (*Source*: Cluff et al. 2013)

and exudative), and DFD (dark purplish red, very firm, and dry). Pork chops classified as RFN have desirable color, and have normal texture with good water-holding capacity. Pork classified as PFN has the desired firmness and good texture but it has an undesirable color. Pork classified as RSE has the desired color of good meat but lacks the desired firmness and water-holding capacity. Pork chops with a classification of PSE have undesirable appearance and lack firmness with excessive drip loss. Similarly, DFD classified pork chops are undesirable as they are very firm with a sticky surface and are very dry. Figure 10.15 shows a set of typical pork images of different quality.

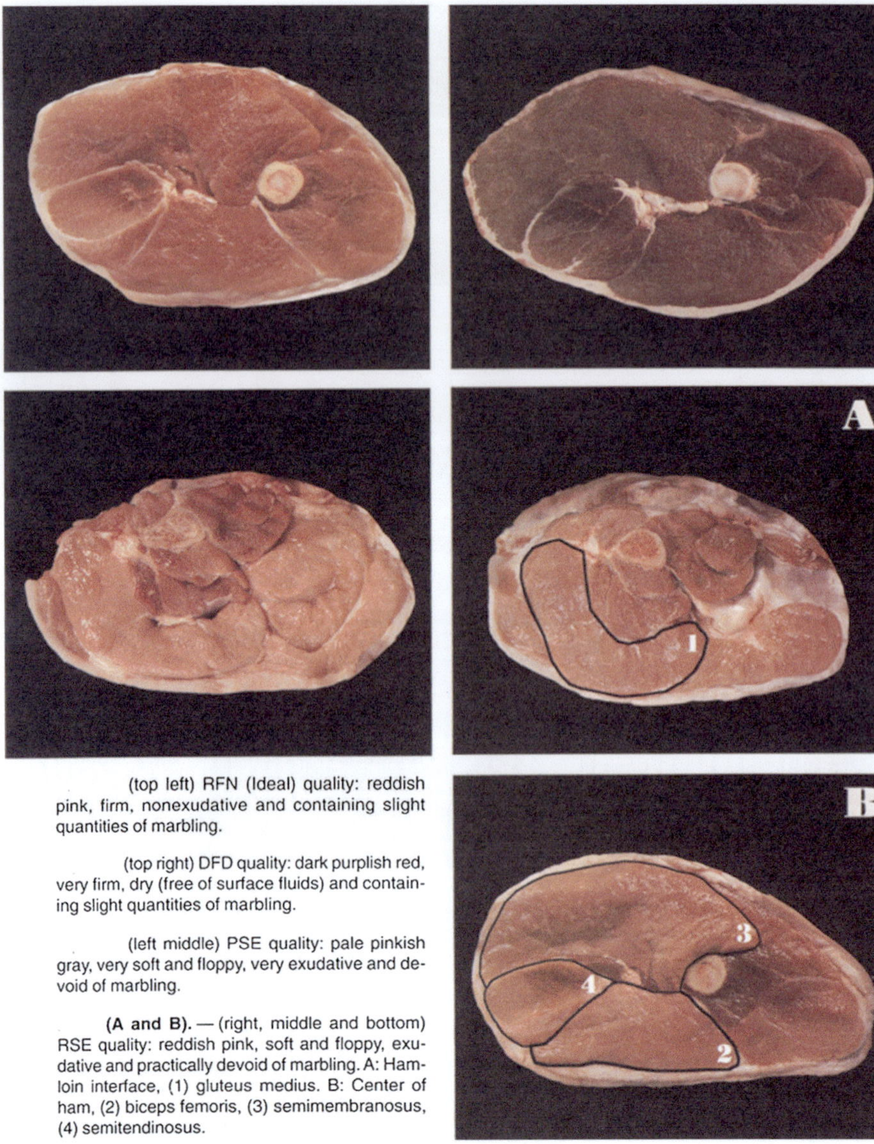

(top left) RFN (Ideal) quality: reddish pink, firm, nonexudative and containing slight quantities of marbling.

(top right) DFD quality: dark purplish red, very firm, dry (free of surface fluids) and containing slight quantities of marbling.

(left middle) PSE quality: pale pinkish gray, very soft and floppy, very exudative and devoid of marbling.

(A and B). — (right, middle and bottom) RSE quality: reddish pink, soft and floppy, exudative and practically devoid of marbling. A: Ham-loin interface, (1) gluteus medius. B: Center of ham, (2) biceps femoris, (3) semimembranosus, (4) semitendinosus.

Fig. 10.15 Pork quality grades (*Source*: Kauffman et al. 1993)

10.3.1 Objective Evaluation of Pork Quality

Many objective measures have been developed to aid in the assessment of quality attributes of pork meat quality. Van Laack et al. (1994) classified these five pork quality classes in terms of L-value. However, they concluded that brightness was not necessarily a reliable predictor of quality. They presented a table of L-values that were associated with quality classes of pork. The measure of acidity or basicity (pH) has been a common measurable quality attribute, but Channon et al. (2000) concluded that pH alone was not sufficient to characterize pork meat quality. They suggested the need for additional measurements of other quality attributes to create a satisfactory model for sorting pork into different eating quality classes. In addition, sophisticated laboratory-based techniques (chemical methods for analyzing water holding capacity, protein, and fat) have been developed, but are not suitable for large volumes of meat and high-speed processing in meat production facilities.

10.3.2 Near Infrared (NIR) Spectroscopy

Spectroscopy deals with the changes in molecular vibrational energies that occur when light interacts with matter, and produces unique absorbance spectra. Visible/Near infrared (Vis-NIR) spectroscopy deals with spectroscopic interactions that occur in the Vis-NIR region (350–2,500 nm) of the electromagnetic spectrum. Xing et al. (2007) performed an investigation using visible spectroscopy (400–700 nm) to classify pork meat into four quality categories namely RFN (red, firm, and non-exudative, RSE (red, soft, and exudative), PFN (pale, firm, and non-exudative), and PSE (pale, soft, and exudative). They were able to successfully separate pale meat from red meat and distinguish between PFN and PSE meat. However, they were not able to successfully classify all four quality groups, although they did report that visible reflectance spectroscopy was able to better classify meat qualities than L^*, a^*, and b^* values. In continuation, Monroy et al. (2010) used a wider spectral range from 350 to 2,500 nm to classify pork quality grades and achieved an overall classification accuracy of 79 %. Figure 10.16 shows reflectance spectra of different pork grades.

10.3.3 Hyperspectral Imaging

Emerging technologies such as hyperspectral imaging include both physical and biochemical features. So, hyperspectral imaging has the potential to measure a number of quality traits. It has been used for predicting quality grades such as RFN, RSE, PFN and PSE, marbling levels, drip-loss, pH, color, and tenderness of pork.

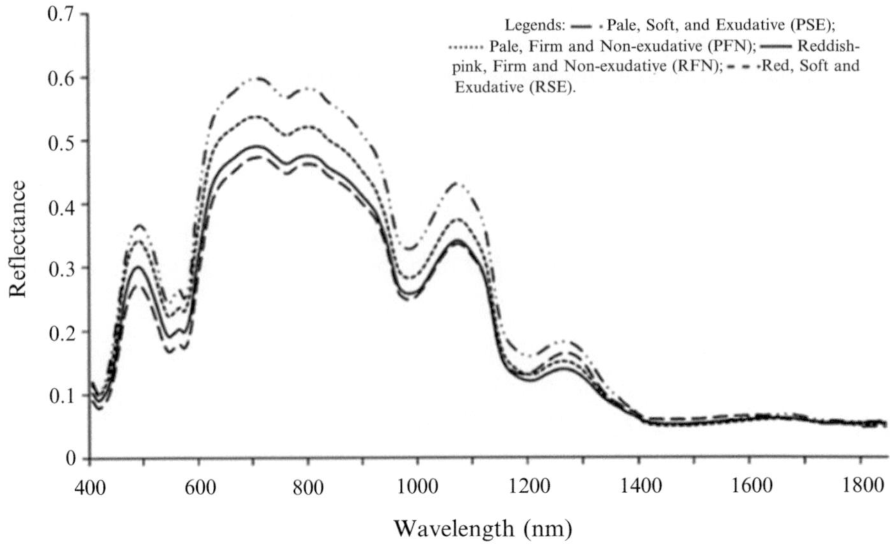

Fig. 10.16 Reflectance spectra of pork samples representing different quality grades (*Source*: Monroy et al. 2010)

10.3.3.1 Marbling Scores

According to the pork marbling standards established by the National Pork Board, there are seven levels—1.0 (Devoid) to 6.0 and 10.0 (Abundant). Figure 10.17 shows the pork marbling standards released by the National Pork. In a study conducted by Qiao et al. (2007a), hyperspectral images of pork loins (n = 40) in the 400–1,000 nm spectral range with a spectral resolution of 2.8 nm were obtained. After analyzing each hyperspectral image band for highest contrast between lean and marbling, they used the band 661 nm for predicting marbling scores. On the 661 nm image, a rectangular region-of-interest (ROI) was defined and a textural feature, angular second moment, was calculated and used to predict marbling scores. They concluded that the angular second moment was able to discriminate all marbling scores except Abundant. They also indicated the need to use large sample size to further validate and improve the marbling prediction.

10.3.3.2 Drip-Loss, pH, and Color

In another study conducted by the same group (Qiao et al. 2007b), hyperspectral images of pork loins were used to predict the drip-loss, pH, and color of pork. On the hyperspectral pork images, a circular ROI covering 10,000 pixels around the image center was defined. Prior to defining the ROI, the images were calibrated for

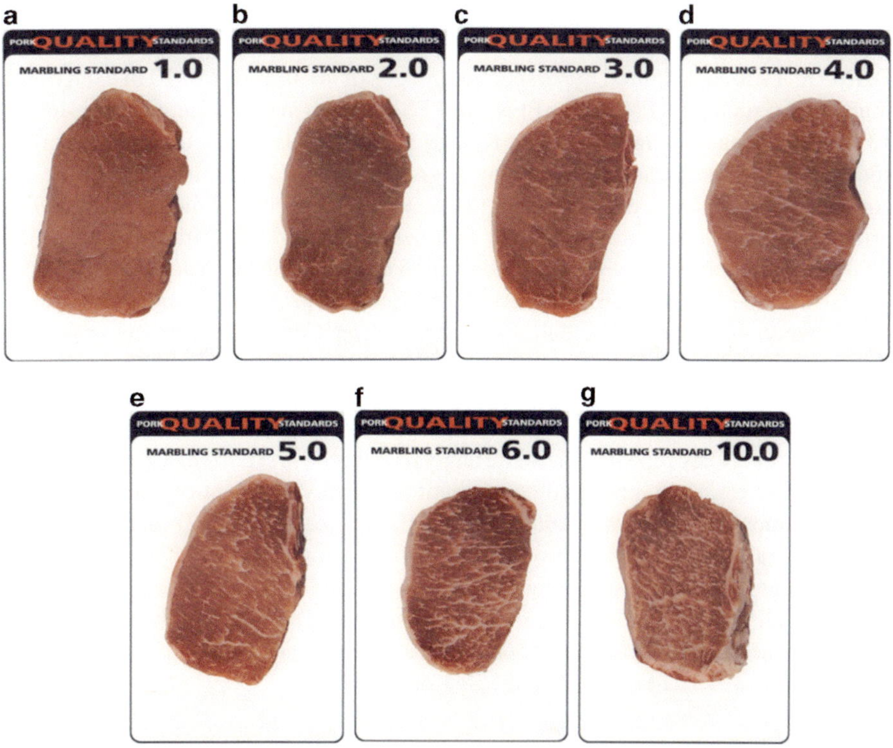

Fig. 10.17 Pork marbling standards published by National Pork Producers Counsil (NPPC) showing various levels of intramuscular fat content: 1% (**a**), 2% (**b**), 3% (**c**), 4% (**d**), 5% (**e**), 6% (**f**), and 10% (**g**).

reflectance using the dark and white reference images. For each band, average intensity of the ROI pixels was calculated. In addition, normalized intensity was calculated by dividing a band intensity by the sum of intensities of all the bands. The raw and normalized intensity values were used separately in neural network models to predict pH, drip-loss, and color of pork. Then the correlations between the intensity values and quality traits were examined in a band-by-band fashion and a set of pertinent bands were identified for predicting each trait: drip loss (459, 618, 655, 685, 755 and 953 nm), pH (494, 571,637, 669, 703 and 978 nm) and color (434, 494, 561, 637, 669 and 703 nm). The correlation coefficient values obtained using the normalized intensity values were 0.77, 0.55, and 0.86 for drip-loss, pH, and color, respectively. The normalized intensity values provided slightly better prediction results than the raw intensity values.

10.3.3.3 Quality Grades

Qiao et al. (2007a) acquired hyperspectral images of 40 pork chops, 10 in each of the four grades such as RFN, RSE, PFN, and PSE, in the spectral range from 400 to 1,000 nm. On each image, a circular ROI of 10,000 pixels was defined and average spectral reflectance was calculated by averaging the ROI pixels. Then principal component analysis was conducted to reduce the spectral dimensionality. The principal component scores were sent to a neural network algorithm to obtain the final classification. This method provided an overall classification accuracy of 75 %, 75 %, and 80 % when using the first 5, 10, and 20 principal component scores, respectively. With the same dataset, incorporating Gabor features increased the accuracy to 84 % in cross-validation (Liu et al. 2010). The same group conducted another study (Qiao et al. 2007c) with 80 pork chops, 20 in each quality grade. They compressed the spectral dimensionality with principal component analysis and step-wise selection. They repeated the same analysis with the first derivative spectra. First derivative followed by principal component analysis yielded an overall classification accuracy of 87.5 %. Additionally, Barbin et al. (2012) acquired NIR hyperspectral images (900–1,700 nm) of 75 pork chops, extracted spectral reflectance values, obtained second derivative spectra, and used them to classify PSE, RFN, and DFD. They achieved an overall classification accuracy of 96 %. They also noted that there were significant reflectance variations among these quality grades at 960, 1,074, 1,124, 1,147, 1,207, and 1,341 nm.

10.3.3.4 Tenderness

Tenderness is an important trait related to consumer satisfaction. Barbin et al. (2011) used a hyperspectral imaging system sensitive in the spectral range from 900 to 1,700 nm to predict sensory tenderness of pork. They obtained hyperspectral images of 30 pork loins and conducted partial least squares regression to predict sensory tenderness and juiciness. They reported an R^2 value of 0.49 and 0.54 for juiciness and tenderness, respectively, in a cross-validation analysis. When they classified samples into two groups based on tenderness namely tender and tough, the R^2 value increased to 0.82. Similarly, when classifying samples based on juiciness as juicy or dry, they achieved an R^2 value of 0.67. In 2013, the same research group conducted a study with 90 pork loins to relate the NIR reflectance measurements (900–1,700 nm) and textural features obtained from discrete wavelet transforms to pork tenderness. They achieved a R^2 values 0.63 and 0.48 when using the NIR reflectance measurements and wavelet textural features separately. When both the measurements were combined, the R^2 value increased to 0.75 (Barbin et al. 2013).

While both the studies mentioned above used reflectance mode of hyperspectral imaging, Tao et al. (2012) used scattering measurements acquired in the 400–1,100 nm range to pork tenderness. After acquiring the scattering signals, they used a three-parameter Lorentzian function to fit the scattering profile. Using the three fitting parameters were *a* (asymptotic value), *b* (peak value), and *c* (full width at half maximum), they developed multiple linear regression models to predict pork tenderness. The parameters combinations *a*, *b*, $(b - a)$, and $(b - a)/c$ yielded a R^2 values of 0.83, 0.86, 0.86, and 0.93, respectively, in cross-validation analysis. They also identified a set of optimal wavelengths: 612, 632, 708, 770, 786, and 814 nm, for the $(b - a)/c$ parameter model.

10.4 Conclusions

Hyperspectral imaging has potential to characterize meat muscle tissue. In beef, tenderness is a major trait. The industry is looking for a non-destructive instrument for forecasting 14-day aged cooked beef tenderness by scanning the exposed ribeye on the hanging carcass in a beef packing plant at 2–3 day postmortem. Because hyperspectral imaging extracts both spatial (structural) and spatial (biochemical) information from the ribeye, it has potential to evaluate muscle structure and biochemical information related to aging. At the laboratory level, the hyperspectral imaging has been shown to forecast beef tenderness. There is a need to build a commercial system that can work in a beef packing plant to acquire the images at line speed and predict beef tenderness real-time. Hyperspectral imaging has also shown to predict pork quality grades, color, water holding capacity and tenderness.

References

Barbin D, Elmasry G, Sun DW, Allen P (2011) Prediction of pork sensory attributes using NIR hyperspectral imaging technique. In: The sixteenth annual biosystems engineering research review, University College Dublin

Barbin D, Elmasry G, Sun DW, Allen P (2012) Near-infrared hyperspectral imaging for grading and classification of pork. Meat Sci 90:259–268

Barbin D, Valous N, Sun DW (2013) Tenderness prediction in porcine longissimus dorsi muscles using instrumental measurements along with NIR hyperspectral and computer vision imagery. Innov Food Sci Emerg Technol 20:335–342

Belk KE, Scanga JA, Wyle AM, Wulf DM, Tatum JD, Reagan JO (2000) The use of video image analysis and instrumentation to predict beef palatability. Proc Recip Meat Conf 53:10–15

Burns DA, Ciurczak EW (eds) (2001) Handbook of near-infrared analysis, 2nd edn. Marcel Dekker, New York

Byrne CE, Downey G, Troy DJ, Buckley DJ (1998) Non-destructive prediction of selected quality attributes of beef by near-infrared reflectance spectroscopy between 750 and 1098 nm. Meat Sci 49(4):399–409

Hildrum KI, Nilsen BN, Mielnik M, Næs T (1994) Prediction of sensory characteristics of beef by near-infrared spectroscopy. Meat Sci 38(1):67–80

Hildrum KI, Isaksson T, Næs T, Nilsen BN, Rodbotten M, Lea P (1995) Near infrared reflectance spectroscopy in the prediction of sensory properties of beef. J Near Infrared Spectrosc 3:81–87

Kauffman RG, Cassens RG, Scherer A, Meeker DL (1993) Variations in pork quality: history, definition, extent, resolution. Swine Health Prod 1:28–34

Konda Naganathan G, Grimes LM, Subbiah J, Calkins CR (2006a) Predicting beef tenderness using hyperspectral imaging. In: 2006 ASABE annual international meeting. Paper No. 063036

Konda Naganathan G, Grimes L, Subbiah J, Calkins C, Samal A (2006b). VNIR imaging for beef tenderness prediction. In: Annual ASABE international meeting, OR. Paper No. 063036

Konda Naganathan G, Grimes LM, Subbiah J, Calkins CR, Samal A, Meyer GE (2008) Visible/near-infrared hyperspectral imaging for beef tenderness prediction. Comput Electron Agric 64(2):225–233

Lin SP, Wang LH, Jacques SL, Tittel FK (1997) Measurement of tissue optical properties by the use of oblique incidence optical fiber reflectometry. Appl Optics 36:136–143

Lin BH, Variyam J, Allshouse J, Cromartie J (2003) Food and agricultural commodity consumption in the United States: looking ahead to 2020. Agr. Econ. Report No. 820. U.S. Department of Agriculture, Economic Research Service, Washington, DC, p 820

Lipsey J (1999) Development of multibreed genetic evaluation. In: Proceedings, the range beef cow symposium XVI

Liu L, Ngadi MO, Prasher SO, Gariépy C (2010) Categorization of pork quality using Gabor filter-based hyperspectral imaging technology. J Food Eng 99:284–293

Liu L, Ngadi MO, Prasher SO, Gariépy C (2012) Objective determination of pork marbling scores using the wide line detector. J Food Eng 110:497–504

McKenna DR, Roebert DL, Bates PK, Schmidt TB, Hale DS, Griffin DB, Savell JW, Brooks JC, Morgan JB, Montgomery TH, Belk KE, Smith GC (2002) National Beef Quality Audit—2000: survey of targeted cattle and carcass characteristics related to quality, quantity, and value of fed steers and heifers. J Anim Sci 80(5):1212–1222

Mitsumoto M, Maeda S, Mitsuhashi T, Ozawa S (1991) Near infrared spectroscopy determination of physical and chemical characteristics in beef cuts. J Food Sci 56:1493–1496

Monroy M, Prasher S, Ngadi MO, Wang N, Karimi Y (2010) Pork meat quality classification using visible/near-infrared spectroscopic data. Biosyst Eng 107:271–276

Naes T, Hildrum KI (1997) Comparison of multivariate calibration and discriminant analysis in evaluating NIR spectroscopy for determination of meat tenderness. Appl Spectrosc 51:350–357

NPB (National pork board) (2002) Pork quality standards. Des Moines

Park B, Chen YR, Hruschka WR, Shackelford SD, Koohmaraie M (1998) Near-infrared reflectance analysis for predicting beef longissimus tenderness. J Anim Sci 76:2115–2120

Peng Y, Wu J (2008) Hyperspectral scattering profiles for prediction of beef tenderness. In: ASABE annual international meeting. ASABE Paper No. 080004

Qiao J, Ngadi MO, Wang N, Gariépy C, Prasher SO (2007a) Pork quality and marbling level assessment using a hyperspectral imaging system. J Food Eng 83:10–16

Qiao J, Wang N, Ngadi MO, Gunenc A, Monroy M, Gariépy C, Prasher SO (2007b) Prediction of drip-loss, pH, and color for pork using a hyperspectral imaging technique. Meat Sci 76:1–8

Qiao J, Wang N, Ngadi MO, Gunenc A, Monroy M, Gariépy C, Prasher SO (2007c) Pork quality classification using a hyperspectral imaging system and neural network. J Food Eng 3(1):6

Shackleford SD, Wheeler TL, Koohmaraie M (2005) On-line classification of US Select beef carcasses for longissimus tenderness using visible and near-infrared reflectance spectroscopy. Meat Sci 69:409–415

Subbiah J (2004) Nondestructive evaluation of beef palatability. Dissertation, Biosystems Engineering, Oklahoma State University

Tao F, Peng Y, Li Y, Chao K, Dhakal S (2012) Simultaneous determination of tenderness and Escherichia coli contamination of pork using hyperspectral scattering technique. Meat Sci 90:851–857

USDA-ERS (2010) Economic Research Service (ERS), U.S. Department of Agriculture (USDA). Food availability (per capita) data system. http://www.ers.usda.gov/Data/FoodConsumption/. Accessed 4 July 2011

USDS-ERS (2011) United States Department of Agriculture, Economic Research Service. http://www.ers.usda.gov/Data/MeatPriceSpreads/. Accessed 4 July 2011

Vote DJ, Belk KE, Tatum JD, Scanga JA, Smith GC (2003) Online prediction of beef tenderness using a computer vision system equipped with a BeefCam module. J Anim Sci 81:457–465

Wang LH, Jacques SL (1995) Use of a laser beam with an oblique angle of incidence to measure the reduced scattering coefficient of a turbid medium. Appl Optics 34:2362–2366

Webb NB, Kahlenberg OJ, Naumann HD (1964) Factors influencing beef tenderness. J Anim Sci 23:1027–1031

Wyle AM, Cannell RC, Belk KE, Goldberg M, Rifle R, Smith GC (1999) An evaluation of the prototype portable HunterLab video imaging system (BeefCam) as a tool to predict tenderness of beef carcasses using objective measure of lean and fat color. Research reports. Department of Animal Science, Colorado State University, Fort Collins

Wyle AM, Vote DJ, Roeber DL, Cannell RC, Belk KE, Scanga JA, Goldberg M, Tatum JD, Smith GC (2003) Effectiveness of the SmartMV prototype BeefCam System to sort beef carcasses into expected palatability groups. J Anim Sci 81:441–448

Xia J, Weaver A, Gerrard DE, Yao G (2006) Monitoring sarcomere structure changes in whole muscle using diffuse light reflectance. J Biomed Opt 11(4):040504

Xia JJ, Berg EP, Lee JW, Yao G (2007) Characterizing beef muscles with optical scattering and absorption coefficients in VIS-NIR region. Meat Sci 75(1):78–83

Chapter 11
Plant Health Detection and Monitoring

Won Suk Lee

11.1 Introduction

Plant health detection and monitoring is one of the main applications of hyperspectral imaging to agriculture, which include detection of water content, nutrient status, and pest damages including disease infections and insect damages. Different ground-based, airborne, and spaceborne sensing systems are used to detect and monitor plant health. Airborne and spaceborne imaging spectroscopy was established in the 1990s (Ustin et al. 2004). These systems basically can measure electromagnetic radiation from different crop plants to determine their status of water, nutrition, and degree of pest infestation using specific spectral signatures of different plant status.

Ustin et al. (2004) provided an excellent overview by describing advances in airborne and spaceborne imaging spectrometers to observe and monitor different processes and properties in ecosystems along with various applications. They reported that detection of photosynthetic pigments (chlorophyll, xanthophylls, carotenes, and algal pigments) in the visible range shows great potential for various applications; in the NIR range, strong absorption exists by water (870 and 1,240 nm) and leaf compounds such as cellulose, lignin, and carbohydrates. In 1,100–2,500 nm, carbon compounds including cellulose, lignin, nitrogen, starches, and sugars exhibit strong absorption. These chemical differences can be used to create a vegetation map with different types of stresses. When plants lose chlorophyll, reflectance in the visible and NIR ranges increases, and the red edge is shifted up to 25 nm toward shorter wavelengths (named as the "blue shift"). Figure 11.1 shows an example of plant pigment extraction using the Airborne Visible/Infrared Imaging Spectrometer (AVIRIS). For plant water content detection, several water

W.S. Lee (✉)
Agricultural and Biological Engineering, University of Florida, Gainesville, FL, USA
e-mail: wslee@ufl.edu

B. Park, R. Lu (eds.), *Hyperspectral Imaging Technology in Food and Agriculture*,
Food Engineering Series, DOI 10.1007/978-1-4939-2836-1_11

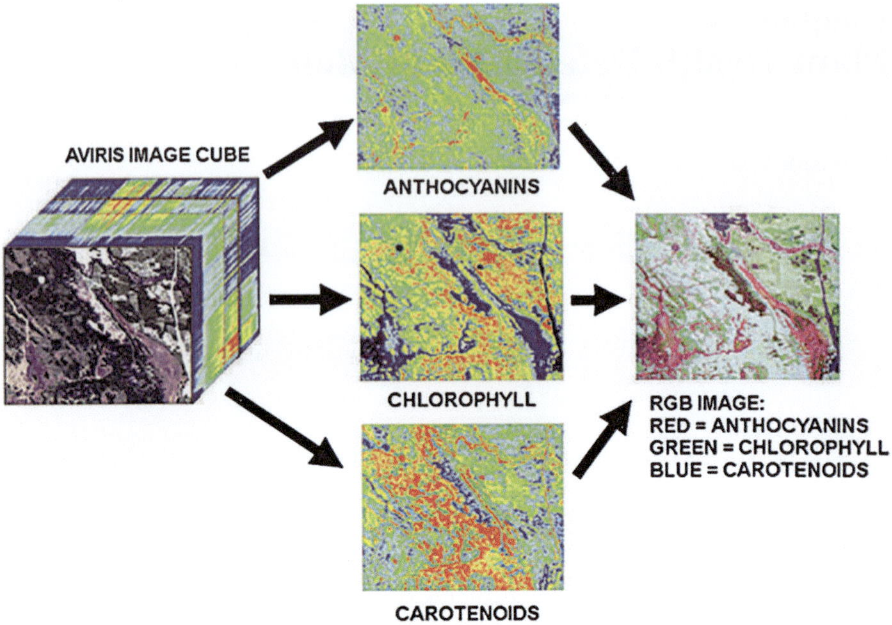

Fig. 11.1 Example of plant pigments extraction from a hyperspectral image *cube* using the Airborne Visible/Infrared Imaging Spectrometer (AVIRIS) (adapted from Ustin et al. 2004)

absorption bands (1,450, 1,940, and 2,500 nm) can be used as well as different band ratios such as the normalized difference water index (NDWI) and plant water index (PWI).

Sankaran et al. (2010) reviewed different techniques for detecting plant diseases. They reported that no commercial sensor exist for tree health conditions, and presented the need for a rapid and reliable sensor system. They reported two different approaches for detecting plant diseases: (1) direct methods including serological and molecular methods, and (2) indirect methods including imaging and spectroscopic techniques, biomarkers using volatile organic metabolites, and plant properties/stress based disease detection. They pointed out one of the challenges in hyperspectral image analysis is to choose proper disease-specific spectral bands and statistical method, and that imaging technique can be used with an autonomous agricultural vehicle for automatic real-time monitoring.

Lee et al. (2010) reviewed plant disease detection methods for specialty crops, and mentioned that any disease causing differences in spectral characteristics can be detected using remote sensing. They mentioned that remote sensing of plant disease could be used more effectively for damage assessment than identifying early disease infection, as early detection would be difficult or impossible in some situations. They also mentioned that more research is needed for developing efficient disease detection algorithm for practical farming operations and distinguishing the diseases from other stresses. The following describes different applications of plant health detection for agricultural crop production.

11.2 Water Status Monitoring

Water plays a very important role for optimal plant growth. With strong water absorption bands in the electromagnetic spectrum, hyperspectral imaging has been used for monitoring plant water status. Allen et al. (1969) defined the equivalent water thickness (EWT) as "the hypothetical thickness of a sheet of liquid water in the target". From this definition, Jacquemoud and Baret (1990) calculated the EWT of plant biomass as:

$$\mathrm{EWT}_{\mathrm{Biomass}}\ (\mathrm{cm}) = \frac{Fresh\ mass\ (\mathrm{cm}^3) - Dry\ mass\ (\mathrm{cm}^3)}{Leaf\ area\ (\mathrm{cm}^2)}$$

Champagne et al. (2003) developed a physical model to estimate the canopy equivalent water thickness from airborne hyperspectral imaging with the Probe-1 sensor using a spectrum matching technique and look-up table approach. They tested the model using various crops (wheat, canola, corn, beans, and peas) by comparing EWT estimation (Fig. 11.2) and actual crop water status, and reported an RMSE of 0.052 cm. They observed that the model was more sensitive to water content in leaf than the water content in the whole plant. Further, Cheng et al. (2006) investigated the relationships between canopy water content and the EWT using three linked leaf and canopy radiative transfer models with different canopy structures: closed, row-structured, and forest canopy architectures. They estimated EWT from the AVIRIS imagery and found good agreement of EWT for agricultural crops with enhanced vegetation index (EVI), however better agreement was observed for conifer forest between EWT and normalized difference water index (NDWI) and shortwave infrared water stress index (SIWSI) than EVI, indicating canopy architecture should be considered when estimating EWT. As a recent study, Kim et al. (2010) utilized spectral indices from hyperspectral images

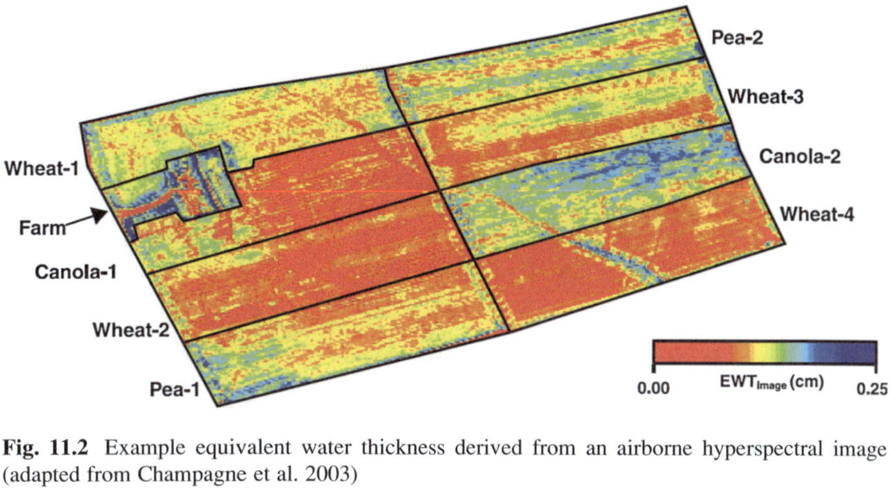

Fig. 11.2 Example equivalent water thickness derived from an airborne hyperspectral image (adapted from Champagne et al. 2003)

to monitor water stress of young apple trees in a greenhouse, and reported red edge normalized difference vegetation index (NDVI) at 705 and 750 nm, and NDVI at 680 and 800 nm yielded the highest correlation with the water stress.

11.3 Nutrient Status Monitoring

Plant nutrient monitoring is also a very important application of hyperspectral imaging for agricultural crop production. Applications include estimation of leaf pigments such as chlorophylls and carotenoids, and nitrogen content.

In early 2000s, Zarco-Tejada et al. (2002) used the fluorescence–reflectance–transmittance (FRT) and PROSPECT leaf models to mimic reflectance which were able to estimate chlorophyll fluorescence and chlorophyll a + b content. They developed a derivative chlorophyll index ($DCI = D_{705}/D_{722}$, where D is derivative of reflectance) based on double peak feature of derivative reflectance, which could be used to estimate plant stress. Further Zarco-Tejada et al. (2003) conducted experiments in controlled environments to observe natural chlorophyll fluorescence emission, and identified that the double-peak feature between 688, 697, and 710 nm of canopy derivative reflectance was due to chlorophyll fluorescence (Fig. 11.3). This double peak feature could be used to monitor plant stress by detecting changes in pigment and canopy structure.

In another study, Haboudane et al. (2004) investigated different vegetation indices (VIs) to green leaf area index, developed new VIs and reported that two newly developed VIs (a modified triangular vegetation index (MTVI2) and a modified chlorophyll absorption ratio index (MCARI2)) worked well to predict green leaf area index (LAI). Blackburn (2007) discussed the importance of leaf pigments such as chlorophylls and carotenoids, different remote sensing platforms for measuring them, various issues in measuring pigments, new methods for

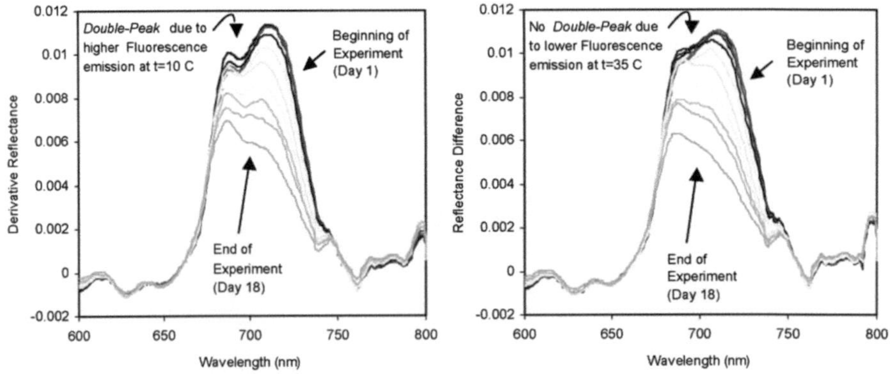

Fig. 11.3 Double-peak feature between 688, 697, and 710 nm of canopy derivative reflectance due to natural chlorophyll fluorescence emission (adapted from Zarco-Tejada et al. 2003)

analyzing hyperspectral data, and emerging applications including determination of crop N requirements, identification of high yielding genotypes, and mapping in-field crop yield variability.

For plant nitrogen status detection, Goel (2003) investigated detection of nitrogen status and weeds in corn and soybean using aerial hyperspectral imaging, and found that 498 and 671 nm were highly related to N levels. They also reported that 701 and 839 nm were found to be important in yield prediction models, and decision trees and artificial neural networks showed strong potential for agricultural remote sensing applications. Min and Lee (2005) investigated reflectance characteristics of citrus leaves of varying nitrogen (N) concentrations, and identified important wavelengths (448, 669, 719, 1,377, 1,773, and 2,231 nm) for N detection using stepwise multiple linear regression (SMLR) and partial least squares (PLS) regression. Further Min et al. (2008) developed a portable hyperspectral sensing system in 620–950 nm and 1,400–2,500 nm to measure citrus nitrogen concentration using detector arrays, linear variable filters, a halogen light, and data acquisition cards. The sensor system had good linearity and stability, and was able to estimate citrus N content with a root mean square difference (RMSD) of 1.69 g/kg.

11.4 Detection of Plant Disease and Insect Damage

Plant disease and insect damage detection are the mostly adopted application for hyperspectral imaging in agriculture. Plant disease detection includes fungal infestation, stem rot disease, yellow rust in winter wheat, rice brown spot disease, citrus canker disease, citrus greening disease (also known as Huanglongbing or HLB), apple scab disease, and diseases in rice, potato, and tomato. Lee et al. (2010) described different measurement techniques of foliar plant diseases based on their infection stages, as shown in Fig. 11.4.

For fungal disease detection, Muhammed (2002) investigated discrimination of healthy and diseased plants, and estimation of level of infection of the fungi

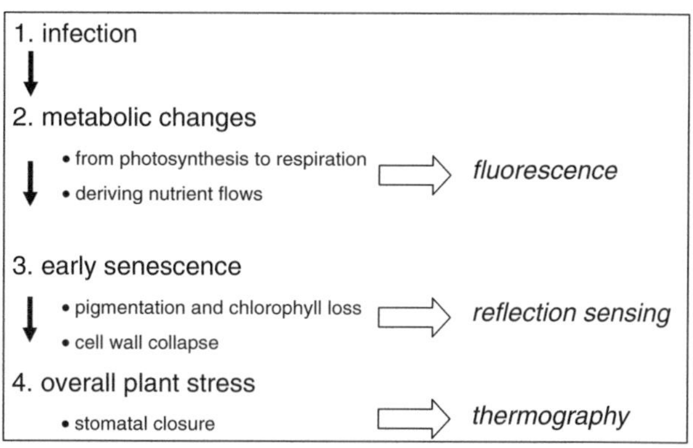

Fig. 11.4 Illustration of different techniques to identify foliar diseases based on infection stages (adapted from Lee et al. 2010)

infested plants through simple steps of pre-processing and nearest neighbor classification, using correlation and sum of squared differences between a reference and unknown spectra. Laudien et al. (2003) studied detection of sugar beet disease by fungal infection using a ground-based hyperspectral spectroradiometer and were able to identify the disease using red edge (a ratio between reflectance at 750 and 700 nm) and modified chlorophyll-absorptions-integral (mCAI) index (an area of the trapeze of reflectance between 545 and 752 nm).

$$\text{mCAI} = \frac{(R545 + R752)}{2} * (752 - 545) - \sum_{R545}^{R752} (R * 1.158)$$

As an example of stem rot disease detection, Vigier et al. (2004) utilized a narrow-band spectrometer to detect sclerotinia stem rot diseases in soybean and reported the red narrow band (R_{675}–R_{685}) was the most contributable to damage estimation.

For yellow rust disease detection for winter wheat, Moshou et al. (2005) investigated fusion of ground based multispectral and hyperspectral fluorescence imaging to detect crop disease (yellow rust in winter wheat) in an early stage before no visible symptoms show, and reported that disease presence could be detected by comparing fluorescence images at 550 and 690 nm. The overall classification error between healthy and diseased plant reduced to 1 % after data fusion using a Self-Organizing Map (SOM) neural network. Huang et al. (2007) evaluated the photochemical reflectance index ($PRI = (R_{531} - R_{570})/(R_{531} + R_{570})$) to quantify yellow rust disease in winter wheat and reported an R^2 of 0.97 between PRI and measured spectral data, and an R^2 of 0.91 between PRI and airborne hyperspectral data.

For the citrus canker disease detection, Qin et al. (2008) utilized principal component analysis (PCA) to detect citrus canker disease from hyperspectral images acquired by a portable hyperspectral imaging unit, and identified four important wavelengths (533, 677, 718, and 858 nm), while reporting an overall 92.7 % detection accuracy. Further Qin et al. (2009) applied spectral information divergence (SID) classification method to detect citrus canker and other citrus surface conditions, and reported an overall accuracy of 96.2 % for citrus canker detection. Balasundaram et al. (2009) identified that wavelengths in 500–800 nm range showed the highest discriminatory power to detect citrus canker disease in grapefruit, and reported that wavelengths above 1,100 nm did not contain significant wavelengths.

Citrus greening disease or Huanglongbing (HLB) is a catastrophic disease in Florida, since there is no cure. The disease was first found in 2005 in Florida and is spread most of Florida and other states in the USA. Currently ground inspection is being conducted to identify infected tree canopies, however it is subjective, time consuming, and labor intensive. Therefore, aerial hyperspectral imaging can be well applied for the HLB detection, which can provide a rapid detection over a wider area. As a first step for the disease detection, Mishra et al. (2007) measured HLB infected tree canopies with a handheld spectrometer (FieldSpec UV/VNIR, Analytical Spectral Devices, Boulder, CO) in 350–2,500 nm, and reported that wavelengths of 530–564, 710–715, 1,041, and 2,014 nm showed higher discriminant capability between healthy and infected canopies. Then, Lee et al. (2008)

compared laboratory and ground measurement spectra of healthy and infected citrus tree canopies to detect the citrus greening disease, and reported that reflectance differences existed at 550 nm for healthy, diseased, and zinc-deficient canopies. HLB infected tree canopy could be detected based on first derivative peak near 730 nm. Also ANOVA analysis showed significant difference was found between healthy and infected tree canopies. However, variability of healthy and HLB infected canopies and georeference errors were main obstacles in detecting the disease. Kumar et al. (2009) investigated detection of the HLB disease using airborne hyperspectral images in 400–1,000 nm. Different algorithms including image-derived spectral library, mixture tuned match filtering (MTMF), spectral angle mapping (SAM), and spectral feature fitting (SFF) were employed. They reported an overall accuracy of 60 % using SAM, largely attributed to the inaccuracies of ground truthing. Further, Kumar et al. (2010) acquired aerial multispectral and hyperspectral images to detect the citrus greening disease, constructed spectral library for healthy and HLB infected canopies, applied SAM, MTMF, and linear spectral unmixing (LSU), and reported detection accuracies of 60–87 %, 73–80 %, and 53–73 % for SAM, MTMF, and LSU, respectively, depending on validation subset images. In a recent study, Li et al. (2011) investigated several classification algorithms for detecting HLB disease from airborne multispectral and hyperspectral images, and reported that the HLB infected canopy showed higher reflectance in the visible range, and that detection accuracies were ranged from 55 to 95 %, however simpler algorithms such as minimum distance and Mahalanobis distance worked better than other algorithms such as SAM or spectral information divergence. Figure 11.5 shows average reflectance spectra of healthy

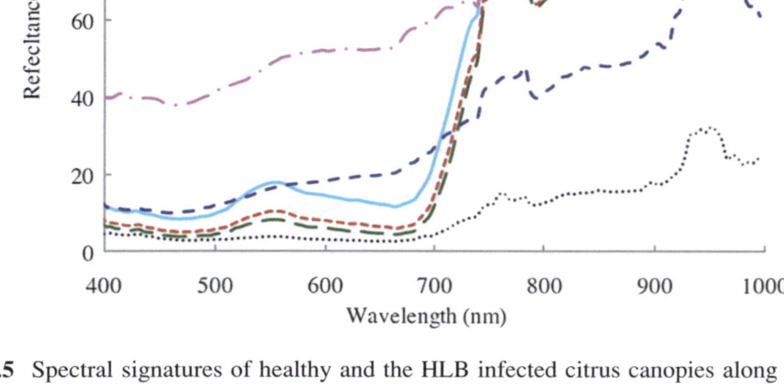

Fig. 11.5 Spectral signatures of healthy and the HLB infected citrus canopies along with other objects found in a citrus grove. Numbers in a parenthesis indicate the number of samples used for averaging (adapted from Li et al. 2011)

and HLB infected citrus canopies, and other objects in a citrus grove which were acquired using an airborne hyperspectral imaging system.

For an application of hyperspectral imaging to apple, Delalieux et al. (2009) detected apple scab disease using hyperspectral images and reported that ratios of reflectance at R_{440}/R_{690} and R_{695}/R_{760} yielded good separation between healthy and more developed infection. For early detection, ratio at water band performed well. They reported 570, 1,460, 1,940, and 2,400 nm were important to identify the disease.

Sugar beet disease could be detected by hyperspectral imaging. Mahlein et al. (2010) investigated detection of sugar beet diseases by using three different VIs: NDVI, anthocyanin reflectance index ($ARI = (1/R_{550}) - (1/R_{700})$), and modified chlorophyll absorption integral ($mCAI = (R_{545} + R_{752})/2 \times (752 - 545) - \sum_{R545}^{R752} (R + 1.423)$) and found that combination of those VIs were able to assess the different diseases for sugar beet. Also Rumpf et al. (2010) adopted support vector machine (SVM) and VIs to detect diseased sugar beet leaves and achieved early disease detection using SVM with a radial basis function. They reported 97 % detection accuracy between healthy and diseased leaves, and 65–90 % for different types and degrees of infections.

Hyperspectral imaging could be applied for disease detection for other crops such as rice, potato, and tomato. For rice disease detection, Liu et al. (2007) utilized stepwise regression, principal component regression (PCR), and PLS to identify severity of rice brown spot disease, and reported root mean square errors of 5.8 %, 13.9 %, and 2.0 % for the three methods, respectively. Liu et al. (2010a) applied neural networks and PCA to detect fungal disease of rice panicles in hyperspectral images in 350–2,500 nm acquired in a laboratory, and reported that four different infection levels were classified with 86–100 % accuracies. Liu et al. (2010b) adopted principal component analysis (PCA) and support vector machine (SVM) to classify healthy and infected rice panicles by rice false smut (*U. virens*), whose hyperspectral images were acquired in the visible and NIR ranges in a laboratory, and reported over 96 % accuracies using the original spectral data, first derivatives, and second derivatives. For potato disease detection, Ray et al. (2011) investigated detection of potato late blight disease using a hand-held spectroradiometer in 325–1,075 nm, utilized stepwise discriminant analysis and different VIs (NDVI, simple ratio (SR), soil adjusted vegetation index (SAVI), and red edge). They reported that 540, 610, 620, 700, 710, 730, 780, and 1,040 nm were optimal wavebands for the disease detection. For tomato, Zhang et al. (2003) employed aerial hyperspectral imaging (AVIRIS) to detect different infection levels of late blight in tomatoes, and observed more difference in reflectance in the NIR range than the visible range. They utilized the minimum noise fraction (MNF) and SAM for the disease detection and reported that severe infection could be identified, however it was not easy to detect less infected plants. They expected that more hyperspectral imaging would be used to detect plant disease on large-scale farms, if proper and fast image processing algorithms were developed. Also Jones et al. (2010) investigated detection of tomato disease using ultra violet, visible, and near infrared (NIR) reflectance spectroscopy, and employed PLS with B-matrix, correlation coefficient, and SMLR. They reported an RMSE of 4.9 % and an R^2 of 0.82 for the best disease prediction model.

Hyperspectral imaging could be applied for detection of insect damage. Carroll et al. (2008) studied different VIs to detect European corn borer infestation using airborne hyperspectral imaging. They reported that better detection could be achieved as the disease progressed more, and that chlorophyll related VIs were more closely related to the disease detection than VIs for anthocyanin and carotenoids. Also Singh et al. (2009) investigated detection of insect damaged wheat kernels using statistical image features and histogram features obtained from images taken at 1,101 and 1,305 nm which were selected from a PCA analysis, and reported that discriminant analyses were able to correctly classify 85–100 % infected wheat kernels.

11.5 Other Applications

Hyperspectral imaging has been applied for other plant health monitoring and detection such as heavy metal contamination, weed, fruit quality, acid rain stress, sprouting, fruit defect, immature fruit detection, and fruit maturity status.

For heavy metal contamination detection, Schuerger et al. (2003) studied the growth of bahiagrass under the stress of heavy metal (zinc) contamination. They utilized four handheld devices for hyperspectral imaging (two imaging systems), laser-induced fluorescence spectroscopy, and laser-induced fluorescent imaging. They reported that NDVI and ratio vegetation index ($RVI = R_{750}/R_{700}$, where R is reflectance) were able to predict chlorophyll concentrations in bahiagrass. Also Wilson et al. (2004) applied support vector classification (SVC), partial least squares (PLS) and logistic discrimination (LD) to classify different levels of heavy metal or petroleum exposed plants, and reported that SVC was better in classifying the images than PLS/LD.

For weed detection, Williams and Hunt (2004) utilized AVIRIS images to distinguish leafy spurge (*Euphorbia esula* L.), a noxious perennial weed, using mixture tuned matched filtering (MTMF) method and reported an overall detection accuracy of 95 %.

As an example of fruit quality detection, Mehl et al. (2002) studied detection of defects for three apple varieties by applying hyperspectral images to identify multispectral bands and to develop a multispectral imaging system. The multispectral imaging system with 705, 460, and 575 nm was able to detect 76–95 % of contaminated apples. Also Lenk et al. (2007) investigated multispectral fluorescence imaging in the blue, green, read, and far red, and reflectance imaging in the green and NIR regions and discussed applications such as detection of fruit quality, photosynthetic activity, disease symptoms, and leaf tissue structure along with different instrumentations. As an another example of quality detection, Xing et al. (2010) investigated detection of sprouted wheat kernels, identified a reflectance ratio of 878 nm over 728 nm as a feasible index to distinguish sprouted kernels, and also reported that four wavelengths identified from PCA loadings

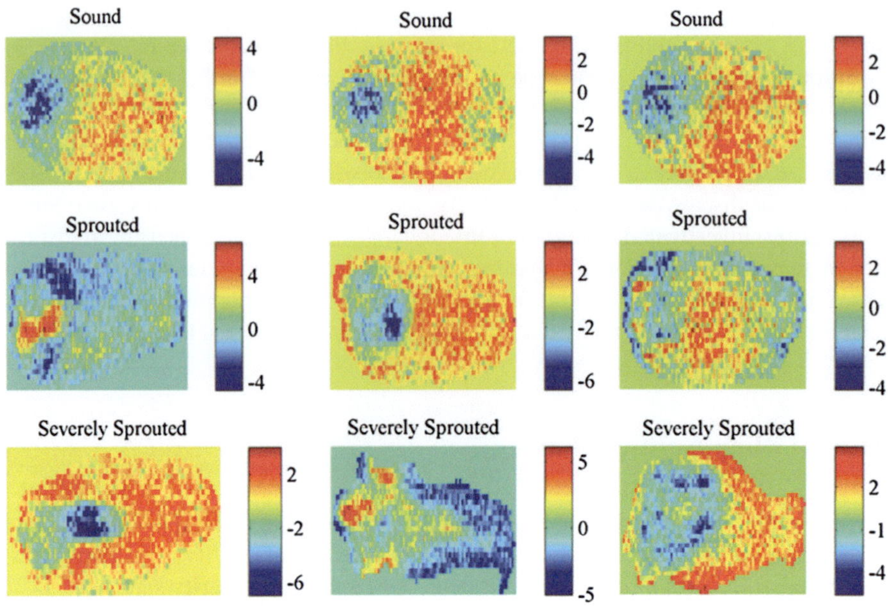

Fig. 11.6 Hyperspectral images of sound and sprouted wheat kernels represented by a principal component score (adapted from Xing et al. 2010)

could be used to evaluate quality of wheat kernels. Figure 11.6 shows principal component score images of sound and sprouted wheat kernels.

Hyperspectral imaging was applied for acid rain stress detection. Song et al. (2008) conducted ground-based hyperspectral imaging to detect acid rain stress on native forest by employing continuum removal, VIs, and PCA, and used two new chlorotic indices ($R_{GY} = R_G/R_Y$ and $R_{GO} = R_G/R_O$ where R_G, R_Y, and R_O are reflectance at green, yellow, and orange wavebands) to explain leaf chlorosis from acid deposition stress.

Immature fruit detection was conducted using hyperspectral imaging. Okamoto and Lee (2009) utilized a hyperspectral imaging system to detect green immature citrus fruit of three different varieties, and reported 80–89 % correct identification of foreground fruit by employing pixel discriminant function and spatial image processing steps.

For fruit maturity detection, Yang et al. (2012) investigated spectral signatures of blueberry fruit and leaves of seven different varieties by measuring their reflectance using a spectrophotometer in 200–2,500 nm, and reported that significant differences were identified among mature fruit, intermediate fruit, immature fruit, light-green leaf and dark-green leaf, enabling monitoring fruit maturity status, as shown in Fig. 11.7. They developed normalized vegetation indices to find significant bands for identifying different maturity status, used classification tree and

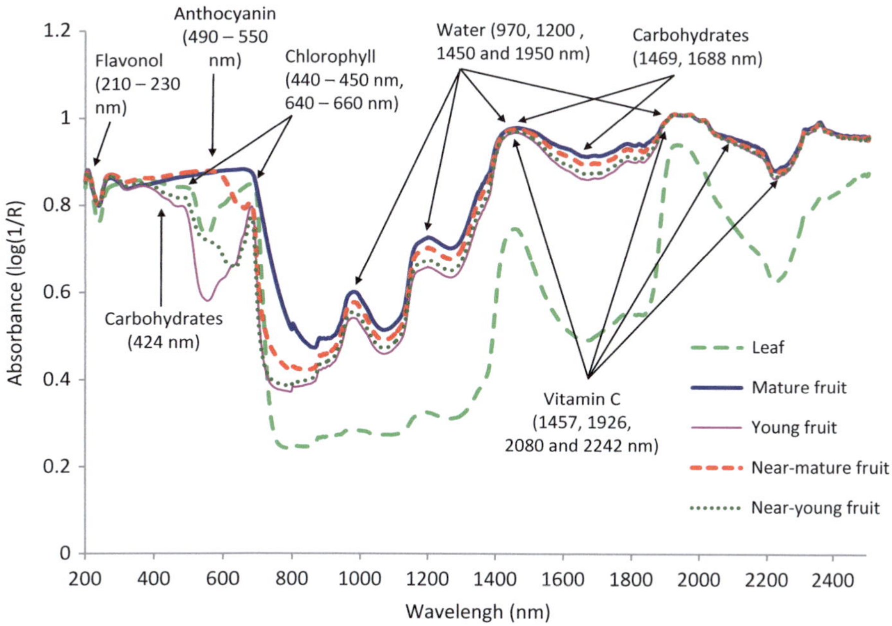

Fig. 11.7 Spectral signatures of different growth stages of blueberry fruit and leaves. All spectra are averages of seven southern highbush varieties (adapted from Yang et al. 2012)

multinomial logistic regression with principal components, and achieved 95–100 % prediction accuracy of fruit maturity status.

In conclusion, hyperspectral imaging provides a wide variety of applications for plant health detection and monitoring in agricultural crop production, and will continue to show great potential in future applications as hyperspectral sensing technology improves and equipment cost decreases.

References

Allen WA, Gausman HW, Richardson AJ, Thomas JR (1969) Interaction of isotropic light with a compact leaf. J Opt Soc Am 58(8):1023–1028

Balasundaram D, Burks TF, Bulanon DM, Schubert T, Lee WS (2009) Spectral reflectance characteristics of citrus canker and other peel conditions of grapefruit. Postharvest Biol Technol 51(2):220–226

Blackburn GA (2007) Hyperspectral remote sensing of plant pigments. J Exp Bot 58(4):855–867

Carroll MW, Glaser JA, Hellmich RL, Hunt TE, Sappington TW, Calvin D, Copenhaver K, Fridgen J (2008) Use of spectral vegetation indices derived from airborne hyperspectral imagery for detection of European corn borer infestation in Iowa corn plots. J Econ Entomol 101(5):1614–1623

Champagne CM, Staenz K, Bannari A, McNairn H, Deguise J-C (2003) Validation of a hyperspectral curve-fitting model for the estimation of plant water content of agricultural canopies. Remote Sens Environ 87:148–160

Cheng Y-B, Zarco-Tejada PJ, Riaño D, Rueda CA, Ustin SL (2006) Estimating vegetation water content with hyperspectral data for different canopy scenarios: relationships between AVIRIS and MODIS indexes. Remote Sens Environ 105:354–366

Delalieux S, Somers B, Verstraeten W, van Aardt JAN, Keulemans W, Coppin P (2009) Hyperspectral indices to diagnose leaf biotic stress of apple plants, considering leaf phenology. Int J Remote Sens 30(8):1887–1912

Goel PK (2003) Hyper-spectral remote sensing for weed and nitrogen stress detection. Ph.D. Dissertation, McGill University, Montreal

Haboudane D, Miller JR, Pattey E, Zarco-Tejada PJ, Strachan I (2004) Hyperspectral vegetation indices and novel algorithms for predicting green LAI of crop canopies: modeling and validation in the context of precision agriculture. Remote Sens Environ 90(3):337–352

Huang W, Lamb DW, Niu Z, Zhang Y, Liu L, Wang J (2007) Identification of yellow rust in wheat using in-situ spectral reflectance measurements and airborne hyperspectral imaging. Precis Agric 8:187–197

Jacquemoud S, Baret F (1990) PROSPECT: a model of leaf optical properties spectra. Remote Sens Environ 34:75–91

Jones CD, Jones JB, Lee WS (2010) Diagnosis of bacterial spot of tomato using spectral signatures. Comput Electron Agric 74(2):329–335

Kim Y, Glenn DM, Park J, Ngugi HK, Lehman BL (2010) Hyperspectral image analysis for plant stress detection. ASABE Paper No. 1009114. ASABE, St. Joseph

Kumar A, Lee WS, Ehsani R, Albrigo LG (2009) Airborne hyperspectral imaging for citrus greening disease detection. In: Proceedings of the 3rd Asian conference on precision agriculture (ACPA), Beijing

Kumar A, Lee WS, Ehsani R, Albrigo LG, Yang C, Mangan RL (2010) Citrus greening disease detection using airborne multispectral and hyperspectral imaging. In: 10th international conference on precision agriculture, Hyatt Regency Tech Center, Denver, 18–21 July 2010

Laudien R, Bareth G, Doluschitz R (2003) Analysis of hyperspectral field data for detection of sugar beet diseases. In: EFITA 2003 conference 5–9, Debrecen

Lee WS, Ehsani R, Albrigo LG (2008) Citrus greening disease (Huanglongbing) detection using aerial hyperspectral imaging. In: Proceedings of the 9th international conference on precision agriculture, Denver, 20–23 July

Lee WS, Alchanatis V, Yang C, Hirafuji M, Moshou D, Li C (2010) Sensing technologies for precision specialty crop production. Comput Electron Agric 74(1):2–33

Lenk S, Chaerle L, Pfündel EE, Langsdorf G, Hagenbeek D, Lichtenthaler HK, van Der Straeten D, Buschmann C (2007) Multispectral fluorescence and reflectance imaging at the leaf level and its possible applications. J Exp Bot 58(4):807–814

Li X, Lee WS, Li M, Ehsani R, Mishra A, Yang C, Mangan R (2011) Comparison of different detection methods for citrus greening disease based on airborne multispectral and hyperspectral imagery. ASABE Paper No. 1110570. ASABE, St. Joseph

Liu Z-Y, Huang J-F, Shi J-J, Tao R-X, Zhou W, Zhang L-L (2007) Characterizing and estimating rice brown spot disease severity using stepwise regression, principal component regression and partial least-square regression. J Zhejiang Univ Sci B 8(10):738–744

Liu Z-Y, Wu H-F, Huang J-F (2010a) Application of neural networks to discriminate fungal infection levels in rice panicles using hyperspectral reflectance and principal components analysis. Comput Electron Agric 72(2):99–106

Liu Z-Y, Shi J-J, Zhang L-W, Huang J-F (2010b) Discrimination of rice panicles by hyperspectral reflectance data based on principal component analysis and support vector classification. J Zhejiang Univ Sci B 11(1):71–78

Mahlein A-K, Steiner U, Dehne HW, Oerke EC (2010) Spectral signatures of sugar beet leaves for the detection and differentiation of diseases. Precis Agric 11:413–431

Mehl PM, Chao K, Kim M, Chen YR (2002) Detection of defects on selected apple cultivars using hyperspectral and multispectral image analysis. Appl Eng Agric 18(2):219–226

Min M, Lee WS (2005) Determination of significant wavelengths and prediction of nitrogen content for orange. Trans ASAE 48(2):455–461

Min M, Lee WS, Burks TF, Jordan JD, Schumann AW, Schueller JK, Xie H (2008) Design of a hyperspectral nitrogen sensing system for citrus. Comput Electron Agric 63(2):215–226

Mishra A, Ehsani R, Albrigo LG, Lee WS (2007) Spectroscopic study to identify citrus greening from other nutrient deficiencies. ASABE Paper No. 073056. ASABE, St. Joseph

Moshou D, Bravo C, Oberti R, West J, Bodria L, McCartney A, Ramon H (2005) Plant disease detection based on data fusion of hyper-spectral and multi-spectral fluorescence imaging using Kohonen maps. Real Time Imag 11:75–83

Muhammed HH (2002) Using hyperspectral reflectance data for discrimination between healthy and diseased plants, and determination of damage level in diseased plants. In: Proceedings of the 31st applied imagery pattern recognition workshop, Washington

Okamoto H, Lee WS (2009) Green citrus detection using hyperspectral imaging. Comput Electron Agric 66(2):201–208

Qin J, Burks TF, Kim MS, Chao K, Ritenour MA (2008) Citrus canker detection using hyperspectral reflectance imaging and PCA-based image classification method. Sens Instrum Food Qual 2:168–177

Qin J, Burks TF, Ritenour MA, Bonn WG (2009) Detection of citrus canker using hyperspectral reflectance imaging with spectral information divergence. J Food Eng 93:183–191

Ray SS, Jain N, Arora R, Chavan S, Panigrahy S (2011) Utility of hyperspectral data for potato late blight disease detection. J Indian Soc Remote Sens 39(2):161–169

Rumpf T, Mahlein A-K, Steiner U, Oerke E-C, Dehne H-W, Plümer L (2010) Early detection and classification of plant diseases with Support Vector Machines based on hyperspectral reflectance. Comput Electron Agric 74(1):91–99

Sankaran S, Mishra A, Ehsani R, Davis C (2010) A review of advanced techniques for detecting plant diseases. Comput Electron Agric 72(1):1–13

Schuerger AC, Capelle GA, Di Benedetto JA, Maoc C, Thai CN, Evans MD, Richards JT, Blank TA, Stryjewski EC (2003) Comparison of two hyperspectral imaging and two laser-induced fluorescence instruments for the detection of zinc stress and chlorophyll concentration in bahia grass (*Paspalum notatum* Flugge.). Remote Sens Environ 84(4):572–588

Singh CB, Jayas DS, Paliwal J, White NDG (2009) Detection of insect-damaged wheat kernels using near-infrared hyperspectral imaging. J Stored Prod Res 45:151–158

Song X, Jiang H, Yu S, Zhou G (2008) Detection of acid rain stress effect on plant using hyperspectral data in Three Gorges region, China. Chin Geogr Sci 18(3):249–254

Ustin SL, Roberts DA, Gamon JA, Asner GP, Green RO (2004) Using imaging spectroscopy to study ecosystem processes and properties. Bioscience 54(6):523–534

Vigier BJ, Pattey E, Strachan IB (2004) Narrowband vegetation indexes and detection of disease damage in soybeans. IEEE Geosci Remote Sens Lett 1(4):255–259

Williams AEP, Hunt ER Jr (2004) Accuracy assessment for detection of leafy spurge with hyperspectral imagery. J Range Manage 57(1):106–112

Wilson MD, Ustin SL, Rocke DM (2004) Classification of contamination in salt marsh plants using hyperspectral reflectance. IEEE Trans Geosci Remote Sens 42(5):1088–1095

Xing J, Symons S, Shahin M, Hatcher D (2010) Detection of sprout damage in Canada Western Red Spring wheat with multiple wavebands using visible/near-infrared hyperspectral imaging. Biosyst Eng 106:188–194

Yang C, Lee WS, Williamson JG (2012) Classification of blueberry fruit and leaves based on spectral signatures. Biosyst Eng 113(4):351–362

Zarco-Tejada PJ, Miller JR, Mohammed GH, Noland TL, Sampson PH (2002) Vegetation stress detection through chlorophyll a+b estimation and fluorescence effects on hyperspectral imagery. J Environ Qual 31:1433–1441

Zarco-Tejada PJ, Pushnik JC, Dobrowski S, Ustin SL (2003) Steady-state chlorophyll a fluorescence detection from canopy derivative reflectance and double-peak red edge effects. Remote Sens Environ 84:283–294

Zhang M, Qin Z, Liu X, Ustin SL (2003) Detection of stress in tomatoes induced by late blight disease in California, USA, using hyperspectral remote sensing. Int J Appl Earth Obs Geoinf 4(4):295–310

Chapter 12
Hyperspectral Imagery for Mapping Crop Yield for Precision Agriculture

Chenghai Yang

12.1 Overview of High Resolution Imagery for Yield Estimation

Airborne multispectral imaging systems provide image data at fine spatial resolutions from less than a meter to a few meters and at up to 12 narrow spectral bands in the visible to middle-infrared region of the spectrum. Airborne multispectral image data have been related to crop yield samples to generate yield maps (Richardson et al. 1990; Yang and Anderson 1999; Shanahan et al. 2001; Leon et al. 2003; Inman et al. 2008). The availability of both yield monitor data and remote sensing imagery allows the relations between yield and image data to be evaluated more thoroughly than the use of limited numbers of yield samples. Many researchers have evaluated the relationships between yield monitor data and airborne multispectral imagery (Senay et al. 1998; Yang et al. 2000; Yang and Everitt 2002; Dobermann and Ping 2004). Based on yield monitor data and multispectral imagery acquired from sorghum fields on multiple dates during a growing season, Yang and Everitt (2002) found that relations of yield to imagery reached the strongest around the peak vegetative development, indicating imagery taken during this period (approximately 1 month) would be the best indicator of yield for grain sorghum.

Imagery from high spatial resolution satellite systems such as IKONOS, QuickBird, and SPOT 5 has opened up new opportunities for mapping within-field variability. IKONOS, launched by Space Imaging (now part of GeoEye) in 1999, provides multispectral data in three visible bands and one near-infrared (NIR)

C. Yang (✉)
Aerial Application Technology Research Unit, U.S. Department of Agriculture, Agricultural Research Service, Southern Plains Agricultural Research Center, 3103 F and B Road, College Station, TX 77845, USA
e-mail: chenghai.yang@ars.usda.gov

B. Park, R. Lu (eds.), *Hyperspectral Imaging Technology in Food and Agriculture*, Food Engineering Series, DOI 10.1007/978-1-4939-2836-1_12

band at 4-m resolution. QuickBird, launched in 2001 by DigitalGlobe, delivers multispectral imagery with 2.4 m (at nadir) or 2.8 m (off nadir) resolution in four spectral bands similar to those of IKONOS. SPOT 5 was launched in 2002 to acquire multispectral data in two visible bands (green and red) and one NIR band at 10-m resolution and one short-wave infrared (SWIR) band at 20-m resolution. These satellite sensors have significantly narrowed the gap in spatial resolution between satellite and airborne imagery. Imagery from these sensors has been evaluated for assessing crop yields (Chang et al. 2003; Dobermann and Ping 2004; Yang et al. 2006, 2009). Newer satellite systems such as GeoEye and WorldView-2 provide multispectral data at even higher spatial resolution and have great potential for this type of applications.

Hyperspectral imaging sensors can collect image data in tens to hundreds of very narrow, continuous spectral bands throughout the visible, NIR, mid-infrared and thermal infrared portions of the spectrum. These systems offer new opportunities for better differentiation and estimation of biophysical attributes for a variety of remote sensing applications. Many commercial airborne hyperspectral sensors such as AVIRIS, CASI, HYDICE, HyMap, ASIA, and HySpex have been developed and used for various remote sensing applications. Advances in charge coupled device (CCD) cameras, frame grabber boards, and modular optical components have also led to developments of low-cost airborne hyperspectral imaging systems from off-the-shelf products (Mao 1999; Yang et al. 2003). Despite significant progress in airborne hyperspectral remote sensing, hyperspectral imagery has not been used as widely as multispectral imagery partially due to higher costs for image acquisition and special needs for handling and processing vast volumes of data.

Hyperspectral imagery provides additional information that multispectral data may have missed. Several researchers have evaluated airborne hyperspectral imagery for estimating crop yields. Goel et al. (2003) examined the potential of 72-band airborne hyperspectral imagery for estimating corn yield and other biophysical parameters and found that yield was significantly related to the image data taken at the tasseling stage. Yang et al. (2004a) applied stepwise regression analysis on grain sorghum yield monitor data and 102-band airborne hyperspectral imagery to identify optimum band combinations for mapping yield variability. They also used principal component analysis and stepwise regression to select the significant principal components to explain the variability in yield. To demonstrate the advantage of narrow hyperspectral bands over broad multispectral bands for yield estimation, Yang et al. (2004b) aggregated hyperspectral bands into Landsat-7 ETM+ sensor's four broad visible and NIR bands and found that the combinations of significant narrow bands explained more variability in cotton yield than the four broad bands. Zarco-Tejada et al. (2005) calculated a number of vegetation indices (VIs) using selected narrow bands from airborne hyperspectral imagery to estimate cotton yield.

Yang et al. (2007, 2008) derived all 5,151 possible narrow-band normalized difference vegetation indices (NDVIs) from 102-band hyperspectral images and related them to yield. They also applied spectral angle mapper and linear spectral unmixing techniques to the 102-band hyperspectral images to generate single-layer spectral angle images and plant cover fraction images, respectively, for mapping the variation in crop yields. Their results showed that the spectral angle images and plant cover fraction images provided better r-values with yield than the majority of the narrow-band NDVIs derived from the hyperspectral images. Ye et al. (2007) used partial least squares (PLS) regression models to predict the yields of citrus trees from their canopy features obtained from airborne hyperspectral imagery as compared with VIs and multiple linear regression models. Their results showed that VIs and multispectral regression models failed to predict citrus yield, but PLS models successfully predicted citrus yield with r-squared values of 0.51–0.90.

The rest of the chapter will describe the methods and procedures for hyperspectral image acquisition and processing and illustrate with an example how airborne hyperspectral imagery can be used for crop yield estimation based on the work by Yang et al. (2004a, 2007, 2008, 2010).

12.2 Hyperspectral Image Acquisition, Processing and Analysis

12.2.1 Image Acquisition

An airborne hyperspectral imaging system described by Yang et al. (2003) was used for image acquisition. The system consists of a digital CCD camera, a prism-grating-prism imaging spectrograph, and a PC equipped with a frame grabbing board and camera utility software. The CCD camera provides 1,280(h) $\times$ 1,024 (v) pixel resolution and 12-bit dynamic range. The imaging spectrograph is attached to the camera via an adapter to disperse radiation into a range of spectral bands. The effective spectral range resulting from this integration is from 457.2 to 921.7 nm. The hyperspectral imaging system is configured with a horizontal binning of 2 and a vertical binning of 8 to capture images with a 640-pixel width and 128 bands at a bandwidth of 3.63 nm.

Hyperspectral image acquisition requires careful planning and preparations. First, optimum exposure time and aperture settings were determined based on various imaging experiments so that the images for all bands were not too dark or saturated for ground cover conditions. These optimum settings were then set and used for all hyperspectral image acquisitions. A Cessna 206 single-engine and a Cessna 404 twin-engine aircraft were used as the platform for image acquisition. The hyperspectral imaging system was mounted on a light aluminum frame along with a three-camera color-infrared (CIR) imaging system. No stabilizer or inertial measurement device (IMU) was used to dampen or measure platform variations, but care was taken to minimize the effects of winds and changes in the aircraft's speed and flight direction. Images were acquired between 11:00 and 15:00 h local time under calm and sunny conditions.

For hyperspectral image acquisition, aircraft speed and flight height have to meet certain conditions to obtain images with square pixels and to avoid forward skip for this type of hyperspectral imaging systems. Based on the camera parameters and requirements for pixel size and ground coverage, the imaging system was flown at a height of 1,680 m above ground level and at a speed of 150 km/h. The aircraft was stabilized at the predetermined flight height, speed, and direction before taking each image and was maintained throughout the scanning process. The swath of the imagery was approximately 840 m and the ground pixel size achieved was 1.3 m.

12.2.2 Image Correction, Rectification and Calibration

The hyperspectral imaging system captures one line image for all the bands at a time and the aircraft serves as a mobile platform to carry out pushbroom scanning in the along-track direction. All the line images captured in a scanning process are combined to form a three-dimensional hyperspectral image. A moving aircraft has six degrees of freedom, that is, speed changes in the along-track direction, movements in the across-track direction, variations in altitude, pitch, roll, and yaw. The pitch refers to the movement of the airplane's nose either up or down. Roll is known as the rising or dipping of the airplane's wing. The yaw allows the airplane to move towards the left or right while in flight. Since no IMU was used to measure platform variations, a reference line approach described by Yang et al. (2003) was used for geometrical correction. A reference line, such as a straight field boundary or a road within the image approximately parallel to the flight line, is first identified and overlaid on the corresponding distorted line on the raw image. Then the distances in pixels between the reference line and the distorted line are determined for each row of the raw image. Finally, each row is shifted in the across-track direction by the number of pixels determined. Although this approach can only correct the distortions due to variations in roll and movements in the across-track direction, it has been used successfully for geometric correction if the variations in the other degrees of freedom are minimal. Besides, this approach requires a reference line approximately parallel to the flight line in the imaging area. This may not be a problem for crop fields because straight field boundaries can be found in most fields.

Geometrically corrected images were then rectified to the Universal Transverse Mercator (UTM), World Geodetic Survey 1984 (WGS-84), coordinate system using a set of ground control points collected with a submeter-accuracy GPS receiver or already georeferenced aerial photographs or airborne images covering the hyperspectral images. The hyperspectral images were generally resampled to 1 m pixel resolution using the nearest-neighbor algorithm during the rectification process. The root mean square (RMS) errors were within 3–4 m for the hyperspectral images based on first order polynomial transformations. If the RMS errors were over 4 m, a rubber sheeting procedure was applied for rectification.

For radiometric calibration of the images, three 8 m by 8 m tarpaulins with nominal reflectance values of 4 %, 32 %, and 48 %, respectively, were placed near the fields during image acquisition. Actual reflectance values from the tarpaulins were measured using a FieldSpec HandHeld spectroradiometer (Analytical Spectral Devices, Inc., Boulder, Colorado), which covers the spectral region of 350–1,050 nm. The rectified hyperspectral images were converted to reflectance based on 128 empirical line calibration equations relating reflectance values to the digital count values extracted from the three tarpaulins in the images. Because the hyperspectral camera had low quantum efficiency near the NIR end of the observed spectrum, the reflectance values for wavelengths greater than 845 nm dropped sharply. In addition, the first few bands in the blue region appeared to be noisy. Therefore, bands 1–5 and 108–128 (a total of 26 bands) were removed from each hyperspectral image and the remaining 102 bands with wavelengths from 475 to 845 nm were used for analysis.

12.2.3 Yield Data Collection

A Yield Monitor 2000 system (Ag Leader Technology, Ames, Iowa) was used to collect yield data from grain sorghum fields. A submeter AgGPS 132 receiver (Trimble Navigation Limited, Sunnyvale, California) was integrated with the yield monitor for position data collection. The yield monitor was calibrated to ensure data accuracy before data collection. Instantaneous yield and GPS data were simultaneously recorded at 1-s intervals. The data were then viewed, cleaned and exported in text files for further analysis using SMS Basic software (Ag Leader Technology, Ames, Iowa).

12.2.4 Calculation of Narrow-Band NDVIs

NDVI is typically formed from a NIR band and a red band. For a hyperspectral image, a large number of NDVIs can be derived. To generalize, a narrow-band NDVI-type index can be formed from any two different bands. Narrow-band NDVIs involving all possible two band combinations can be calculated using the following formula:

$$\mathrm{NDVI}_{i,j} = \frac{R_i - R_j}{R_i + R_j} \tag{12.1}$$

where R_i is the reflectance for band i, $i = 1, 2, \ldots, n-1$, $j = i+1, \ldots, n$, and $n =$ number of bands. The images from the airborne hyperspectral imaging system contained 102 usable spectral bands. Thus the number of NDVIs that can be derived from a 102-band hyperspectral image is $102!/(100!2!) = 5{,}151$.

12.2.5 Linear Spectral Unmixing

When a pixel is composed of one single surface component such as pure plants or pure soil, the spectrum from the pure pixel can be regarded as the signature for the component. However, if a pixel contains two or more components, the spectrum from the mixed pixel is a spectrally mixed response of all the components represented by the pixel. A hyperspectral image can be viewed as a collection of individual band images, and each image pixel contains a spectrum of reflectance values for all the wavebands in the image. These spectra can be regarded as the signatures of ground components such as crop plants or soil, provided that a component occupies the whole pixel. Spectra from mixed pixels can be analyzed with linear spectral unmixing, which models each spectrum in a pixel as a linear combination of a finite number of spectrally pure spectra of the components in the image, weighted by their fractional abundances (Adams et al. 1986). The unique ground components are referred to as endmembers with their unique spectra as endmember spectra. A simple linear spectral unmixing model has the following form:

$$y_i = \sum_{j=1}^{m} a_{ij}x_j + \varepsilon_i, \quad i = 1, 2, \ldots, n, \tag{12.2}$$

where

y_i is the measured reflectance in band i for a pixel,
a_{ij} is the known or measured reflectance in band i for endmember j,
x_j is the unknown cover fraction or abundance for endmember j,
ε_i is the residual between measured and modeled reflectance for band i,
m is the number of endmembers, and
n is the number of spectral bands.

Equation 12.2 is referred to as the unconstrained linear spectral unmixing model. For fully constrained linear spectral unmixing, the following additional conditions should be satisfied: (a) the abundance sum-to-one constraint, $\sum_{j=1}^{m} x_j = 1$ and (b) the abundance non-negativity constraint, $x_j \geq 0, \ j = 1, 2, \ldots, m$. Assuming that the endmembers are not linearly dependent, the fractions can be determined from the data.

The fractions determined by linear spectral unmixing might be preferred to band ratios and NDVIs for identifying spectrally defined components as it uses all the bands in the data (Bateson and Curtiss 1996). When linear spectral unmixing is applied to an image, it produces a suite of images of the fractions, one for each endmember in the model. Each fraction image shows the spatial distribution of the spectrally defined component as a NDVI image does. Different types of linear spectral unmixing have been used with multispectral and hyperspectral imagery for mapping the distributions of geological materials and vegetation types (Adams et al. 1986, 1995; Roberts et al. 1998; Lobell and Asner 2004). As VIs are indicators of plant vigor and canopy cover, the fractional abundance of crop plants determined from linear spectral unmixing is a more direct measure of plant cover and provides a more intuitive link between the image data and ground observations.

Linear spectral unmixing analysis requires the spectra of the known endmembers. They can be obtained directly from the image, measured on the ground or derived from a spectral library. In our studies, crop plants and bare soil were selected as the relevant endmembers. A pair of plant and soil spectra was extracted from each image to represent pure and healthy plants and bare soil, and was used as endmember spectra for spectral unmixing analysis for each field. To obtain pure spectra for crop plants, 100 pixels that had a bright red color on a CIR image (corresponding to healthy plants and high yielding areas) were first identified from each image. Similarly, 100 pixels that contained pure bare soil were identified from each image. The endmember spectra for plants and soil for each image were then obtained by averaging the spectra of the respective training pixels from that image. Both unconstrained and constrained linear spectral unmixing models were applied to each image, and four abundance images (two unconstrained and two constrained) were generated for each image.

12.2.6 Statistical Analysis

Considering the coarse yield data resolution and positional errors, the hyperspectral, NDVI, and abundance images were aggregated from 1-m pixels to 9-m pixels to be comparable with the effective cutting width of the harvester. The yield values were similarly averaged from the data points within each larger pixel area. Correlation coefficients with yield were calculated for the 5,151 NDVIs and the four abundance images for each hyperspectral image. Stepwise regression was used to relate yield to the 102 bands and to identify the significant bands for each image.

12.3 Relationship Between Yield and Airborne Hyperspectral Imagery

12.3.1 Hyperspectral Imagery and Yield Monitor Data

Figure 12.1 shows three band images (NIR, red and green) and their CIR composite extracted from a 102-band hyperspectral image for a 14-ha grain sorghum field (26°28′ 55″ N, 98°02′28″ W) in south Texas. The center wavelengths of the green, red, and NIR bands were 560.6, 629.6, and 829.2 nm, respectively. The hyperspectral image was taken around the peak plant development for the crop. Both the individual band images and the CIR image reveal distinct spatial patterns of plant growth. On the NIR band image, healthy plants have a light gray color,

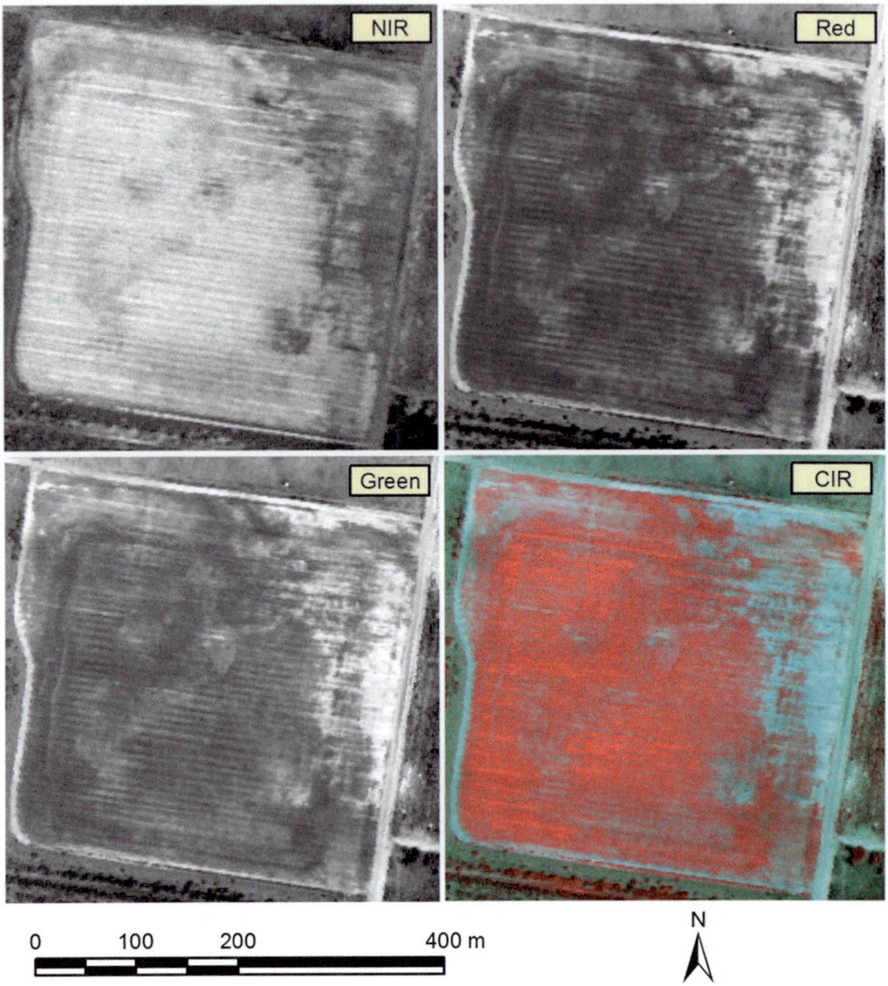

Fig. 12.1 *Black*-and-*white* images for a near-infrared (NIR) band, a *red band*, and a *green band* and their color-infrared (CIR) composite extracted from a 102-band hyperspectral image for a 14-ha grain sorghum field in south Texas

while areas with more bare soil exposure have a dark gray tone. On the red and green band images, healthy plants have a dark gray color, while areas with more bare soil exposure have a light gray tone. On the CIR image, healthy plants have a reddish response, while the problem areas exhibit a bluish color. The problem areas in the field were mainly due to the very sandy soil. Plants in those areas had poor stand and low canopy cover because of the low water and nutrient holding capability of the sandy soil.

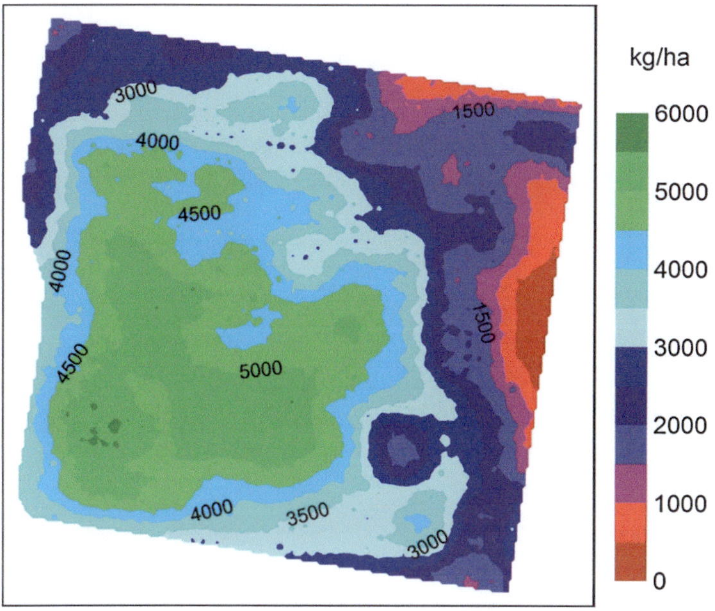

Fig. 12.2 Yield map generated from yield monitor data for a grain sorghum field

Figure 12.2 presents the yield map generated from yield monitor data for the field. Yield varied from essentially 0 to approximately 6,000 kg/ha with a mean of 3,440 kg/ha and a standard deviation of 1,480 kg/ha, indicating there existed large variability in yield within the field.

12.3.2 Yield Correlations with Narrow-Band NDVIs

Figure 12.3 shows a NDVI map derived from a NIR band (800 nm) and a red band (668 nm) for the field. NDVI ranges from 0.1 to 0.8 with the low NDVI values corresponding to low yield and high NDVI values to high yield. The correlation coefficient between yield and the NDVI was 0.83.

Figure 12.4 shows a contour map of absolute r-values between yield and all 5,151 possible NDVIs derived from the 102-band image for the field. The contour map is symmetrical about the diagonal line and clearly illustrates the r-value distributions for all the band pairs. The absolute r-values vary from essentially 0 for the band pair of 778 and 822 nm to 0.88 for the band pair of 742 and 789 nm. The median of the r-values is 0.82, indicating 50 % of the NDVIs have r-values higher than 0.82. The r-values are generally high (>0.825) when one band has wavelengths below 730 nm and the other band has wavelengths above 730 nm for this particular field. However, the best r-values (>0.85) occur when one band has

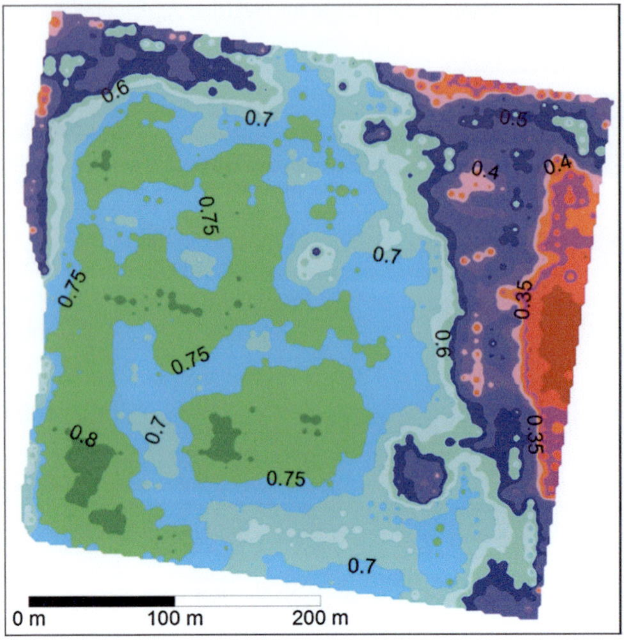

Fig. 12.3 NDVI map derived from a NIR band (800 nm) and a red band (668 nm) from a 102-band hyperspectral image for a grain sorghum field

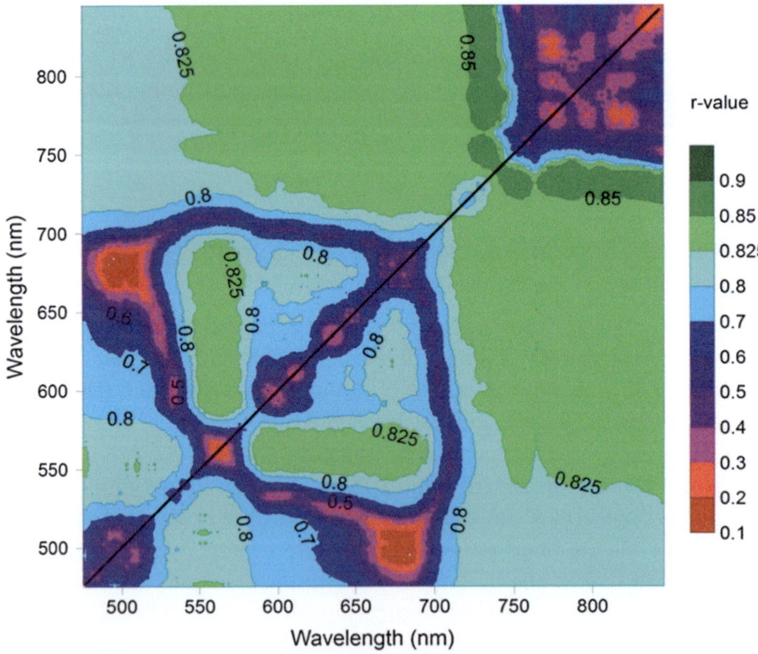

Fig. 12.4 Contour map showing absolute correlation coefficients between yield and all possible narrow-band NDVIs derived from a 102-band airborne hyperspectral image for a grain sorghum field. When band i = band j, $NDVI_{ij} = 0$ and r values do not exist (shown by the *diagonal line*)

wavelengths between 730 and 750 nm and the other band has wavelengths above 760 nm. Also large *r*-values (>0.825) occur when one band in a pair has wavelengths between 550 and 575 nm and the other has wavelengths between 575 and 690 nm. Based on the contour map of *r*-values, better NDVI images are more likely to be obtained by selecting one band in the visible region and the other in the NIR region.

It should be noted that the best NDVI identified for this field may not be the best for another field because of different field conditions. For example, Yang et al. (2008) found that the center wavelengths that resulted in the highest r-value for another field was 543 and 728 nm. Nevertheless, a NDVI calculated from a visible band and a NIR band can generally be a better representation than that derived from two visible bands or two NIR bands despite the fact that band pairs from either visible or NIR bands can result in better r-values.

12.3.3 Yield Relation with Plant Abundance

Table 12.1 gives the univariate statistics of unconstrained and constrained plant and soil abundance fractions derived from the hyperspectral image for the field. Ideally, abundance fraction values should be within the 0–1 range, but in unconstrained fraction images they can be negative or exceed 1. For example, the unconstrained plant abundance ranges from −0.15 to 1.01 and the unconstrained soil abundance ranges from 0.02 to 1.16 for the field. This is because spectral unmixing results can be affected by the purity of the endmembers and the number of endmembers. The linearity assumption of linear spectral unmixing is at best an approximation of the generally non-linear mixing of surface components. The fully constrained fractions have values in the range of 0–1. Figure 12.5 presents a constrained plant abundance fraction image derived from the hyperspectral image for the field. Red areas have small plant abundance values and represent pixels with large soil exposure and sparse plant cover. Conversely, green areas indicate large plant abundance values and represent pixels with dense plant cover.

Table 12.1 Univariate statistics of unconstrained and constrained plant and soil abundance fractions derived from a 102-band airborne hyperspectral image for a grain sorghum field based on a pair of plant and soil endmember spectra extracted from the image

Endmember fraction	Mean	Standard deviation	Minimum	Maximum
UPF[a]	0.63	0.28	−0.15	1.01
USF	0.32	0.25	0.02	1.16
CPF	0.64	0.26	0.00	1.00
CSF	0.36	0.26	0.00	1.00

[a]*UPF* unconstrained plant fraction, *USF* unconstrained soil fraction, *CPF* constrained plant fraction with the sum-to-one and positivity constraints, *CSF* constrained soil fraction with the sum-to-one and positivity constraints

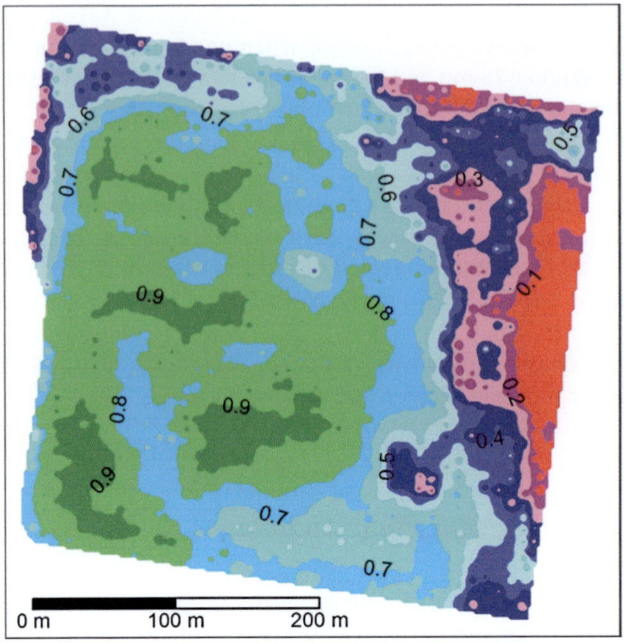

Fig. 12.5 Constrained plant abundance fraction images derived from a 102-band airborne hyperspectral image for a grain sorghum field based on a pair of plant and soil endmember spectra extracted from the image

The mean unconstrained plant and soil abundance fractions are 0.63 and 0.32, respectively, indicating mean plant canopy cover was approximately 63 % at the time of the image acquisition. Although the unconstrained model does not force the endmember abundance fractions to sum to 1, the sum (0.95) is close to 1, indicating the unconstrained two-endmember linear unmixing model is appropriate for characterizing plant and soil cover in the images. The mean constrained plant and soil abundance fractions are 0.64 and 0.36, respectively, with a sum of 1 as expected.

Yield is positively related to unconstrained and constrained plant abundance fractions, and negatively related to the unconstrained and constrained soil abundance fractions. Unconstrained plant abundance fractions have slightly stronger correlations with yield than the unconstrained soil abundance fractions, whereas constrained plant and soil abundance fractions have identical absolute correlations because they sum to unity. The correlation coefficient for the unconstrained plant abundance fraction is 0.85, whereas the r-value for the unconstrained soil abundance fraction is −0.82 for the field. The *r*-values for the fully constrained plant and soil abundance fractions are both 0.85.

The best NDVI has a larger r-value (0.88) with yield than the best abundance fraction for this field. Nevertheless, the best abundance fraction-based *r*-value (0.85) is better than 97.1 % of the 5,151 NDVI-based *r*-values for the field. If the objective of a study is to determine the best correlation based on actual yield data, all possible narrow-band NDVIs could be derived to identify the best NDVI. However, if the objective is to generate a spectral map from a hyperspectral image to characterize the spatial variation in yield without knowing the actual yield, an unconstrained or constrained plant fraction image based on a pair of plant and soil spectra will be a better choice. This is because a NDVI image uses only two narrow bands, whereas a plant fraction image is based on all bands in the image. Moreover, the best NDVI can only be identified if the yield is known and all possible narrow-band NDVIs are calculated. On the other hand, the plant fraction image can be generated using all the bands and a pair of plant and soil endmember spectra without the need to know the actual yield.

12.3.4 Stepwise Regression of Yield with Image Bands

Table 12.2 summarizes the stepwise regression statistics for multiple linear models relating grain yield to the 102 bands. Seven of the 102 bands were identified to be significant in the final regression equation. The best single band was 782 nm and explained 71 % of variability in yield. The best two-band combination was 738 and 782 nm and the R^2-value increased by 7.6 % with the addition of the second band. The best three-band combination was 713, 731, and 782 nm and explained 80.4 % of the variability. The R^2-values only improved by 2 % with the addition of the third band and subsequent significant bands contributed even less. All seven significant bands explained about 82 % of the variability in yield. It should be noted that these optimum bands were only the best for the image and yield data from which they were derived and might not be the best for different datasets. For example, in the same study, Yang et al. (2004a) identified four significant bands for another grain sorghum field, which were completely different from the seven significant bands identified for the 14-ha field.

Table 12.2 Stepwise regression results for relating grain yield to 102 bands of a hyperspectral image for a grain sorghum field

No. of bands	Significant wavelength[a] (nm)	Model R^2
1	782	0.709
2	738, 782	0.785
3	713, 731, 782	0.804
4	481, 713, 731, 782	0.808
5	481, 543, 713, 731, 782	0.817
6	481, 543, 713, 731, 735, 782	0.819
7	481, 543, 713, 731, 735, 771, 818	0.824

[a]Stepwise regression was applied to a full linear model with 102 variables (wavebands). The best fitting one-, two-, . . ., and seven-variable models and all the variables remaining in the models were significant at the 0.0001 level. No other variable could be added to the seven-variable model and remain significant at the 0.0001 level

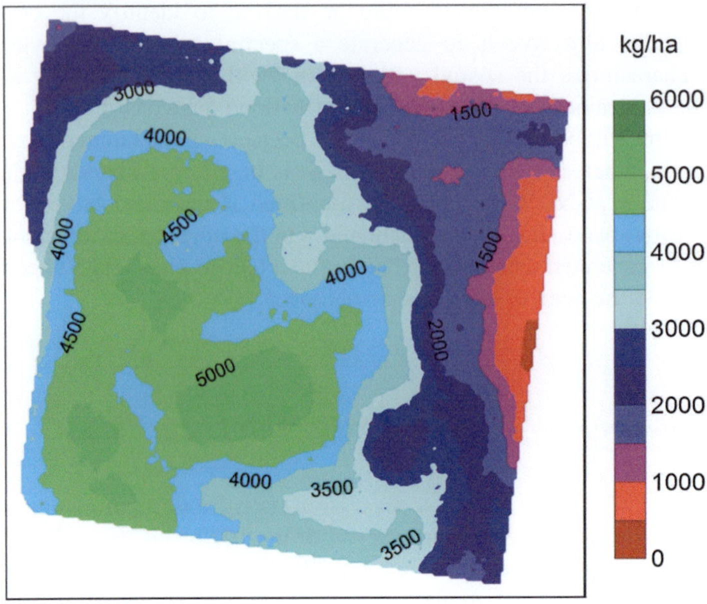

Fig. 12.6 Yield map generated based on seven significant bands in a 102-band hyperspectral image for a 14-ha grain sorghum field

Figure 12.6 presents a yield map generated from the hyperspectral image with the regression equation relating yield to the seven significant bands. The spatial pattern displayed on the yield map is similar to that from yield monitor data (Fig. 12.2). Regression analysis can be used to identify the best NDVI and the best combinations of bands for estimating yield if yield data samples are available. However, if yield data are not available, either a NDVI image derived from a NIR band and a visible band or a plant abundance image derived from the all the bands can convert a hyperspectral image to a single-layer image to represent relative yield.

12.4 Summary

This chapter has provided an overview of high resolution remote sensing imagery for mapping crop yield variability, described the methods and procedures for hyperspectral image acquisition, processing and analysis, and illustrated how airborne hyperspectral imagery can be used for crop yield estimation based on narrow-band NDVIs, stepwise regression, and linear spectral unmixing.

Research has demonstrated that high resolution remote sensing imagery can be useful for estimating and mapping within-field crop yield variability. Both multispectral and hyperspectral images can be used to determine the spatial patterns in plant growth and yield before harvest. Nevertheless, hyperspectral imagery has the potential to provide additional information that multispectral data may have missed. Depending on the availability of yield sampling data, hyperspectral imagery can be used to generate absolute or relative yield maps to characterize spatial variations in crop yields. These maps are important for site-specific crop management in precision agriculture. As more high resolution multispectral and hyperspectral imagery, including the high resolution multispectral imagery from newly launched satellite sensors (i.e., GeoEye-1 and WorldView-2), is becoming more available and less expensive, more research is needed to compare different types of image data and analysis techniques for yield estimation and other precision agriculture applications for different crops and growing conditions.

References

Adams JB, Smith MO, Johnson PE (1986) Spectral mixture modeling: a new analysis of rock and soil types at the Viking Lander 1 site. J Geophys Res 91:8098–8112

Adams JB, Sabol DE, Kapos V, Filho RA, Roberts DA, Smith MO, Gillespie AR (1995) Classification of multispectral images based on fractions of endmembers: application to land-cover change in the Brazilian Amazon. Remote Sens Environ 52:137–154

Bateson A, Curtiss B (1996) A method for manual endmember and spectral unmixing selection. Remote Sens Environ 55:229–243

Chang J, Clay DE, Dalsted K, Clay S, O'Neill M (2003) Corn (*Zea mays* L.) yield prediction using multispectral and multidate reflectance. Agron J 95:1447–1453

Dobermann A, Ping JL (2004) Geostatistical integration of yield monitor data and remote sensing improves yield maps. Agron J 96:285–297

Goel PK, Prasher SO, Landry JA, Patel RM, Viau AA, Miller JR (2003) Estimation of crop biophysical parameters through airborne and field hyperspectral remote sensing. Trans ASAE 46(4):1235–1246

Inman D, Khosla R, Reich R, Westfal DG (2008) Normalized difference vegetation index and soil color-based management zones in irrigated maize. Agron J 100:60–66

Leon CT, Shaw DR, Cox MS, Abshire MJ, Ward B, Wardlaw MC, Watson C (2003) Utility of remote sensing in predicting crop and soil characteristics. Precis Agric 4(4):359–384

Lobell DB, Asner GP (2004) Cropland distributions from temporal unmixing of MODIS data. Remote Sens Environ 93:412–422

Mao C (1999) Hyperspectral imaging systems with digital CCD cameras for both airborne and laboratory application. In: Tueller PT (ed) Proceedings of the 17th Biennial Workshop on color photography and videography in resource assessment, Reno, NV, 5–7 May 1999. American Society of Photogrammetry and Remote Sensing, Bethesda, pp 31–40

Richardson AJ, Heilman MD, Escobar DE (1990) Estimating grain sorghum yield from video and reflectance based PVI measurements at peak canopy development. J Imag Technol 16 (3):104–109

Roberts DA, Gardner M, Church R, Ustin S, Scheer G, Green RO (1998) Mapping chaparral in the Santa Monica Mountains using multiple endmember spectral mixture models. Remote Sens Environ 65:267–279

Senay GB, Ward AD, Lyon JG, Fausey NR, Nokes SE (1998) Manipulation of high spatial resolution aircraft remote sensing data for use in site-specific farming. Trans ASAE 41 (2):489–495

Shanahan JF, Schepers JS, Francis DD, Varvel GE, Wilhelm WW, Tringe JM, Schlemmer MR, Major DJ (2001) Use of remote sensing imagery to estimate corn grain yield. Agron J 93:583–589

Yang C, Anderson GL (1999) Airborne videography to identify spatial plant growth variability for grain sorghum. Precis Agric 1(1):67–79

Yang C, Everitt JH (2002) Relationships between yield monitor data and airborne multidate multispectral digital imagery for grain sorghum. Precis Agric 3(4):373–388

Yang C, Everitt JH, Bradford JM, Escobar DE (2000) Mapping grain sorghum growth and yield variations using airborne multispectral digital imagery. Trans ASAE 43(6):1927–1938

Yang C, Everitt JH, Davis MR, Mao C (2003) A CCD camera-based hyperspectral imaging system for stationary and airborne applications. Geocarto Int J 18(2):71–80

Yang C, Everitt JH, Bradford JM (2004a) Airborne hyperspectral imagery and yield monitor data for estimating grain sorghum yield variability. Trans ASAE 47(3):915–924

Yang C, Everitt JH, Bradford JM, Murden D (2004b) Airborne hyperspectral imagery and yield monitor data for mapping cotton yield variability. Precis Agric 5(5):445–461

Yang C, Everitt JH, Bradford JM (2006) Evaluating high resolution QuickBird satellite imagery for estimating cotton yield. Trans ASAE 49(5):1599–1606

Yang C, Everitt JH, Bradford JM (2007) Airborne hyperspectral imagery and linear spectral unmixing for mapping variation in crop yield. Precis Agric 8(6):279–296

Yang C, Everitt JH, Bradford JM (2008) Yield estimation from hyperspectral imagery using Spectral Angle Mapper (SAM). Trans ASABE 51(2):729–737

Yang C, Everitt JH, Bradford JM (2009) Evaluating SPOT 5 multispectral imagery for crop yield estimation. Precis Agric 10(4):292–303

Yang C, Everitt JH, Du Q (2010) Applying linear spectral unmixing to airborne hyperspectral imagery for mapping yield variability in grain sorghum and cotton fields. J Appl Remote Sens 4:041887

Ye X, Sakai K, Manago M, Asada S, Sasao A (2007) Prediction of citrus yield from airborne hyperspectral imagery. Precis Agric 8:111–125

Zarco-Tejada PJ, Ustin SL, Whiting ML (2005) Temporal and spatial relationships between within-field yield variability in cotton and high-spatial hyperspectral remote sensing imagery. Agron J 97:641–653

Chapter 13
Real-Time Hyperspectral Imaging for Food Safety

Bosoon Park and Seung-Chul Yoon

13.1 Introduction

Hyperspectral imaging (HSI) is an emerging platform technology that integrates conventional imaging and spectroscopy to attain both spatial and spectral information from an object. Although HSI was originally developed for remote sensing, it has recently emerged as a powerful process analytical tool for non-destructive food analysis (Gowen et al. 2007). This technique is capable of providing an absolute radiometric measurement over a contiguous spectral range for every pixel of an image. Thus, data from a hyperspectral image contain two–dimensional spatial information, as well as spectral information. These data are considered as a three–dimensional hypercube, or datacube that can provide physical and/or chemical information on an object or material under test. This information includes physical and geometric observations of size, orientation, shape, color, and texture, in addition to chemical/molecular information such as water, fat, proteins, and other hydrogen–bonded constituents. Since hyperspectral imaging has been developed as a powerful technique in earth remote sensing during past decades, this technique continues to be utilized in medical, biological, agricultural, and industrial areas.

Although hyperspectral imaging technology was mainly used in the field of remote sensing, recently hyperspectral imaging technology is emerging on food quality (Kim et al. 2001; Ariana et al. 2006; Lu and Peng 2006; Nicolai et al. 2006; El Masry et al. 2007; Qiao et al. 2007) and safety (Park et al. 2002; Kim et al. 2004; Yoon et al. 2010) evaluation with a unique platform that integrates spectral as well

B. Park (✉)
U.S. Department of Agriculture, Agricultural Research Service, Athens, GA, USA
e-mail: bosoon.park@ars.usda.gov

S.-C. Yoon
U.S. Department of Agriculture, Agricultural Research Service, U.S. National Poultry Research Center, Athens, GA 30605, USA
e-mail: seungchul.yoon@ars.usda.gov

B. Park, R. Lu (eds.), *Hyperspectral Imaging Technology in Food and Agriculture*,
Food Engineering Series, DOI 10.1007/978-1-4939-2836-1_13

as spatial information of the scene. Although hyperspectral imaging technology has the potential for food safety inspection and quality control, there exist several limitations to widely adoption in the food industry. The most limiting factors are a relatively longer time necessary for data acquisition, processing, and costs. Since process control in food industry is crucial to maintain quality and safety, it would be beneficial to have an instrument that is able to detect directly the spatial distribution of various chemical and biological components from a surface in real-time. To accomplish this, a real-time hyperspectral scanning method is needed. However, the challenge with hyperspectral cameras is that the datacubes are too large to process in real-time. Therefore, minimizing the datacube by eliminating redundant data from the chemometric model, which is applicable directly every time the camera collects spectra is needed for real-time hyperspectral imaging platform (Kester et al. 2011).

For biological hyperspectral imaging, Fletcher-Holmes and Harvey (2005) demonstrated a new, biologically inspired approach, in which a compact hyperspectral fovea is embedded within a conventional panchromatic periphery. The system enables hyperspectral imaging to be applied only to small regions of interest previously identified using the panchromatic periphery with an intelligent scanning system, resulted in improving the efficiency for object recognition. A Real-time hyperspectral imaging endoscope has also been developed for resolving a vasculature pattern of the lower lip, while simultaneously detecting oxy-hemoglobin based on an image mapping technique capable of addressing parallel high throughput nature, which enables the system to operate at the rate of 5.2 frames per second with a datacube size of $350 \times 350 \times 48$. The most challenging obstacle for real-time hyperspectral imaging is how to record a three-dimensional datacube with a conventional two-dimensional detector array, and how to most efficiently transmit the spectral datacube through the information bottleneck constituted by the detector's limited space (Fletcher-Holmes and Harvey 2005). For another real-time hyperspectral application, the Dark HORSE 1 (Hyperspectral Overhead Reconnaissance and Surveillance Experiment 1) flight tests have demonstrated autonomous, real-time visible hyperspectral detection of military ground targets, with real-time cuing of a high-resolution framing camera (Stellman et al. 2000).

13.2 Imaging Platform for Real-Time Applications

Prior to real-time hyperspectral imaging technology, more adequate real-time processing was possible with a multispectral imaging system with pre-selected and pre-assembled platforms such as a common aperture camera (Park et al. 2006). For instance, it was needed to develop an online real-time poultry inspection system to detect condemned birds due to either feces or diseases, in order to increase production yield by eliminating or re-washing birds. Researchers have studied hyperspectral imaging methods for fecal detection and found three key wavelengths of 517, 565 and 802 nm, for maximizing detection accuracy on surface fecal contamination (Park et al. 2002).

The detection algorithm for fecal matter with selected band ratios using 567-nm/517-nm, followed by an advanced fecal detection algorithm using the additional band ratios of 802-nm/517-nm was developed to increase the overall performance of fecal detection by reducing false positive errors (Park et al. 2006). However, the development of real-time multispectral imaging systems with selected bands poses a challenge to both researchers and real-time system designers, because the hyperspectral imaging system platform used for research was not appropriate for high-speed real-time multispectral imaging applications. In the past, a real-time imaging system using a common aperture camera for pre-selected bands was developed (Park et al. 2004). However, the disadvantage of a common aperture multispectral imaging platform is that once the optic lens with selected bands is assembled, it is impossible to replace them with other bands, without replacing all optical components, which are expensive. Therefore, the development of a real-time hyperspectral imaging platform will benefit to researchers and food industry.

13.2.1 Real-Time Multispectral Imaging

Real-time multispectral imaging technology is a candidate for food safety inspection in food industry. For example, fecal contamination on broiler carcasses is the primary avenue for contamination by foodborne pathogens, because pathogens may reside in fecal materials potentially found in both the gastrointestinal tract and exterior surfaces of the poultry carcasses. During slaughter and processing, the edible portions of the carcass can be contaminated with bacteria causing illness in human. For science-based food safety inspection in the food industry, researchers have developed multispectral imaging technology for online contaminant detection during food processing (Windham et al. 2003a, b; Park et al. 2004). Based on the principles and technologies from real-time multispectral imaging, the researchers developed the common aperture multispectral imaging system (Park et al. 2004) and portable multispectral imaging system (Kise et al. 2007) for online applications of fecal contaminant detection. After several upgrades of the system hardware and software, the real-time online multispectral imaging system (Park et al. 2007a) was developed and tested at commercial poultry processing plants. The prototype real-time multispectral imaging system utilized a common aperture camera, with three selected optical filters that were pre-installed into the camera assembly.

Pilot-scale tests demonstrated the potential of multispectral imaging system at poultry processing plants for detection of fecal and ingesta contamination with commercial processing speed of 140 birds per minute. However, the challenges for real-time imaging system development are to eliminate false positive errors caused by variability of carcasses and instability of detectors. To eliminate false positive errors, both hardware and software solutions are needed. For hardware solutions, an additional optical interference filter such as 802 nm can be added (Heitschmidt et al. 2007), which can remove most false positive errors caused by the carcasses cuticles. However, adding the filter to existing common aperture camera assembly is not cost-effective and has additional technical difficulties. Therefore, although their limitation for improving performance to eliminate false errors, various algorithms including dynamic thresh-holding methods (Park et al. 2005) such as fisher linear discriminant analysis (Park et al. 2007b), kernel density estimation (Yoon et al. 2007), and textural analysis of hyperspectral images (Park et al. 2008) were examined for software solutions. The implication of those algorithms was not easy for real-time in-line system because of limited image processing time for the task. Recently, an adaptive image processing method has been tested with a success to reduce false positive errors, particularly cuticle filtering (Park et al. 2009). This image processing method for real-time application included band ratio, binning, cuticle filtering, median filtering as well as morphological image processing. In-plant tests have also demonstrated the real-time multispectral imaging system performed with over 91 % detection accuracy and 3.3 % false positive errors at the processing speed of 150 birds per minute. Thus, the performance of a real-time on-line imaging system for contaminant detection on food can be improved by adopting a high quality camera, which has a higher signal-to-noise ratio, such as line-scan real-time hyperspectral imaging system.

13.2.2 Real-Time Hyperspectral Imaging

Researchers have developed a spectral line-scan imaging system for high-speed non-destructive inspection of wholesome broilers (Chao et al. 2007, 2010), and a real-time online multispectral imaging system for fecal contaminant detection, during poultry processing (Park et al. 2004). Due to the similarity of technology and industry demands, a common platform imaging system that has the ability of detecting both wholesomeness and contaminants (Park et al. 2011), benefits the poultry industry in terms of food safety inspection and quality control. In doing those two different tasks, a line-scan hyperspectral imaging platform can be used for systemic diseased carcass inspection and contaminant detection without any system hardware modification required at poultry processing plants. Thus, the advantage of a real-time hyperspectral imaging system is that the system is able to conduct multiple tasks with simple appropriate image processing modules (or system software) that have already been implemented in the system. Therefore, a line-scan hyperspectral imaging system with appropriate image processing algorithms has the potential for real-time applications in the food industry.

In general, a real-time hyperspectral imaging system consists of a line-scan hyperspectral camera; including an imaging spectrograph, a camera sensor, an objective lens, lighting sources (tungsten-halogen or LED, depending on band selection), power supplies, and a computer for camera control and image acquisition. The line-scan real-time hyperspectral imaging system performed with data binning utilizes the unique feature of the electron-multiplying charge coupled device (EMCCD) sensor for random access to user-defined areas on the CCD sensor, and multitasking software that is crucial for system customization for various applications. Both hardware and software of the system are major considerations for designing and implementing real-time hyperspectral and multispectral imaging system. For system hardware, the challenge is to determine whether a given hardware platform could produce quality images and quantity data that are good enough for real-time applications. In contrast, for system software, the challenge is determining how to implement algorithms developed at off-line computing environments for application in a real-time on-line version.

Although real-time multispectral imaging solutions based on line-scan hyperspectral imaging platforms are not quite common, line-scan based hyperspectral imaging systems have been used for several applications; such as industrial polymer sorting (Leitner et al. 2003), cellulose-based material sorting (Tatzer et al. 2005), apple sorting (Kim et al. 2007; Noh and Lu 2007), diseased chicken sorting (Chao et al. 2007) and contaminant detection for poultry carcasses (Park et al. 2011). When real-time hyperspectral imaging systems are developed, several steps such as real-time application requirements, design methodology and strategies, hardware platforms, software architecture, and algorithm implementations need to be prepared.

13.3 High-Speed Hyperspectral Imaging System

The transportable hyperspectral imaging system can be designed and assembled for food industry application as shown in Fig. 13.1. Specifically, for food production inspection applications, a pushbroom hyperspectral imaging instrument operates similar to a line scanner, producing spectral responses aligned with the target objects to be scanned.

13.3.1 Real-Time Hyperspectral Imaging Platform

The common platform for real-time hyperspectral imaging to scan moving objects is a pushbroom hyperspectral imager for real-time monitoring objects moving in the shackles or conveyer belts during processing. As shown in Fig. 13.1, the

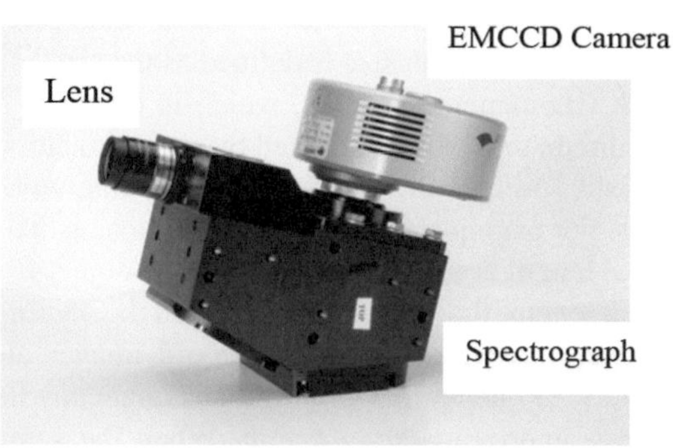

Fig. 13.1 Line-scan hyperspectral camera attached with spectrograph

hyperspectral imager can be consisted of a spectrograph (HyperSpec VNIR, Headwall Photonics, Fitchburg, MA, USA), an electron-multiplying charge coupled device (EMCCD) detector (Luca-R, Andor Technology, Belfast, UK) and an objective lens (CNG f-1.4/12 mm, Schneider Optics, Hauppauge, NY, USA). The spectrograph and the detector are usually pre-aligned by the spectrograph manufacture for the end-users without any further adjustment. In order to construct a two-dimensional scan area, samples moving in shackles need to be scanned while the imaging system is stationary. The hyperspectral imager is installed on top of a monopod to scan surface of objects hung in moving shackles line by line. As a result, the long side of the spectrograph's entrance slit is perpendicular to the movement direction of shackles, and running parallel to the scanning direction so that lights passed through the line slit are dispersed onto the image sensor, at which one axis of the two-dimensional sensor array is the spatial dimension and the other axis is the spectral dimension. An industrial portable computer is usually used for camera control and software operation in field applications.

The spectrograph operates in the visible and near-infrared (VNIR) wavelength range between 400 and 1,000 nm, and short wavelength infrared (SWIR) is also available up to 2,500 nm depending on the applications. The spectrograph is based on holographic diffraction gratings and aberration-corrected to minimize spatial and spectral image distortion due to smile and keystone effects. Therefore, no additional fixation for aberration problems is needed to increase the spatial and spectral image fidelity. The entrance slit (or thin-line opening) of the spectrograph can be 18 mm high and 40 μm wide, resulting in 4 nm spectral resolution (measured as full width at half maximum, FWHM) of the spectrograph. The EMCCD detector is a monochrome megapixel camera and thermoelectrically cooled (−20 °C). The EMCCD camera has a USB 2.0 port and 1004 × 1002 pixels (8 × 8 μm square pixels on an 8 × 8 mm sensor) with 14-bit digitization. The frame rate of the camera is 12.4 frames/s at the full-frame resolution. The pixel readout rate of the camera is 13.5 MHz. The lens mounted in front of the spectrograph's entrance slit is a compact C-mount lens with 12-mm focal length, and suitable for 1/2- and 2/3-in. imager.

13.3.2 Line Scan Rate of Hyperspectral Imager

The minimum scan size is defined as the size of the smallest area that a pixel on the EMCCD camera can map, which is different from a sample size. Therefore, the minimum scan size is affected by several factors such as the focal length of the lens, the slit width of the spectrograph, and the working distance to an object from the lens. For example, if the working distance is fixed with 48 cm, the lens with 12-mm focal length is able to fully sample the object from top to bottom. When the 12-mm focal length, the 40-μm slit width and the 48 cm working distance are used, the field of view (FOV) of the hyperspectral imager along with the scan line is approximately 32 cm. In this case, the instantaneous field of view (IFOV) across the scan line is approximately 1.6 mm. When the samples are not moving during image acquisition, the area scanned by the imager is long narrow rectangle with the dimension 1.6 mm (width) × 32 cm (height). Hence, one pixel on the EMCCD is mapped to an area with the dimension 1.6 mm (width) × 0.3 mm (height) that is also the minimum scan size, which has an aspect ratio of 5 and minimum scan area of 0.48 mm^2 that is mapped to a square pixel on the imager.

When the lateral movement of objects in the shackles is considered, a scanning rate is an important parameter for system design, to prevent a under sampling problem. For instance, if the minimal scan size is already known from a sample, and the minimum scanning rate is 95 scans per sample (i.e. 1 scan per every 1.6 mm distance) at 140 samples per minute from 15.24 cm/1.6 mm, and the number of scans is less than 95, the area becomes under sampled (scanning less than target sample). In practice, the over sampling is usually recommended because it will increase the detectability of small features by increasing spatial resolution. Usually an optimum scanning rate is an additional 25 % more than actual scan, considering the size. If 25 % additional scans are added, the number of scans becomes 119 per sample, so that the line-scan rate is controlled between 95 and 119 scans per sample. Based on the optimum scan rate, the system can be designed to meet this line-scan rate requirement by reducing exposure time, binning, image processing algorithm and multitasking strategy.

13.3.3 Frame Rate of Hyperspectral Imager

The frame rate of a hyperspectral imager is defined as the number of CCD image acquisition per second that includes time required for exposure, readout, data transfer, as well as any delay. For instance, at the processing speed of 140 samples per minute (SPM), the total processing time required is 4.5 ms per line that is equivalent to 222 line scan per second. Hence, the frame rate of the camera must at least support the minimum frame rate 222 Hz (frames/s) to avoid an under sampling problem. Since the EMCCD camera has a 12.4 Hz frame rate, image binning is required to achieve the desired frame rate. With each line scan without binning, the data size is approximately 1-mega pixels (1,004 × 1,002). If the full spatial dimension (1,004 pixels) of the CCD is binned down to half resolution (502 pixels) and the spectral dimension is binned with three selected non-contiguous bands using a random track mode at λ_1 nm, λ_2 nm and λ_3 nm, the frame rate will be met as required.

Through the software development, any spectral bands can be selected at random processing locations for a spectral dimension, along with their bandwidths. After binning, the spectral image resolution becomes 3 (spectral) × 502 (spatial) per scan. If the width of sample to be scanned is assumed as N pixels, the size of final hypercube becomes N (sample width) × 512 (sample height) × 3 (spectral bands). Based on this approach, the frame rate is 286 Hz with 3.5 ms per line, for 180 SPM and 317 Hz for 200 SPM. In this case, the data transfer rate; i.e. the amount of data transferred by a 3-band images per sample will be 16 Mbps with the 286 Hz and 28 Mbps with the 500 Hz frame rate, respectively.

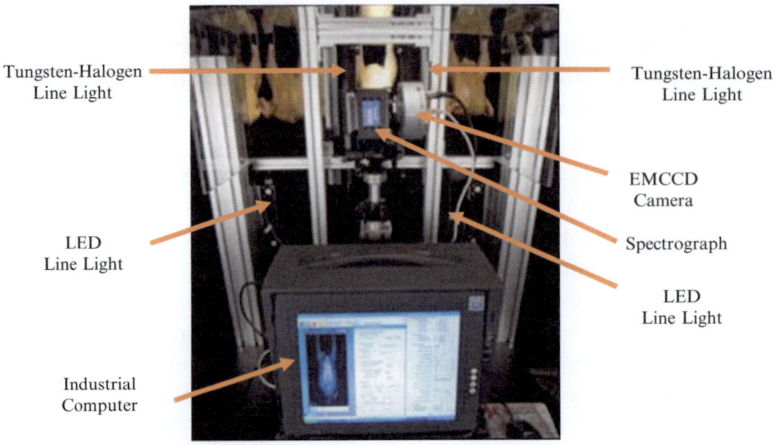

Fig. 13.2 Transportable real-time hyperspectral imaging system for poultry safety inspection

13.3.4 Lighting System of Hyperspectral Imager

Tungsten-halogen line-light is a resource for the transportable real-time hyperspectral imaging system as shown in Fig. 13.2. Tungsten halogen lighting has been proven to be a reliable light source for illumination at the visible-near infrared spectral range. White light-emitting diode (LED) may be an alternative lighting resource, although white LED lighting does not produce light energy in the near-infrared (NIR) spectral range, so additional NIR LED lighting is required for LED illumination at the NIR spectral range, if necessary. The tungsten halogen light source for a transportable imaging system uses two 150-W fiber optic line illumination modules (Fiber Lite A-240L and A-240P, Dolan-Jenner Industries, Boxborough, MA, USA) focusing the lines to be scanned.

Figure 13.3 shows an industrial scale line-scan hyperspectral imaging system setup to acquire real-time on-line image data for contaminate detection on broiler carcasses, with a commercial line speed. The components of imaging system except lighting and monitor were enclosed for running in harsh food processing environment. Two high intensity LED line lights (LL6212, Advanced Illumination, Rochester, VT) are attached to the system for distributing uniform intensity for quality image acquisition. Each line light had 12 LED elements. The optimal working distance of lights is between 15 and 60 cm.

Figure 13.4 shows the devices inside the enclosure including EMCCD camera (Luca, Andor Technology Inc. CT, USA), which has the detector of 1,004 × 1,002 pixels, 8 μm pixel size and 12.4 fps coupled with a 14-bit digitizer board at a 13.5 MHz readout rate. An imaging spectrograph (400–1,000 nm, Hyperspect-VNIR, Headwall Photonics Inc., Fitchburg, MA) and C-mount lens (1.4/23 mm, Schneider, Germany) in the enclosure is attached to the EMCCD camera.

Fig. 13.3 Real-time high-speed hyperspectral imaging system with LED lights

Fig. 13.4 Enclosed line-scan hyperspectral camera with spectrograph

13.4 Real-Time Hyperspectral Image Processing

Hyperspectral imaging for practical applications requires real-time processing of large data volumes recorded by a hyperspectral imager. The use of graphics processing units (GPU) is a candidate for real-time processing (Tarabalka et al. 2009). In particular, real-time processing with GPU is effective for a hyperspectral anomaly detection algorithm, based on normal mixture modeling of the background spectral distribution, a computationally demanding task, relevant to target detection and numerous other applications. Overall, the GPU implementation runs significantly faster, particularly for highly data-parallelizable and arithmetically intensive algorithms. Detection results on the actual data set demonstrate that the total speedup provided by the GPU enables real-time anomaly detection with normal mixture models for airborne hyperspectral imager with high spatial and spectral resolution.

Real-time image processing requires high computational and input/output (I/O) throughputs obtained by an optoelectronic system. A novel architecture that uses focal-plane optoelectronic-area I/O with a fine-grain, low-memory, single-instruction-multiple-data (SIMD) processor array is presented as an efficient computational solution for real-time hyperspectral image processing. The focal-plane SIMD architecture is capable of supporting real-time performance with sustained operation throughputs between 500 and 1,500 giga-operations per second (Chai et al. 2000). A linearly constrained minimum variance (LCMV) beam forming an approach to real-time processing algorithms has been presented for target detection and classification in hyperspectral imagery. The idea is to design a finite impulse response (FIR) filter to pass through the targets using a set of linear constraints, while also minimizing the variance resulting from unknown signal sources (Chang et al. 2001). Real-time enhancement of local anomalies in images of the over flown scene can be presented to the operator to facilitate the decision making process. Within this framework, one of the ultimate research interests is the design of complexity-aware anomaly detection algorithm architectures capable of assuring real-time processing with a fast decision making algorithm (Acito et al. 2013). A real-time, parallel version of optical real-time adaptive spectral identification algorithm is available for adaptive, accommodating changing lighting conditions and terrain for ground based real-time analysis and visualization system (Bowles et al. 1997). An effective approach to speed up real-time implementation can be conducted by using a small portion of pixels in reduced computational complexity, with a simplified hardware for remotely detection of objects using hyperspectral imagery (Du and Nekovei 2009).

A constrained linear discriminant analysis (CLDA) approach to hyperspectral image detection and classification as well as its real-time implementation is also available. The basic idea of CLDA is to design an optimal transformation matrix that can maximize the ratio of inter-class distance to intra-class distance, while imposing the constraint that is different class centers, after transformation, along with different directions can be better separated. The CLDA approach is useful for both detection and classification problems. With a CLDA the real-time implementation can be developed to meet the requirement of on-line image analysis when the immediate data assessment is critical (Du and Ren 2003). For real-time processing of hyperspectral images, two artificial intelligence algorithms for target detection have been studied. Both algorithms are based on the application of artificial neural networks to the hyperspectral data. The first neural network algorithm is applied at the level of individual pixels within an image, and the second algorithm is a multi-resolution based approach to scale invariant target identification using hierarchical artificial neural network architecture (Heras et al. 2011). Two optimized implementations of the principal component analysis (PCA) algorithm, primarily targeted on spectral image analysis in real-time were reported. The implementations are evaluated and compared with a multithreaded C implementation, compiled by an optimizing compiler and the results show speed-ups of approximately ten times that allows for using PCA on RGB and spectral images in real-time (Josth et al. 2012). The development of new real-time hyperspectral endoscope (called the image mapping spectroscopy endoscope) based on the image mapping technique has been reported. The parallel high throughput nature of this technique enables the device to operate at 5.2 frames per second with a datacube size of $350 \times 350 \times 48$. Using this technology tissues *in vivo* are imaged for resolving a vasculature pattern of the lower lip, while simultaneously detecting oxy-hemoglobin (Kester et al. 2011). In recent years, several efforts have been directed towards the incorporation of high-performance computing (HPC) systems and architectures in real-time remote sensing research. With the aim of providing an overview of parallel and distributed systems for remote sensing applications, three HPC-based paradigms for efficient implementation of the Pixel Purity Index (PPI) algorithm have been explored. Several different parallel programming techniques are used to improve the performance of the PPI on a variety of parallel platforms, including a set of message passing interface (MPI)-based implementations (Plaza et al. 2010).

Hyperspectral image analysis algorithms exhibit inherent parallelism at multiple levels, and map nicely on high performance systems such as massively parallel clusters and networks of computers. An exciting new development in this field is the emergence of programmable graphics hardware. Setoain et al. (2008) investigated graphics processing units (GPU)-based implementations of a morphological endmember extraction algorithm, which is used for joint spatial/spectral techniques for hyperspectral analysis. Both endmember extraction accuracy and parallel efficiency are quantitatively compared and assessed in terms of implementing hyperspectral imaging algorithms on commodity graphics hardware. For real-time visible hyperspectral detection of military ground targets with real-time cuing of a high-resolution framing camera, the system hardware components with an emphasis on the visible hyperspectral sensor and the real-time processor were developed. The results from field experiments are described along with an analysis of the collected data and demonstrate the improved performance obtained by operating two detection algorithms simultaneously (Stellman et al. 2000). N-finder algorithm (N-FINDR) (Wu et al. 2010) is also widely used for endmember extraction in hyperspectral imagery. When it comes to practical implementation, four major obstacles including the number of endmembers, initialization of N-FINDR, dimensionality reduction, and computational cost need to be considered. Thus, real-time image data processing is a popular research area for hyperspectral remote sensing. In particular, target detection surveillance, which is demanding real-time or at least near real-time processing. However, the massive amount of hyperspectral image data limits the processing speed so that a strategy of spatial-spectral information extraction (SSIE) that is composed of band selection and sample covariance matrix estimation is needed for fast hyperspectral image processing (Zhang et al. 2012). Band selection utilizes the high-spectral correlation in spectral image, while sample covariance matrix estimation fully utilizes the high-spatial correlation in hyperspectral image. In order to overcome the inconsistent and irreproducible shortage of random distribution, a scalar method can be used to select sample pixels and implemented target detection algorithm based on the SSIE strategy on the hardware of a digital signal processor.

13.4.1 Software Architecture for Real-Time Hyperspectral Imaging

Using a random track mode, selected wavelengths at λ_1, $\lambda_{2,}$ and λ_3 nm are transferred from the camera buffer to the computer memory. Real-time image processing is possible by adopting a ping-pong memory technique as well as a circular buffer technique embedded in the software. The ping-pong memory architecture enables the operation of image acquisition and processing simultaneously. Multithreading of Microsoft Visual C++ 6.0 (Microsoft Foundation Class or MFC) can be used to implement a ping-pong memory. A thread is assigned to an image acquisition module while another thread process a newly acquired image from a circular buffer. There is delay of one object at the beginning, until first acquisition begins. After acquisition is started, the processing module processes the previous object while the acquisition module acquires the following object. A circular memory buffer allows an access to data previously scanned up to 8–10 objects while acquiring new data. The size of the circular buffer is 1,004 × 1,004 × M, where M is the number of random tracks (i.e. number of bands selected). The OpenCV™ that is a library of open source C/C++ programming functions for real-time computer vision can be used for implementing basic image processing operations such as median filter.

13.4.2 Hyperspectral Image Calibration

Spectral calibration is usually completed by manufactures of the hyperspectral imager nowadays. For a calibration task, dark current and 99 % reflectance panels (SRT-99-120, Spectralon, North Sutton, NH, USA) are usually imaged to calibrate the measured images to obtain percent reflectance. For data analysis, regions of interest (ROI) are manually obtained as a ground-truth, and used for evaluating performance of image processing algorithms.

13.4.3 Software for Real-Time Image Acquisition

Software for camera control and image acquisition can be developed using a software development kit (SDK) provided by the camera manufacturer with the Microsoft Visual Basic programming environment. The spectral information of each pixel is collected in the vertical (traverse) direction of the EMCCD. Although push-broom-based line-scan method acquires a full range of hyperspectral images, multispectral images with a few selected wavelengths can be acquired for the real-time in-line applications. The spectral interval and resolution are determined by vertical binning in a multispectral mode. For real-time applications with multispectral mode, the vertical pixel shift rate of the EMCCD is one of the limiting factors along with exposure time and the number of lines scanned for the sample.

13.4.4 Image-Based Polling Algorithm

An image-based polling algorithm can be developed to determine the start and end position of the object. The algorithm is based on monitoring scanned lines to find if the object enters or leaves the field of view. The fundamental assumption for polling is that there must be a gap or space between the objects. The contacts between neighboring objects are usually observed at one end and the other end of objects in moving shackles. The scheme of polling algorithm can be developed as follows. One selected band is usually used for the polling algorithm. After reflectance calibration is completed, a pixel position is selected from the current scanned line. A position between middle and top of the field of view, preferably 40 % down from the top is selected. Then, reflectance calibration values of several pixels upward and down from the selected pixel position are examined to poll the number of dark background pixels within the range previously selected. A threshold value is used to determine the pixel as background or not. If the piece of line examined is greater than the threshold value, the polling for the start of an object ended, followed by imaging the object. The polling for the end of the object is opposite, i.e. if the area of location examined becomes dark from bright, the polling for the end of the object ended, then finish to image the object. The determination of bright or dark is based on the count of bright or dark pixels within the search limit from the position with appropriated calibration.

13.4.5 Hyperspectral Image Processing Algorithm for Contaminant Detection

The fundamental fecal detection algorithm is based on the band ratio algorithm of λ_2/λ_1. The enhanced fecal detection algorithm is the dual band ratio algorithm with λ_2/λ_1 and λ_3/λ_1 followed by a threshold value. The dual band ratio algorithm performs to remove false positive errors, especially cuticles on samples. A calibrated λ_1 band image is used for background masking and polling. Both mask and polling threshold is 1.5 % of reflectance. Various techniques can be applied and compared in terms of eliminating or reducing false errors that are major cause of false positives in the two-band imaging system. In case of poultry contaminant detection application, the algorithms could be two-band ratio (λ_2/λ_1) with a threshold without further filtering, three-band ratios (λ_2/λ_1) and (λ_3/λ_1) with two selected thresholds without further filtering, and two-band ratio (λ_2/λ_1) with software cuticle removal filter with threshold of 40 % reflectance applied to λ_1 without further filtering. For optimum algorithms to detect contaminants, both basic and enhanced band-ratio algorithms can be applied and compared for hyperspectral imaging systems in a real-time multispectral imaging mode.

13.5 Performance of Real-Time Hyperspectral Imaging System

13.5.1 Speed Performance

In order to measure the frame rates of the imaging system as a function of exposure time, the following variables are considered: a trigger mode (internal or external), a random track mode (two bands or three bands) from the camera manufacture's software. After the frame transfer mode is turned on, two random tracks (i.e. spectral bands) are selected from the bands corresponding to λ_1 and λ_2 or three tracks are selected from the bands at λ_1, λ_2 and λ_3. The bandwidth for the two bands is just one track; whereas the bandwidths for three bands are 13, 13, and 26 tracks centered at λ_1, λ_2 and λ_3, respectively. Several experiments need to be conducted for understanding the frame rate limit of the camera system and finding an exposure time range meeting the minimum frame rate requirement such as 222 Hz at 140 SPM and 286 Hz at 180 SPM. With an external trigger mode, pulses generated by a 4 MHz sweep function generator (4003A, B&K Precision, Yorba Linda, CA, USA) can be used with a singe continuous image acquisition mode. The actual frame rate is automatically computed by software and a number is read at arbitrary time, because the frame rate is almost constant over the time unless the operator interrupts the image acquisition process.

As shown in Table 13.1, any exposure time below 100 μs exceeds the minimum requirement of frame rate (>222 Hz). The external triggering has faster frame rate than internal triggering. The minimum frame rate requirement of 286 Hz for 180 SPM may not be met if internal triggering is used. However, if 1×2 spatial binning is used for spatial domain, it is expected to achieve the frame rate of 286 Hz.

Additional experiments are conducted to evaluate the performance speed of the software with a higher processing speed of 200 SPM. The software for contaminant detection is used for grabbing and processing images with a multitasking environment. Both image grabbing and processing time are measured and recorded in a text file as output. For this task, the grabbing time refers to the time for which the software grabbing module spent to transform a three dimensional data cube of N (width) × 512 (height) × 3 (bands/tracks), where N is 128 or 118 scans per sample. If the software grabs more than 95 lines per sample, and processes within 300 ms, the software meets the requirement of minimum speed of 200 SPM. If the software continuously runs for 1 h, the number of datacubes created is 12,078 that is equivalent to the speed of approximately 201 SPM.

Table 13.1 Relationship between exposure time and frame rate of real-time hyperspectral image acquisition

		Frame rate (Hz)			
		External trigger		Internal trigger	
		Two bands	Three bands	Two bands	Three bands
Exposure time (μs)	10	518.1	395.3	269.5	232.0
	40	510.2	390.6	267.4	230.4
	50	507.6	389.1	266.7	229.8
	100	495.0	381.7	263.2	227.3
	1,000	342.5	212.8	284.1	188.7

Since the grabbing time for 128 lines is greater than 300 ms and there is a delay for image grabbing, race condition occurs and causes incorrect and unexpected results. On the other hand, grabbing 118 lines per sample requires 282 ms. In this case, the processing time is approximately between 47 and 86 ms, resulting in more than 200 ms for other tasks. For this processing, the CPU usage is approximately 6–11 % with memory usage of 92 KB without any software issues such as memory leaks. Thus, the software can handle 200 SPM processing speed with high resolution images of 118 (width) × 512 (height) × 3 (bands).

13.5.2 Algorithms for Removing False Errors

Figure 13.5 demonstrates the real-time image processing for cuticle removal with three algorithms discussed in previous section, and Table 13.2 summarizes the performance of the algorithms. The dual band ratio algorithm using three bands can remove the cuticles completely with only 2 pixels remaining as shown in Fig. 13.5a, b. The software filtering is able to remove most cuticles except boundary pixels that are darker than pixels from samples. The false positives (red) except the cuticles (cyan) are along with the boundaries of poultry carcasses. These isolated pixels along the boundaries of samples can be removed by a morphological erosion algorithm such as the median filter that is effective to remove most small false positive pixels. Optical density of cuticles affects mostly the camera, in addition to the sample size, i.e. the thicker the cuticles are, the more false positives. The three-band algorithm performs well for removing cuticles with post-processing filtering as shown in Table 13.2. To visualize false positive errors removal, scatter plots are used to analyze data distribution of cuticles in two-dimensional space.

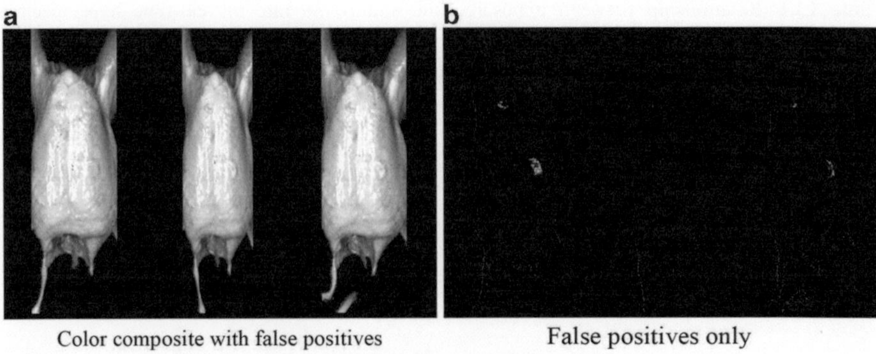

Fig. 13.5 Comparison of three algorithms for cuticle detection using two bands, three bands, and two bands with software cuticle filtering from *left* to *right*, respectively. *Note*: the *cyan* pixels are false positives from cuticles and the *red* pixels are false positives from boundaries. (**a**) Color composite with false positives; (**b**) False positives only

Table 13.2 Comparison of three algorithms performance

	False positives (pixels)	Cuticles (pixels)	Cuticle rate among false positives (%)
Two bands	575	225	39.1
Three bands	154	2	1.3
Two bands with software filtering	517	66	12.8

13.5.3 Distribution Analysis of False Positive Errors

In order to find out how the false positive errors are distributed statistically, the scatter plots are used to visualize the statistical distributions of false positives. For scatter plots, all 293 pixels from cuticles on the images from three samples in Fig. 13.5 are assigned to one group and 953 pixels from false positives are assigned to the other group. Figure 13.6 demonstrates a scatter plot of λ_1 (517 nm) and λ_3 (802 nm) bands. As shown in the figure, cuticles and most false positives can be removed by selecting decision boundary of 1.5 threshold. The cuticle pixels (represented with 'red') are distributed relatively evenly from 0.35 to 0.5. However, the pixels below 0.15 shown at λ_1 (517 nm) are not removed by this decision boundary. Most of these false positive pixels are from background and body contours, so erosion of the sample masking image is effective for removing false positive errors caused by the boundary of samples. Alternatively, a modified decision boundary (piecewise linear) by adding a line as shown in Fig. 13.6 and median filtering can remove the remaining false positives.

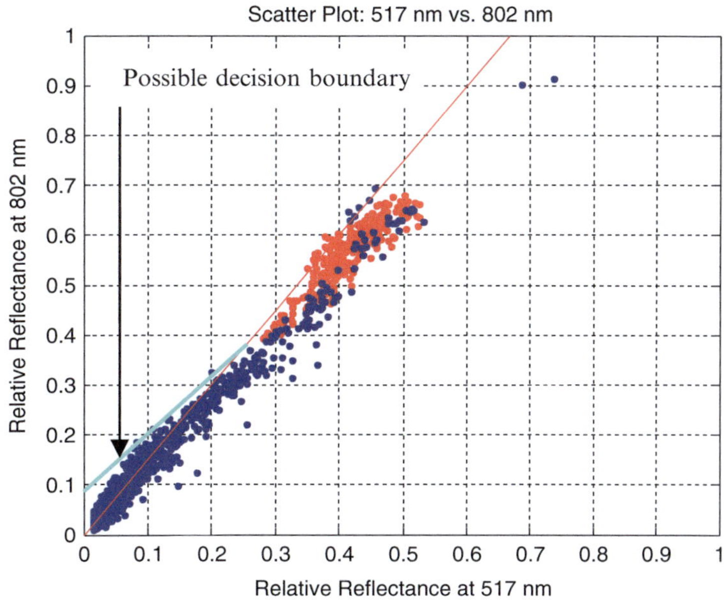

Fig. 13.6 2-D scatter plot of false positives from cuticles (*red*) and background, body contours (*blue*) found in the 517 and 802-nm bands. *Note*: the *red line* is a decision boundary determined by the threshold of 1.5

13.5.4 Algorithms for Contaminant Detection

Figure 13.7 demonstrates a mosaic of pseudo-color composite images generated from 3-band images acquired by the real-time hyperspectral imaging system. Twelve birds are used to examine the efficacy of both fundamental and enhanced algorithms for fecal detection on poultry carcasses. The system parameters used for the experiment are the dimension of 118 (width) × 512 (height) × 3 (bands), the exposure time of 1.68 ms and the EM gain of four to collect the spectral images of three bands from λ_1 (517 nm, 13 bandwidths, BW), λ_2 (565 nm, 9 BW) and λ_3 (802 nm, 21 BW), respectively. Sample contaminants for testing algorithms include duodenum, ceca, colon and ingesta collected from poultry carcass intestines. The images at the top row in Fig. 13.7 are acquired with 180 SPM and the images at the bottom are acquired with 140 SPM. As seen in the figure, the image differences of bird widths are caused mainly by different processing line speeds of shackles (180 vs. 140 SPM). The image polling algorithm performs well, and the processing time of the software meets the requirement. The images are overlaid by the pixels in the regions of interest (ROI) that are color coded as green or red. The sizes of ROI from skin, feces, and cuticles are 6115, 3098 and 293 pixels, respectively.

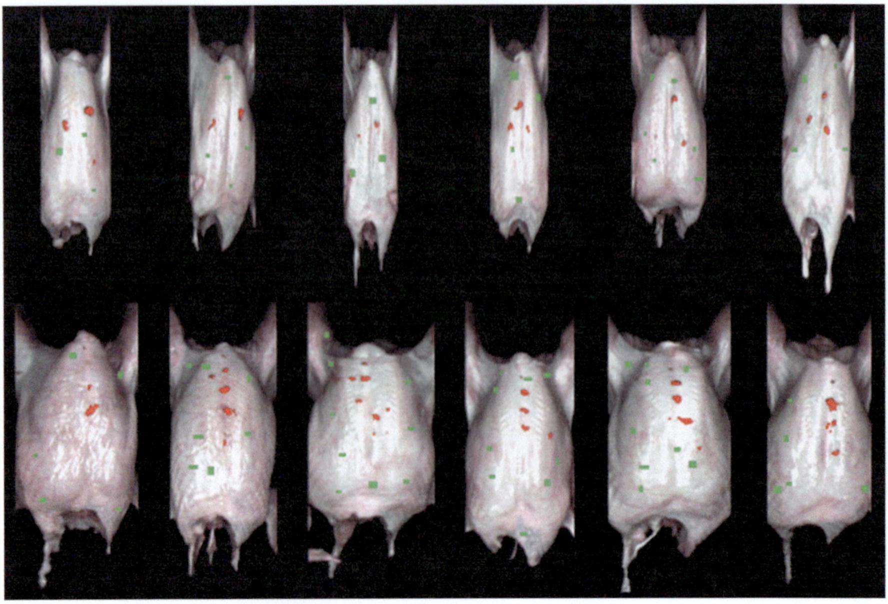

Fig. 13.7 Mosaic images of contaminated birds with feces. *Note*: region of interest (ROI) represents feces (*red*) and clean surface (*green*)

Figure 13.8 demonstrates 2-D scatter plots of the ROIs acquired from the samples in Fig. 13.7. The data from the ROIs are based on two wavelength combinations of λ_1 and λ_2 (517 vs. 565 nm) and λ_1 and λ_3 (517 vs. 802 nm). In addition, the ROIs used for cuticle-skin distribution analysis are added to the ROIs data for Figs. 13.8a, b. The scatter plots in Figs. 13.8c, d are acquired from the ROI pixels after additional image processing with a 3×3 morphological erosion filter applied to the ROI pixels in Figs. 13.8a, b. All pixels above the lines in the scatter plots are considered as feces, although some false positives are included, especially in Figs. 13.8a, b. The eliminated false positives in Figs. 13.8c, d are the pixels from cuticles as well as dark background, which have reflectance values less than 0.2. As shown in Figs. 13.8c, d, the morphological erosion algorithm can remove true fecal pixels, as well as false positives. Most of false positives other than cuticles are observed from the boundaries of true fecal ROIs. Due to the nature of erosion algorithm, most of eliminated pixels along the fecal ROI boundaries are spectrally mixed. Therefore, the advanced fecal detection algorithm using three bands can separate cuticles and feces using the ratio of λ_3 and λ_1 (802 vs. 517 nm) spectral images with the threshold value of 1.5. However, when the pixels are linked to optically less opaque feces such as duodenums or small size of feces, the advanced fecal detection algorithm may also remove mixed pixels around feces. Thus, boundary pixels of small fecal spots are prone to the thresholding operation of the second band-ratio algorithm using λ_3 (802 nm), because median filter possibly wipes out the remaining small pixels.

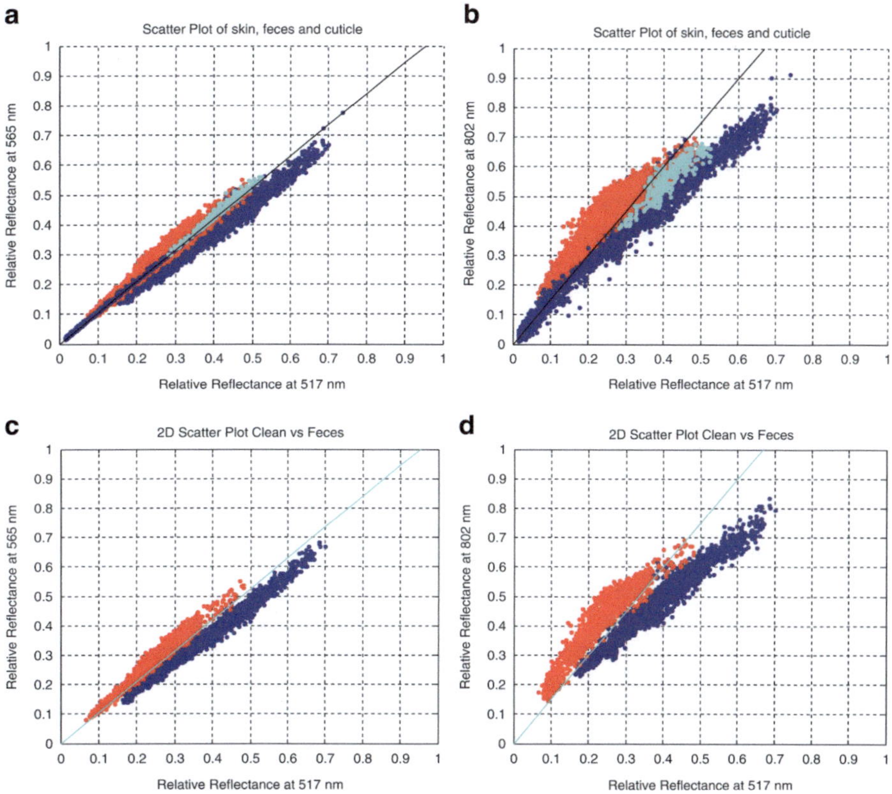

Fig. 13.8 Scatter plots of skin (*blue*), feces (*red*) and cuticles (*cyan*). *Note*: decision boundary lines are overlaid with band-ratio thresholds (**a**) 1.05 and (**b**) 1.5; and the scatter plots in (**c**) and (**d**) are acquired from ROIs with 3×3 morphological erosion algorithms. (**a**) 517 vs. 565 nm, (**b**) 517 vs. 802 nm, (**c**) Eroded ROIs: 517 vs. 565 nm, (**d**) Eroded ROIs: 517 vs. 802 nm

13.6 Consideration of Real-Time Hyperspectral Imaging Applications

For real-time hyperspectral imaging applications, it is important to understand real-time operation for both hardware and software. For the hardware point of view, the challenge is to determine whether a given hardware platform can produce the data with quality and quantity required by real-time applications. For the software, the challenge is to implement the algorithms developed at non-real-time computing environments into a real-time version.

From the initial feasibility test for common platform imaging technology, a high-speed line-scan hyperspectral imaging system can be used for the application of online detection. Among the several issues need to be addressed for field applications of a real-time system, some observations about the imagery from initial tests in the laboratory include system calibration, as usual for the case of most instruments. Although large contaminants appear to be detected at uncalibrated image ratio of two bands of λ_2 (565 nm) and λ_1 (517 nm), about 30 % of the calibration images are saturated at the λ_1 (517 nm) band. Therefore, more precise calibration methods are needed for line-scan image acquisition, because it is not easy to assess the performance of standard threshold values for the uncalibrated imagery. If spatial resolution is very low across the sample, false negative errors could increase. In order to eliminate both false positive and false negative errors, the optimum spatial resolution needs to be determined for line-scan camera setting throughout the image acquisition process.

13.6.1 Calibration Issues of Line-Scan Camera

Based on the modification of image analysis software with Interactive Data Language (IDL) programming environment, on-line reflectance and dark current images are used to calibrate the composite images of samples. Since there are differences between λ_1 (517 nm) and λ_2 (565 nm) in dark current data and considerable inconsistency from top to bottom of the sensor, calibration is necessary step for real-time hyperspectral imaging system for on-line applications. For instance, the image could be saturated at λ_1 (517 nm) band so that this results in false positives, especially in the poorly lit and saturated areas. Sometimes there is significant shadowing and also tapering off of the light that follows the curvature of the samples, resulting in false positives along the edges of the sample such as wings and legs of poultry carcasses. Therefore, an accurate calibration procedure for real-time imaging needs to be carefully taken care of for better detection accuracy.

13.6.2 Issues of Real-Time Hyperspectral Imaging System for Contaminant Detection

In order to accomplish contaminant detection for a real-time online mode, several factors need to be considered and resolved prior to implementing at commercial processing plants. For the specification of sample measurement, the minimum weight of fecal matter correlated with microbial counts need to be determined. A second factor to consider is the detectable minimal pixel size of fecal contaminates. According to the observation from spot size studies (Windham et al. 2005), the detectable fecal size is approximately 10 pixels, which can be generated from a 5 mg cecal sample. Although the optimum camera setting is dependent on specifications of the primary detecting module such as a camera sensor and corresponding optics, the parameter setting of hyperspectral camera can be 125 ms of exposure time, with 27 in. working distance, 4×2 binning for 380 lines scan that covers the whole poultry carcass, for example, with 23-mm lens.

Lighting is another key subsystem for quality image acquisition, specifically the high-speed line-scan hyperspectral imaging system. Both tungsten halogen (TH) and light-emitting device (LED) are good candidates, because of their characteristics of high excitation intensity at 517, 565 and 802 nm, which are adequate for fecal contaminant detection. For three-band spectral image data acquisition, the spectral range of the hyperspectral imager needs to be optimized at the wavelengths of 512–522, 560–570, and 729–812 nm for a line-scan hyperspectral imaging system with the slit width less than 40 μm.

13.6.3 Real-Time Fecal Contaminant Detection with Different Lighting Sources

Figure 13.9 demonstrates pseudo-color composite images with contaminants and corresponding binary images that indicate contaminants tested with pilot-scaled online processing, with a real-time in-line hyperspectral imaging system. In this test, spectral images selected with three bands at 517, 565 and 600 nm (or 802 nm) are used to create color composite images. Since the 802 nm band validated false positive removal from the study of optical filter-based common aperture camera research (Heitschmidt et al. 2007), the 802 nm band form line-scan hyperspectral imaging system is tested to determine if similar results in terms of eliminating false positive errors caused by cuticles could be obtained. From this study, 600 nm is an alternative instead of 802 nm, which indicates that the tungsten halogen line lights can be simply used for lighting sources instead of additional 802 nm LED light. Yet, more lighting tests are needed to validate which lighting system is optimum for a line-scan hyperspectral imaging system for quality image acquisition and performance for other contaminant detection. For example, various contaminants from poultry carcasses such as ingesta, colon and ceca are detected, except duodenum that is not distinct in the composite color images. The size of contaminant from colon is approximately 12 mm^2, and corresponding image size of the same spot is 15 pixels with 600 nm band, but only 7 pixels with 802 nm band. Thus, same size of contaminants is detected differently depending on lighting sources. Although the results with 600 nm are better than 802 nm in this sample, the overall performance of 600 nm is similar to that of 802 nm, because the dominant algorithm is still based on the image ratio from 517 to 565 nm bands. The white spots in Figs. 13.9b, c demonstrates fecal and ingesta contaminants and a yellow marker that indicates ground truth for contaminated samples.

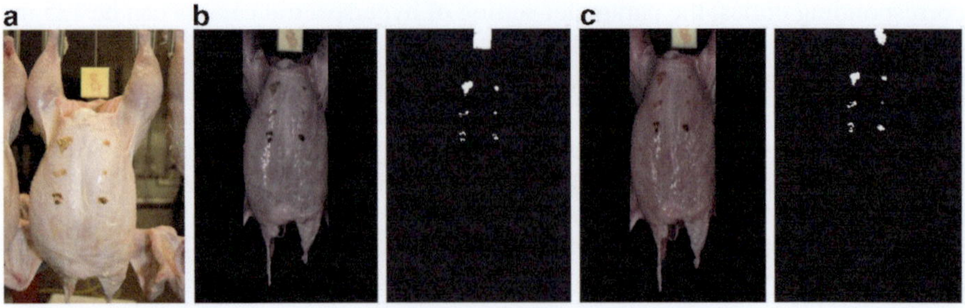

Fig. 13.9 Test results for fecal contaminant detection with line-scan hyperspectral imaging system; (**a**) pseudo-color composite image with various contaminants of duodenum, cecum, colon and ingesta; (**b**) three bands of 517, 565 and 600 nm; (**c**) three bands of 517, 565 and 802. *Note*: Third band was selected for false positive removal and yellow marker was used for ground truth indicating a contaminated carcass

13.6.4 Technical Requirement of In-Line Hyperspectral Imaging System for Contaminant Detection

In order to meet the high throughput across processing lines consistent with food safety inspection regulation, in-line fecal inspection instruments need functional and technical specification with enhancements for real-time processing at commercial poultry processing plants. With high speed data acquisition and real-time image processing algorithms, the in-line hyperspectral imaging system is able to inspect poultry at least 140 birds per minute with fecal contaminants of larger than 3-mm spot size, which is corresponding with approximately about 5 mg fecal mass. The imaging system includes a CCD detector, spectrograph, lens, power supply, electronic control system, thermal control system, and enclosure to protect electronic modules in the harsh environment of a poultry slaughterhouse. The recommended camera lens is 25-mm, f/2 and illumination system includes LED light with 24 VDC if all excitation spectral bands meet the minimum requirement (517, 565, and 600 nm for fecal contaminants) as a lighting source. The spectral range of the spectrograph is between 450 and 900 nm with a 6-nm spectral resolution. Overall image distortion must be less than 0.1 pixel for smile and keystone effect and less than 5 % for vignetting for hyperspectral imager. The CCD detector has non-cooling with 1004 × 1002 pixel format, 8-mm square pixel size, 8 × 8 mm focal plane array (FPA), custom defined binning option, minimum frame speed of 12 fps and USB2 interface functionality. The software for initial testing is based on Visual Basic, but C++ programming environment is feasible for a common platform system configuration for universal purpose.

13.7 Concluding Remarks

The high-speed line-scan hyperspectral imaging system has the potential for real-time online inspection during food processing. Specifically, this technology benefits the poultry industry for fecal contaminant as well as unwholesomeness detection. To improve detection accuracy and performance, fully calibrated (spatially and spectrally) images need to be acquired for further image processing. In addition, either external or internal mode for triggering imagers is required for real-time implementation with appropriate image processing methods. Software triggering methods based on the shape of an object are candidates for online application. In doing this, the entire processing speed from line-scan image acquisition until analysis to identify contaminant spots should meet the industry requirement of 140 birds per minute. For quality image acquisition in real-time applications, a lighting system, either tungsten halogen or LED (LED preferred for industry application) is crucial. In conclusion, high-speed line-scan hyperspectral imaging system can be used for food safety inspections such as fecal contaminant and unwholesomeness detection as a common platform imaging system along with other quality inspections based on industry standards. For advances in the real-time hyperspectral imaging platform, more research needs to be done to fully validate performance of the system to meet industry requirements. For example, in the case of fecal contaminant detection, the ground truths for high detection accuracy with minimum false positive errors need to be validated for commercial implication. For this task, several markers as ground truths based on confirmation by human inspector or other instrumental methods can be tested, and on automatic link with ground truths using a CCD color camera as a reference can be further investigated.

References

Acito N, Matteoli S, Diani M, Corsini G (2013) Complexity-aware algorithm architecture for real-time enhancement of local anomalies in hyperspectral images. J Real-Time Image Proc 8:1–16

Ariana DR, Lu R, Guyer DE (2006) Near-infrared hyperspectral reflectance imaging for detection of bruises on pickling cucumbers. Comput Electron Agric 53:60–70

Bowles JH, Antoniades JA, Baumback MM, Grossmann JM, Haas D, Palmadesso PJ, Stracka J (1997) Real-time analysis of hyperspectral data sets using NRL's ORASIS algorithm. Proc SPIE 3118:38. doi:10.1117/12.283841

Chai SM, Gentile A, Lugo-Beauchamp WE, Fonseca J, Cruz-Rivera JL, Wills DS (2000) Focal-plane processing architectures for real-time hyperspectral image processing. Appl Opt 39:835–849

Chang CI, Ren H, Chiang SS (2001) Real-time processing algorithms for target detection and classification in hyperspectral imagery. IEEE Trans Geosci Rem Sens 39:760–768

Chao K, Yang CC, Chen YR, Kim MS, Chan DE (2007) Hyperspectral-multispectral line-scan imaging system for automated poultry carcass inspection applications for food safety. Poult Sci 86:2450–2460

Chao K, Yang CC, Kim MS (2010) Spectral line-scan imaging system for high-speed non-destructive wholesomeness inspection of broilers. Trends Food Sci Technol 21:129–137
Du Q, Nekovei R (2009) Fast real-time onboard processing of hyperspectral imagery for detection and classification. J Real-Time Image Proc 4:273–286
Du Q, Ren H (2003) Real-time constrained linear discriminant analysis to target detection and classification in hyperspectral imagery. Pattern Recognition 36:1–12
El Masry G, Wang N, El Sayed A, Ngadi N (2007) Hyperspectral imaging for nondestructive determination of some quality attributes for strawberry. J Food Eng 81:98–107
Fletcher-Holmes DW, Harvey AR (2005) Real-time imaging with a hyperspectral fovea. J Opt A Pure Appl Opt 7:S298–S302
Gowen AA, O'Donnell CP, Cullen PJ, Downey G, Frias JM (2007) Hyperspectral imaging—an emerging process analytical tool for food quality and safety control. Trends Food Sci Technol 18:590–598
Heitschmidt GW, Park B, Lawrence KC, Windham WR, Smith DP (2007) Improved hyperspectral imaging system for fecal detection on poultry carcasses. Trans ASABE 50:1427–1432
Heras DB, Arguello F, Gomez JL, Becerra JA, Duro RJ (2011) Towards real-time hyperspectral image processing, a GP-GPU implementation of target identification. In: 2011 I.E. 6th international conference on intelligent data acquisition and advanced computing systems (IDAACS), vol 1, pp 316–321
Josth R, Antikainen J, Havel J, Herout A, Zemcik P, Hauta-Kasari M (2012) Real-time PCA calculation for spectral imaging (using SIMD and GP-GPU). J Real-Time Image Proc 7:95–103
Kester RT, Bedark N, Gao L, Tkaczyk TS (2011) Real-time snapshot hyperspectral imaging endoscope. J Biomed Opt 16:056005. doi:10.1117/1.3574756
Kim I, Kim MS, Chen YR, Kong SG (2004) Detection of skin tumors on chicken carcasses using hyperspectral fluorescence imaging. Trans ASABE 47:1785–1792
Kim MS, Chen YR, Mehl PM (2001) Hyperspectral reflectance and fluorescence imaging system for food quality and safety. Trans ASABE 44:721–729
Kim MS, Lee K, Chao K, Lefcourt AM, Jun W, Chan DE (2007) Multispectral line-scan imaging system for simultaneous fluorescence and reflectance measurements of apples: multitask apple inspection system. Sens Instrumen Food Qual 2:123–129
Kise M, Park B, Lawrence KC, Windham WR (2007) Design and calibration of a dual-band imaging system. Sens Instrumen Food Qual 1:113–121
Leitner R, Mairer H, Kercek A (2003) Real-time classification of polymers with NIR spectral imaging and blob analysis. Real-Time Imaging 9:245–251
Lu R, Peng Y (2006) Hyperspectral scattering for assessing peach fruit firmness. Biosystems Eng 93:161–171
Nicolai BM, Lotze E, Peirs A, Scheerlinck N, Theron KI (2006) Non-destructive measurement of bitter pit in apple fruit using NIR hyperspectral imaging. Postharvest Biol Technol 40:1–6
Noh H, Lu R (2007) Hyperspectral laser-induced fluorescence imaging for assessing apple quality. Postharvest Biol Technol 43:193–201
Park B, Lawrence KC, Windham WR, Buhr RJ (2002) Hyperspectral imaging for detecting fecal and ingesta contaminants on poultry carcasses. Trans ASABE 45:2017–2026
Park B, Lawrence KC, Windham WR, Smith DP (2004) Multispectral imaging system for fecal and ingesta detection on poultry carcasses. J Food Process Eng 27:311–327
Park B, Yoon SC, Lawrence KC, Windham WR (2005) Dynamic threshold method or improving contaminant detection accuracy with hyperspectral images. ASAE paper no. 053071, St. Joseph, MI
Park B, Lawrence KC, Windham WR, Snead MP (2006) Real-time multispectral imaging application for poultry safety inspection. In: Electron imaging SPIE 6070-7, pp 1–10
Park B, Kise M, Lawrence KC, Windham WR, Smith DP (2007a) Real-time multispectral imaging system for online poultry fecal inspection using unified modeling language. Sens Instrumen Food Qual 1:45–54

Park B, Yoon SC, Lawrence KC, Windham WR (2007b) Fisher linear discriminant analysis for improving fecal detection accuracy with hyperspectral images. Trans ASABE 50:2275–2283

Park B, Kise M, Windham WR, Lawrence KC, Yoon SC (2008) Textural analysis of hyperspectral images for improving contaminant detection accuracy. Sens Instrumen Food Qual 2:208–214

Park B, Yoon SC, Kise M, Lawrence KC, Windham WR (2009) Adaptive image processing methods for improving contaminant detection accuracy on poultry carcasses. Trans ASABE 52:999–1008

Park B, Yoon SC, Windham WR, Lawrence KC, Kim M, Chao K (2011) Line-scan hyperspectral imaging for real-time in-line poultry fecal detection. Sens Instrumen Food Qual 5:25–32

Plaza A, Plaza J, Vegas H (2010) Improving the performance of hyperspectral image and signal processing algorithms using parallel, distributed and specialized hardware-based systems. J Sign Process Syst 61:293–315

Qiao J, Ngadi M, Wang N, Garlepy C, Prasher S (2007) Pork quality and marbling level assessment using a hyperspectral imaging system. J Food Eng 83:10–16

Setoain J, Tirado F, Tenllado C, Prieto M (2008) Real-time onboard hyperspectral image processing using programmable graphics hardware. In: Plaza AJ, Chang CI (eds) High performance computing in remote sensing, Chap 18, pp 411–451

Stellman CM, Hazel GG, Bucholtz F, Michalowicz JV, Stocker A, Schaaf W (2000) Real-time hyperspectral detection and cuing. Optical Eng 39:1928–1935

Tarabalka Y, Haavardshlom TV, Kasen I, Skauli T (2009) Real-time anomaly detection in hyperspectral images using multivariate normal mixture models and GPU processing. J Real-Time Image Proc 4:287–300

Tatzer P, Wolf M, Panner T (2005) Industrial application for inline material sorting using hyperspectral imaging in the NIR range. Real-Time Imaging 11:99–107

Yoon SC, Lawrence KC, Park B, Windham WR (2007) Optimization of fecal detection using hyperspectral imaging and kernel density estimation. Trans ASABE 50:1063–1071

Yoon SC, Lawrence KC, Line JE, Siragusa GR, Feldner PW, Park B, Windham WR (2010) Detection of Campylobacter colonies using hyperspectral imaging. Sens Instrumen Food Qual 4:35–49

Windham WR, Lawrence KC, Park B, Martinez LA, Lanoue MA, Smith DA, Heitschmidt GW, Poole GH (2003a) Method andarticlet system for contaminant detection during food processing. U.S. Patent No. 6,587,575

Windham WR, Lawrence KC, Park B, Buhr RJ (2003b) Visible/NIR spectroscopy for characterizing fecal contamination of chicken carcasses. Trans ASAE 46:747–751

Windham WR, Heitschmidt GW, Smith DP, Berrang ME (2005) Detection of ingesta on pre-chilled broiler carcasses by hyperspectral imaging. Int J Poultry Sci 4:959–964

Wu CC, Chen HM, Chang CI (2010) Real-time N-finder processing algorithms for hyperspectral imagery. J Real-Time Image Proc 7:105–129

Zhang B, Yang W, Gao L, Chen D (2012) Real-time target detection in hyperspectral images based on spatial-spectral information extraction. EURASIP J Adv Sign Proc 2012:142

Chapter 14
LCTF Hyperspectral Imaging for Vegetable Quality Evaluation

Changying Li and Weilin Wang

Abstract This chapter discusses the liquid crystal tunable filter (LCTF)-based hyperspectral imaging technology and its application in vegetable quality inspection by using onion as a case study. A brief overview is provided on using the destructive and nondestructive methods for vegetable quality measurement. A detailed description of the LCTF technology, including system components and calibration, is presented. Two examples are given on using the LCTF technology for onion quality evaluation: one is to detect sour skin disease on the onion surface, and the other is to predict onion internal quality (soluble solid content and dry matter content) using the LCTF system. A brief conclusion is provided at the end of the chapter.

14.1 Introduction

It has been well recognized that increasing fresh vegetables in diet could prevent noncommunicable diseases such as cancer and cardiovascular diseases which contribute to two thirds of all deaths in the world (Ezzati and Riboli 2012). Partly driven by the health benefits of fresh vegetables, the production of vegetables in the world has almost quadrupled in the past four decades (Food and Agriculture Organization of the United Nations 2012). In the past 3 years (2009–2011), the production value of American fresh vegetables has consistently exceeded $10.5 billion (USDA-NASS 2012). The top three most valuable fresh vegetables are lettuce, tomato, and onions, which together account for about 40 % of the total farm-gate value of the U.S fresh vegetables. As consumption rises, consumers are more concerned about the quality of vegetable products, which makes the

C. Li (✉)
College of Engineering, University of Georgia, 712F Boyd Graduate Studies Research Center, Athens, GA 30602, USA
e-mail: cyli@uga.edu

W. Wang
College of Engineering, University of Georgia, 701 Boyd Graduate Studies Research Center, Athens, GA 30602, USA

Monsanto Company, 800 North Lindbergh, St. Louis, MO 63167, USA
e-mail: wweilin@uga.edu; weilin.wang@monsanto.com

B. Park, R. Lu (eds.), *Hyperspectral Imaging Technology in Food and Agriculture*, Food Engineering Series, DOI 10.1007/978-1-4939-2836-1_14

marketing even more competitive. As for fresh vegetables, five general quality factors are appearance, flavor, texture, nutritive values, and defect factors, which largely determine consumer acceptability and the value of the product.

In the postharvest handling system of vegetables, quality inspection and sorting plays a central role and virtually affects all stakeholders. Consumers demand high-quality product and would pay premium prices for products with uniform size and appearance. Any surface blemishes, diseases, or internal defects would reduce consumers' satisfaction. For growers, defective products may result in discounting or even rejection of the entire batch of produce, which is a significant loss. For processors, latent damage or internal defects detected by consumers could be economically devastating and may also have a long-lasting effect on brand reputation. For packers, storing blemished or defective produce wastes valuable storage space and decreases profit margins. Thus, maintaining high-quality products is crucial to the economic survival of the vegetable industry. To ensure high quality, defective products must be separated and removed from the wholesome ones by postharvest sorting and packing.

As one of the major fresh vegetables worldwide, the onion (*Allium cepa* L.) has been cultivated for both culinary and medicinal use for thousands of years on earth. Currently, it is grown in more than 20 states in the United States. The annual farm-gate value of the onion in the U.S. reached over $1 billion in 2003 and 2005 and has been consistently exceeding $800 million in the past 5 years (USDA-NASS 2006; National Onion Association 2008; USDA-NASS 2012). Onion consumption has been growing continuously, due in part to their well documented health benefits and the booming American fast food industry.

At present, only a few onion packinghouses are equipped with machine vision systems to sort onions by size or outside appearance. In most onion packinghouses in the U.S., onions are evaluated for external quality (such as blemishes or surface split) only by human visual inspection. Human visual inspection varies from person to person and is subject to error due to human fatigue. In addition, human inspectors are unable to evaluate internal quality (such as dry matter content) and defects (such as center rot). Human inspectors are not only ineffective in some cases and prone to error, but also costly. Nearly 50 % of onion packinghouse operational cost is related to human labor and management cost (personal communication). This labor-intensive handling system becomes more problematic in light of the looming labor shortage across the U.S.

This chapter discusses the liquid crystal tunable filter (LCTF)-based hyperspectral imaging (HSI) technology and its application in vegetable quality inspection with onion as a case study. A brief overview is provided on destructive and nondestructive methods for vegetable quality measurement. A detailed description of LCTF technology, including system components and calibration, is presented. Two examples are given on using LCTF technology for onion quality evaluation: one is to detect sour skin disease on the onion surface, and the other is to predict onion internal quality (soluble solid content and dry matter content) using the LCTF system. A brief conclusion is provided at the end of the chapter.

14.2 Onion Quality and Evaluation Methods

14.2.1 Onion Quality Factors

The quality aspects of onions can be classified into external and internal factors. The important external quality factors of onions include: size, shape, color, uniformity, and outside defects. The key internal quality factors are firmness, dry matter content, soluble solid content, and internal defects (internal rottenness, void, etc.). Since external quality inspection has been the target of current sorting technologies, whereas internal quality has not been as well investigated, this section only focuses on the discussion of those internal quality factors.

Roughly 13 % of the total onion production in the U.S., accounting for more than $100 million revenues, goes to dehydration and processed market (USDA-NASS 2012). In the dehydration process, higher profit is achieved by using onions with high percentage of dry matter. To provide onions with high dry matter, onion breeders need tools to rapidly screen onion cultivars with high dry matter content. Although there exist automated systems (electronic or vision) to measure onion size and weight, it is still difficult to nondestructively estimate onion internal quality.

Currently the internal qualities of onions (and most vegetables) are measured by destructively testing random samples. For instance, the firmness of most fruits and vegetables is usually measured by the Magness-Taylor (MT) method, in which a steel probe is penetrated into the fruit or vegetable at certain depth (e.g., 9 mm) with certain loading speed (e.g. 2 mm/s). The maximum force recorded during this process is used as a measure of firmness for the product.

Soluble solid content, another important internal quality property of most fruits and vegetables, is usually measured by a refractometer with automatic temperature correction. One or two drops of juice are extracted from a vegetable sample, and then the juice is spread onto the glass slide of the refractometer to read soluble solid content (SSC) values in°Brix.

Dry matter (DM) content is usually measured by heating the vegetable sample in a forced draft oven at 75 °C for 12 h. The dry matter content is calculated as the ratio between the weight of the vegetable samples after and before the oven drying.

Apparently, all these methods require destroying the vegetable sample. There are two main drawbacks of the destructive testing approach: (1) the tested samples cannot be used or sold after testing; (2) it can only test a limited number of samples and leaves many defective products undetected. In contrast, nondestructive testing methods can potentially measure each and every vegetable sample without damaging the product. Therefore, it would be of great value if a rapid and nondestructive method is available to evaluate onion internal quality such as dry matter, SSC, and firmness.

14.2.2 Onion Postharvest Diseases

Almost 60 % of non-processed onions in the nation are put in storage and consumed weeks or months later to extend the season and capitalize on a more favorable market window (Burden 2008; National Onion Association 2008). Normally, onions can be stored in a cold, dry, and well ventilated room for several months with a marketable quality. However, fungal and bacterial diseases affect stored onions and cause substantial losses in storage. Outbreaks of these fungal and bacterial diseases are usually caused by a few damaged and infected onions which eventually spread the pathogen and spoil nearby wholesome onions in storage. Due to the lack of detection methods, onion handlers are unaware of the presence of these diseases in the early stage until the onions exhibit visual symptoms that make them unsalable at the end of the storage period. For instance, Botrytis neck rot (caused by *Botrytis allii*), a virtually undetectable fungal disease, can cause as high as 50–70 % storage losses in some years (Ceponis et al. 1986; Boyhan and Torrance 2002). Another disease, known as sour skin, caused by bacteria *Burkholderia cepacia* (Burkholder 1950), is one of the most serious onion diseases that can affect most onion varieties (Schwartz and Mohan 2008). Not surprisingly, both fungal and bacterial diseases are identified by stakeholders as two of the most serious problems that are afflicting the onion industry. However, it is difficult to detect bacteria or fungi induced diseases in onions by human visual inspection because the symptoms of the onion diseases usually remain latent until the environment becomes favorable (Schwartz and Mohan 2008). To control fungal and bacterial diseases and to reduce massive storage losses, a more effective nondestructive sensing method is in demand.

14.2.3 Nondestructive Sensing Methods

Nondestructive measurement is a technique to obtain the quality information without changing the physical and chemical properties of that product (Shewfelt and Prussia 1993; Florkowski et al. 2009). There are several nondestructive techniques that have been investigated for vegetable quality and safety sensing, such as NIR spectroscopy (Osborne and Fearn 1986), X-ray imaging, magnetic resonance imaging (Cho et al. 1990), machine vision (Liao et al. 1994; Tao 1998), and electronic nose (Li et al. 2009). For instance, X-ray imaging has been studied for detecting internal defects in Vidalia sweet onions, including voids and foreign inclusions (Shahin et al. 2002). Li et al. (2009) reported a method of using the gas sensor array to detect the sour skin disease in onions by measuring the headspace gas. It achieved 85 % correct classification rate when six gas sensors were used. Although this technology showed promise to detect sour skin, the gas sensors are more suitable in confined environment than on packing lines.

Among various nondestructive sensing methods, the machine vision technology has been most extensively studied and successfully applied in the postharvest handling sector to inspect the quality of fruits and vegetables, such as size, shape, volume, color, or texture (Cubero et al. 2011). Machine vision technology, however, cannot detect the internal quality or latent diseases of the fresh produce.

Near infrared (NIR) spectroscopy was first used in food and pharmaceutical industry in the mid of 1960s (Williams and Norris 2001; Reich 2005), and it has been increasingly used in nondestructive measurement of internal qualities of fruits and vegetables in the past two decades (Nicolaï et al. 2007). NIR spectroscopy has been particularly successful in estimating soluble solids and dry matter content of fruits and vegetables due to the absorption of sugar and water in the near infrared spectral region. For instance, NIR spectroscopy has shown promising results in predicting sugar content in apples (Lu et al. 2000; Park et al. 2003), cantaloupes (Dull et al. 1989), papayas (Birth et al. 1984), fresh prunes (Slaughter et al. 2003), tomatoes (Slaughter et al. 1996), and sweet cherries (Lu 2001). It has also demonstrated success in dry weight prediction of kiwifruit (Slaughter and Crisosto 1998), and potato tubers (Dull et al. 1989), and onions (Birth et al. 1985).

However, traditional NIR spectrometry can only collect spectra from one spatial point or small area at a time, which may not be representative of the entire sample given the spatial variation within a vegetable sample. This drawback could be overcome by hyperspectral imaging which combines the strengths of machine vision (with spatial information) and spectroscopy (with spectral information) (Lu 2003). The three-dimensional hyperspectral image (two-dimensional spatial and one-dimensional spectral information) not only provides physical and geometric properties of an object such as size, shape, color and texture, but also provides chemical and molecular characteristics such as sugar, protein, and other hydrogen-bonded components (van de Broek et al. 1995; Kazemi et al. 2005). In the past 10 years, extensive studies have been done to apply the hyperspectral imaging technique for quality evaluation of fruits, vegetables, and nuts (Kim et al. 2001; Lorente et al. 2012; Park et al. 2001, 2002; Qin and Lu 2005; Ariana et al. 2006; ElMasry et al. 2007; Jiang et al. 2007; ElMasry et al. 2008). Most of the applications were for qualitative classification or detection of certain defects on the fresh produce. A handful of studies investigated internal quality estimation of strawberry, grape skin, and banana using the hyperspectral imaging technique (ElMasry et al. 2007; Fernandes et al. 2011; Rajkumar et al. 2012). Lu's group explored using the scattering profile of the diffuse reflectance from the hyperspectral images to predict firmness and sugar content of fruits (Lu 2004; Lu and Peng 2006).

Push-broom-based HSI systems were used in most of the above cited literature, while the electronic tunable filter-based HSI systems have been studied and used in only a few cases. For instance, an LCTF-based spectral imaging system (650–1050 nm) was developed to study plant health (Evans et al. 1998); a similar HSI system (460–1020 nm) was used to detect the rottenness in mandarins (Gómez-Sanchis et al. 2008). A group of studies explored using the LCTF-based HSI system for wheat kernel quality (such as constituents, hardness, color, insect-damage) inspection (Cogdill et al. 2004; Archibald et al. 1999; Williams et al. 2009; Mahesh et al. 2008; Singh et al. 2009). The LCTF-based HSI system, however, has its unique advantages over the push-broom-based HSI system, which will be introduced in detail in the following sections.

14.3 LCTF-Based Hyperspectral Imaging

A liquid crystal tunable filter is a multistage Lyot-Ohman type polarization interference filter, which has a stack of polarizers, birefringent elements, and electronically tunable liquid crystal (LC) wave plates (Gat 2000). Since the filtering elements of the LCTF have selective transmissions, only a narrow bandpass of light can pass through the LCTF and the other light is eliminated. By applying different electric fields to control the retardances of LC elements, the bandpass of the LCTF could be tuned to the desired spectral region.

As a solid-state electronically tunable bandpass filter, LCTF is a major type of electronic filter used for hyperspectral imaging due to its superior image quality. Compared to the latest line-scan HSI systems, LCTF-based HSI systems are often less competitive in terms of transmittance throughput and spectral resolution. However, the LCTF-based HSI system also has its unique advantages over the line-scan HSI system because the former:

1. is a natural extension of the multispectral imaging system, which provides versatility to the system to be used for either hyperspectral or multispectral imaging applications;
2. has an area field of view (FOV), whereas the line-scan HSI system can only see one line of the test object at a time;
3. can access wavelength bands randomly and quickly, and thus have superiority for instantaneous imaging applications requiring selective spectral information;
4. has configuration parameters (i.e. the exposure time of the camera) that are dynamic and adjustable over each spectral band during scans, while a line-scan system often has to keep its parameter settings constant during scanning;
5. is easier to be integrated with other systems since they don't rely on any moving mechanical part such as a linear conveyor.

Particularly, the LCTF-based hyperspectral imaging system is appealing to some research labs since it could be built upon an existing camera system by including an LCTF unit and an extended data acquisition program, which provides a cost-efficient alternative for investing a completely new HSI system.

14.3.1 Major Components of the LCTF-Based Hyperspectral Imager

Developing an LCTF-based spectral imaging system requires a complicated process of system design, integration, and calibration. In system design, the key task is to select proper components for the hyperspectral imager, which mainly consists of LCTF, lens, and camera (Fig. 14.1).

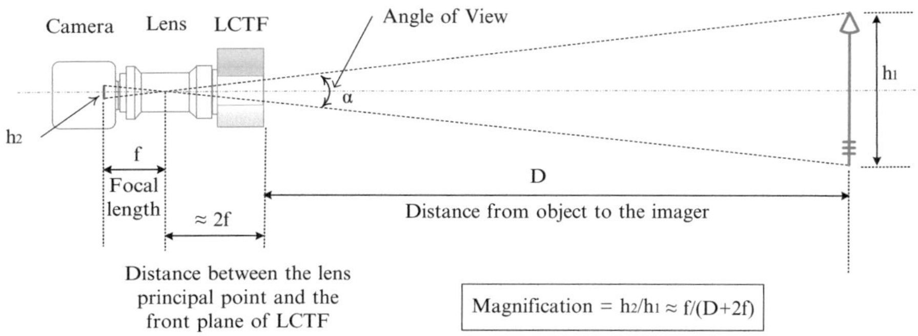

Fig. 14.1 The schematic of the LCTF-based spectral imager. Reproduced with permission from Wang et al. (2012); ©2012 Elsevier

14.3.1.1 Detector

LCTF-based hyperspectral imager requires an area detector. The criteria of selecting a detector for an LCTF-based hyperspectral imager are mostly similar to those for other types of HSI systems. In the Vis-NIR spectral region (400–900 nm), CCD cameras are dominantly used. One particular concern is that the detector should have high sensitivity since LCTF would block more than 90 % of the light. Thus, regular CCD cameras may not be sufficient in low lighting circumstances. In recent years, high performance CCD cameras such as electron multiplying CCD (EMCCD) were increasingly employed in reported HSI applications (Park et al. 2012; Kim et al. 2011; Yoon et al. 2011). For NIR HSI imaging, two types of high performance photodiode detectors are generally used: the indium gallium arsenide (InGaAs) sensor and the mercury cadmium telluride (HgCdTe) sensor. Both the InGaAs and HgCdTe sensors have high quantum efficiency in the NIR spectral region. The HgCdTe detector covers a broad spectral region of 1000–12,000 nm, but it's quite expensive and needs to be operated at a high temperature. In comparison, the InGaAs detector has a limited detection range (900–1700 nm) while the cost is much lower than the HgCdTe detector.

14.3.1.2 LCTF

The tunable spectral range and the aperture size are two fundamental parameters of the LCTF. Generally, due to the low optical throughput of the LCTF, a large aperture LCTF is preferred. Two other important parameters for the LCTF are the tuning speed and the average bandwidth, which often vary across wavelengths. Moreover, the angle of view (AOV) of the LCTF might alter the AOV of the hyperspectral imager. Overall, the off-the-shelf products of LCTF are still quite limited, and users might have to select the optimal one based on the availability.

14.3.1.3 Lens

The lens of the hyperspectral imager is a critical component that directly determines several essential parameters of the spectral imager, such as the focal length (f), angle of view (AOV), field of view (FOV), and the magnification (M). Similar to the conventional camera system, these parameters can be estimated by using the pinhole model of the camera (Fig. 14.1). However, it should be aware that the AOV of the hyperspectral imager could also be affected by the AOV of the LCTF. A couple of other things should be considered in the selection of the lens: a fast lens (with large aperture) is often preferred since the LCTF absorbs a great amount of light, and the lens should have proper coating to enhance the throughput and reduce the image distortion in the desired spectral region.

14.3.1.4 Layout of the Hyperspectral Imager

There are two common layouts (Fig. 14.2) for the LCTF-based hyperspectral imager: (I) the lens is placed between the LCTF and the camera, and (II) the lens is put in front of the spectral imager. The assembly layout of the hyperspectral imager (the camera, the LCTF, and the lens) should be determined before selecting a lens, since different assembly layouts would have different requirements of the lens.

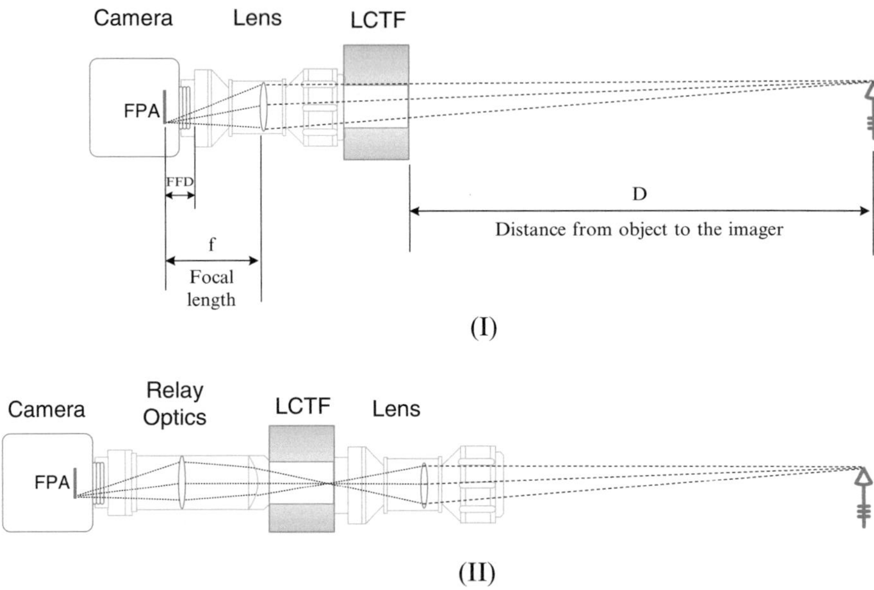

Fig. 14.2 Two common layouts for the LCTF-based spectral imager: (**a**) the lens is placed between the LCTF and the camera, and (**b**) the lens is in front of the LCTF, and relay optics is used to focus the image of the object on the camera. Reproduced with permission from Wang et al. (2012); ©2012 Elsevier

For the layout I, due to the long and narrow light-path inside the LCTF, the light could be easily blocked by the filter, resulting in reduced light at the edge of the lens (mechanical vignetting). As a result, the edges of the captured images could be blurry. A lens with a large *f* could mitigate the vignetting problem, but it would require a longer object-to-camera distance due to the smaller angle of view. Therefore, in some applications, the *f* of the lens would be a trade-off between minimizing image vignetting and maximizing the AOV.

For the layout II, it's necessary to use a lens of a long flange focal distance (FFD) to focus the object on the focal plane arrays (FPA) of the camera. The FFD of a lens refers to the distance between the lens rear flange and the focal plane of the camera. Currently, the thickness of the LCTF is often greater than the FFD of most off-the-shelf lenses. Thus, this layout often requires a custom relay optics to refocus the light to the FPA of the detector, which would greatly increase the complexity, size, and design cost of the spectral imager.

14.3.1.5 Illumination

LCTF-based hyperspectral imaging systems require area illumination. A good area illumination for the hyperspectral imaging should provide stable spectral output and uniform lighting in the illuminated area. Tungsten halogen lamps and LEDs are two most common light sources used for the hyperspectral imaging system (Lawrence et al. 2007). Besides high performance lighting bulbs, stabilized DC power supply is often desired to keep the spectral output of the lamps stable. Similar to conventional machine vision systems, light uniformity can be enhanced by using multiple lamps and arranging them with suitable geometry, and by applying optic diffusers, reflectors, etc. In many applications, chambers with high reflectance coating were also used to enhance the light uniformity.

14.3.1.6 Data Acquisition Program

Another major effort of integrating the LCTF-based HSI system is data acquisition program, which synchronizes the camera and the LCTF to collect 2-D images sequentially, and constructs the 3-D hyperspectral image cube ((x,y) in spatial and λ in spectral axis) in the meantime. To develop data acquisition program for an HSI system, many programming languages could be used, such as C++ (Yoon et al. 2011), Microsoft Visual Basic (Kim et al. 2001), and LabVIEW (Wang et al. 2012). The selection of the programming language depends on many factors, such as the developer's expertise and the availability of the hardware drivers.

The key function of the data acquisition program is to construct the 3-D image cube in the format that could be recognized by other HSI data analysis programs. Currently, there are three common formats to encode a spectral image: band

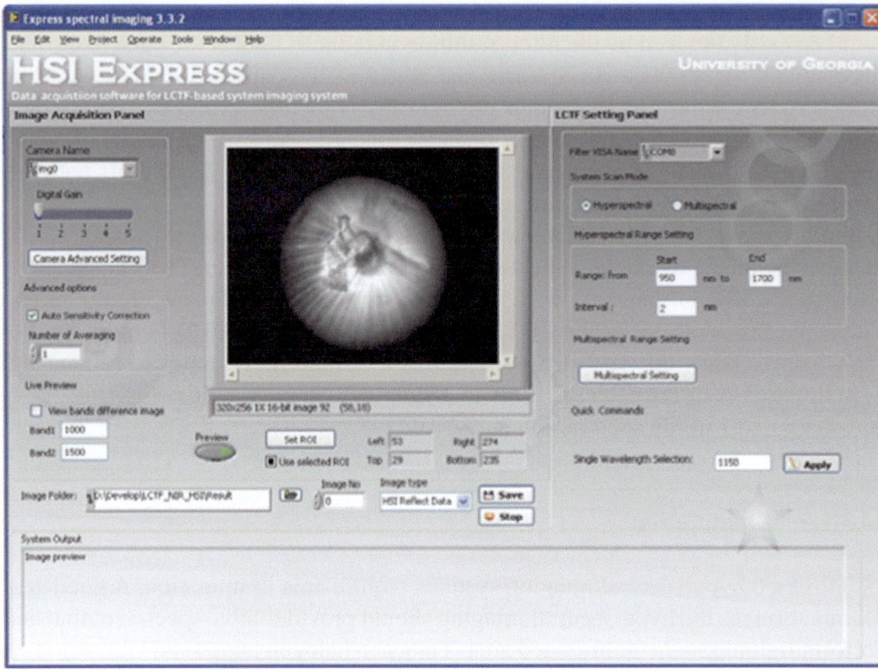

Fig. 14.3 An LabVIEW data acquisition program for an LCTF-based spectral imaging system

interleaved by pixel format (BIP), band interleaved by line format (BIL), and band sequential format (BSQ). The BIP format first stores the spectrum of the first pixel across all bands, and then saves the spectrum of the next pixel in succession until the last pixel. The BIL format uses line as the processing unit: save the first line of the image at the first band, and then iteratively process the same line of the image at the following bands. Then, it saves the remaining lines for all bands successively. The BSQ format stores 2-D spatial images band by band in a sequential order. Although all three formats are inter-convertible, for the LCTF-based spectral imaging system, it's most efficient to encode the spectral images in BSQ format.

The data acquisition program should be user-friendly and reliable. If needed, data pre-processing, post-processing, and data analysis functions could also be included. Figure 14.3 illustrates an LabVIEW data acquisition program for an LCTF-based spectral imaging system (Wang et al. 2012), which was designed to acquire either hyperspectral or multispectral images over the spectral range of 900–1700 nm. The program also provides certain advanced functions to enhance the usability of the system, such as the spectral sensitivity correction, noise reduction, selection of region of interest, ensemble averaging (co-adding), and band ratio image, etc.

14.3.2 System Calibration

Similar to other types of spectral imaging system, an LCTF-based hyperspectral imaging system should be calibrated in both spatial and spectral domains. The main aspects of an HSI system that should be calibrated include: the spectral accuracy, resolution, sensitivity, linearity, stability of the system, the spatial resolution, field of view, spatial pattern noise, and image distortion, etc. (Lawrence et al. 2003; Lu and Chen 1998; Wang et al. 2012).

14.3.2.1 Calibration in Spectral Domain

- Spectral sensitivity

  For an HSI system, its spectral output should be similar at the wavelength bands where the object has identical spectral characteristics. In reality, the system's spectral sensitivity varies across the spectral bands because each component of the system responds differently at different wavelengths. For the LCTF-based spectral imaging system, its spectral sensitivity is determined by the spectral output of the light source, the transmittance of the LCTF, the transmittance of the lens, and the sensitivity of the camera. Since calibrating each unit individually could be formidable, the spectral sensitivity correction of the HSI system is often conducted by considering the HSI system as an integral unit. For the line-scan HSI system, the spectral sensitivity correction mainly counts on the flat-field correction. For the LCTF-based HSI system, in addition to the flat-field correction in the image post-processing stage, the correction spectral sensitivity could be conducted by adjusting the camera's exposure time and digital gains during the data acquisition.
- Linearity

  The calibration of the system's spectral linearity is often carried out by measuring multi-step contrast standards. A multi-step contrast target often consists of multiple side-by-side sub-panels, which have different levels of known reflectance rates (in percentage). These sub-panels and a white reference target should be scanned under the same circumstances (lighting, position, temperature, etc.). Then, the reflectance values of the sub-panels can be converted to relative values (the system's responses to these targets) using flat-field corrections. The linearity of the HSI system can be evaluated by testing the linear relationship between the measured reflectance values and the actual values of the targets band by band.
- Stability

  The stability of an LCTF-based HSI system can be evaluated by the same methods used for other spectroscopic/imaging systems: repeatedly measuring a target of standard reflectance material (i.e., a certified white reference panel or a Teflon board) within a certain period. Then the variances could be evaluated in the spectra point by point, or in the images band by band. If the system tends to be unreasonably unstable, it's critical to identify the reason. There are several factors that could greatly affect the stability of an LCTF-based HSI system: the stability of the power source, the performance of the illumination unit, and the operation temperature of the LCTF/the camera. Sometimes, improper processing algorithm in software could also affect the stability of the system.

- Denoising

  There are many methods that can be applied to reduce the spectral noises of an HSI system. The most common one is ensemble averaging (co-adding), which simply averages multiple scans so that the random noises in scans could counteract each other. In the LCTF-based hyperspectral imaging, this could be done in either spatial domain (by scanning and averaging multiple images at each band) or spectral domain (by binning the pixels after scanning).

14.3.2.2 Calibration in Spatial Domain

- Angle of view (AOV)/Field of view (FOV)

  The FOV/AOV of the LCTF-based hyperspectral imaging system are affected by multiple components (the LCTF, the lens, and camera). Thus, for an LCTF-based hyperspectral imaging system, instead of calculating based on the parameters of the components, it is better to measure its effective FOV as a function of distance and then calculate the AOV of the system.

- Spatial pattern noises

  Due to the non-uniformity of the illumination and the noises in the FPA of the detector, the HSI system often has significant responsivity variations in the spatial domain, known as "pattern noise". In spectral imaging, this kind of noise is often corrected in the image post-processing stage by the flat-field correction, which converts raw spectral images to percentage spectral images using the "white-reference" and "dark" images (Lu and Chen 1998).

- Spatial resolution

  The spatial resolution of this system can be easily measured by using the standard image resolution targets such as USAF 1951 and NBS 1963A resolution targets. However, the focus point of the LCTF-based HSI system could vary at different wavelengths. The spatial resolution in the defocused images could be much lower than that of in the well-focused images. Thus, it is necessary to measure the spatial resolution of the system across the wavelength bands.

- Image distortion and shift

  In an LCTF-based HSI system, the image distortion/shift in the spectral images is mainly caused by the chromatic aberration and geometric distortion of the lens. To calibrate these errors in hyperspectral images, a simple approach is to extend the conventional lens distortion correction method in the 2-D image to the 3-D hyperspectral image. Usually, a target with certain pattern(s) should be scanned from different perspectives, and then the lens distortion can be measured by using certain algorithms band by band. A number of geometric controlled points in the target could be used to evaluate the image shift by comparing the pixel positions of these points at different wavelengths.

14.4 Applications of LCTF-Based Hyperspectral Imaging to Fresh Vegetables

14.4.1 Detecting Sour Skin in Onions

Sour skin (*Burkholderia cepacia*) is a major bacterial disease in onion postharvest. The exposure of onions to sour skin infection is particularly dangerous in storage rooms since the pathogen will spread gradually and affect other clean onions, resulting in substantial storage losses. Moreover, some strains of *Burkholderia cepacia* (*B. cepacia*) are human pathogens, which were deemed as leading causes for death in individuals with cystic fibrosis (Holmes et al. 1998). Thus, it's important to identify and eliminate sour skin-infected onions on the onion sorting lines so that *B. cepacia* could not enter the storage room and not been consumed by human. Conventional automatic classification systems for vegetables, however, are not capable to screen sour skin-infected onions from clean ones due to the complicated presence of the outer dry skins on onions.

Wang et al. (2012) reported an application of detecting sour skin-infected onions by LCTF-based spectral imaging. The study demonstrates the efficacy of using LCTF-based spectral imaging to develop classification models for the quality inspection of vegetables.

14.4.1.1 NIR Hyperspectral Imaging System

An LCTF-based NIR spectral imaging system (Wang et al. 2012) was integrated to acquire hyperspectral reflectance images of onions in the spectral region of 950–1650 nm. The system mainly consisted of the following hardware components:

- An LCTF (LNIR 20-HC-20, Cambridge Research & Instrumentation, Cambridge, MA, USA), which can select a narrow bandpass of light from 850–1800 nm, with 20 nm bandwidths on average.
- An indium gallium arsenide (InGaAs) camera (SU320KTS-1.7RT, Goodrich, Sensors Unlimited, Inc, Princeton, NJ, USA).The camera has 320×256 pixels focal plane array (FPA) with 25 μm pitches, with a maximum speed of 60 fps and 12-bit digital output.
- A near-infrared lens (SOLO 50, Goodrich, Sensors Unlimited, Inc, Princeton, NJ, USA). The lens (50 mm focal length, f/1.4 aperture) has high throughput in the spectral region of 900–1700 nm.
- Four 12-V 50 watt DC quartz halogen lamps (S4121, Superior Lighting, Fort Lauderdale, FL, USA) for providing a NIR lighting source. Ground glass diffusers were applied to increase the uniformity of the lighting.
- A Camera Link frame grabber (NI PCI-1426, National Instruments, Austin, TX, USA) for acquiring images from the InGaAs camera.

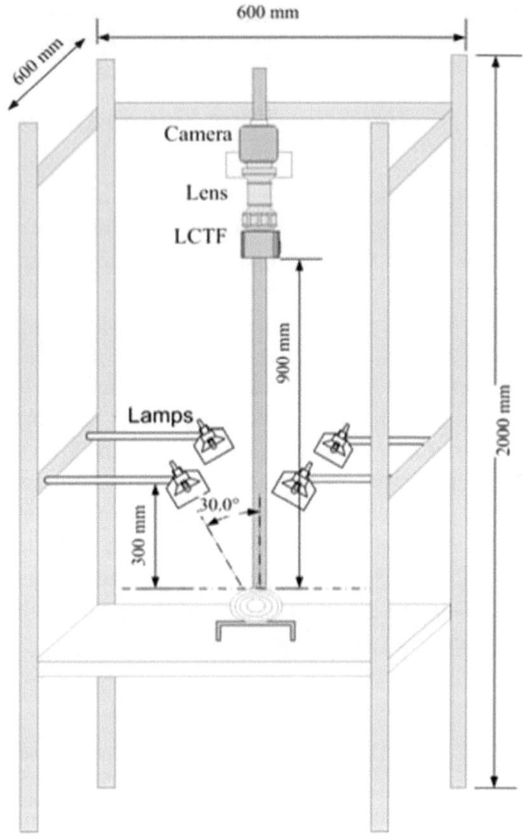

Fig. 14.4 The hardware layout the LCTF-based hyperspectral imaging system for detecting sour skin in onions. Reproduced with permission from Wang et al. (2012); ©2012 Elsevier

Figure 14.4 shows the hardware layout of the LCTF-based hyperspectral imaging system. The system was enclosed in a 600 × 600× mm (L×W×H) aluminum chamber. The chamber was covered by black cloth to avoid the outside ambient light. An image acquisition program was developed in LabVIEW graphical programming language (National Instruments, Austin, TX, USA) and was installed on a desktop computer (Intel Duo processor E8200 and 4 GB RAM). This system was calibrated in both spatial and spectral domains, the detail procedure and techniques for calibrating the system are described in Wang et al. (2012).

14.4.1.2 Sample Preparation and Image Acquisition

In total, 75 medium/jumbo sweet onions (c.v. Savannah Sweet) were used. Onions were harvested from the state of Georgia, USA in May 2010. All onions were manually selected and inspected to be clean. The selected onions were stored individually in plastic bags and labelled by sequential integer numbers written on the bags. The onions were randomly divided to two groups: 70 samples (group 1) and 5 control samples (group 2). A suspension of *B. cepacia* was prepared in sterilized tap water as sour skin inoculum. *B. cepacia* was originally isolated from sour skin-infected onions harvested in Georgia and cultured on potato dextrose agar medium.

On day 0, all 75 onion samples were scanned in 950–1650 nm (2 nm intervals) using the LCTF-based HSI system. Each onion was scanned three times and the averaged hyperspectral image was used. After scanning, the 70 onions in group 1 were inoculated with the *B. cepacia* inoculum. The 5 onions in group 2 were inoculated with sterilized tap water as control samples. Inoculated onions were stored in plastic bags individually to avoid cross contamination. All samples were placed in an incubator at $30 \pm 1\,^{\circ}\text{C}$ and 80 % relative humidity. This inoculation/incubation procedure intended to mimic the natural process of sour skin infection, which could result in the early sour skin symptoms on onion surface in 4–5 days.

On the fifth day after the inoculation (DAI), all onion samples were scanned again using the LCTF-based HSI system, with the same configurations on day 0. After all scans, onions were cut in half from the neck to root to confirm the real infection areas. In total, 150 hyperspectral images were collected from 75 onion samples on 2 days. All hyperspectral images were converted to relative images using flat-field corrections.

14.4.1.3 Wavelength Selection

In non-destructive quality inspection of vegetables, scanning time is often desired to be short. Thus, for many applications of using hyperspectral imaging for quality inspection of vegetables, wavelength selection should be conducted to identify the key wavelengths contributing to the classification. In the other words, the wavelengths that have no or small discrimination power should be eliminated to improve the efficiency of the data processing.

In this application, 5 onions inoculated with *B. cepacia* were selected for wavelength selection. The hyperspectral images of the selected onions on day 0 were used to extract the spectra of healthy onion and the ones on 5 DAI were used to acquire the spectra of sour skin-infected onions. Four regions of interest (ROIs) (6×6 pixels) were manually selected from the neck area and shoulder bulb area on each onion. Spectra were extracted from the selected ROIs using the ENVI software (ITT Visual Information Solutions, Boulder, CO, USA), respectively. The extracted reflectance spectra were averaged and converted to absorption spectra by using log(1/R) (Fig. 14.5), in which R refers to the mean reflectance spectrum of that type of onion tissue.

When onions were sour skin-infected, the absorbance spectrum of neck tissues was about 26–62 % higher than that of the healthy ones in 950–1650 nm. The largest disparity occurred in the spectral region of 1410–1460 nm, which mostly caused by the difference of the moisture content in healthy and diseased neck tissues. Based on the direct observations on these spectra, the neck tissues are more indicative to sour skin.

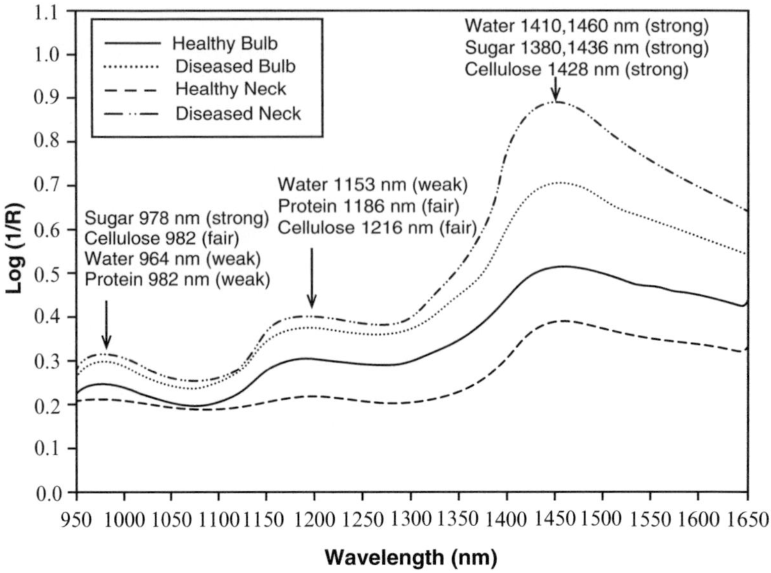

Fig. 14.5 The representative spectra for disease-free (healthy) bulb tissues and neck tissues, sour skin-infected (diseased) bulb tissues and neck tissues of onions. Reproduced with permission from Wang et al. (2012); ©2012 Elsevier

To determine a minimum number of key wavelengths, principal component analysis (PCA) was used to analyze the spectral data. PCA is a classic eigenvector-based algorithm for feature selection and dimension reduction, in which the variance of the original data set are explained by a number of principal components (PCs). The loadings (coefficients) of PCs determine the weights (importance) of original variables in the PCs.

In this application, PCAs were conducted on the datasets of the extracted spectra of the tissues on the onion neck area and the tissues on the bulb area, respectively. For the neck tissues, the first and second PC accounted for 83.53 % and 15.12 % variance of the dataset, respectively. For the bulb tissues, the first and second PC represented 95.26 % and 4.24 % variance of the dataset, respectively. That means, for either PCA model, the majority of the variance can be described by the first two PCs. Therefore, wavelength selection was conducted based on the loading values of the first and second PC (Fig. 14.8). As a common practice, the maxima and minima of their PC loadings were identified as the potential key spectral bands. As a result, two pairs of important wavelengths (1070 and 1420 nm) and (1070 and 1400 nm) were identified from the PCA models on the spectra of onion bulb tissues (Fig. 14.8a) and the spectra of onion neck tissues (Fig. 14.8b), respectively.

The wavelength pair of 1070 and 1400 nm was finally selected for the following justifications: (1) from the direct observations in Fig. 14.5, the neck tissue is more indicative; (2) the spectral characteristics of onion tissues at wavelengths 1400 and

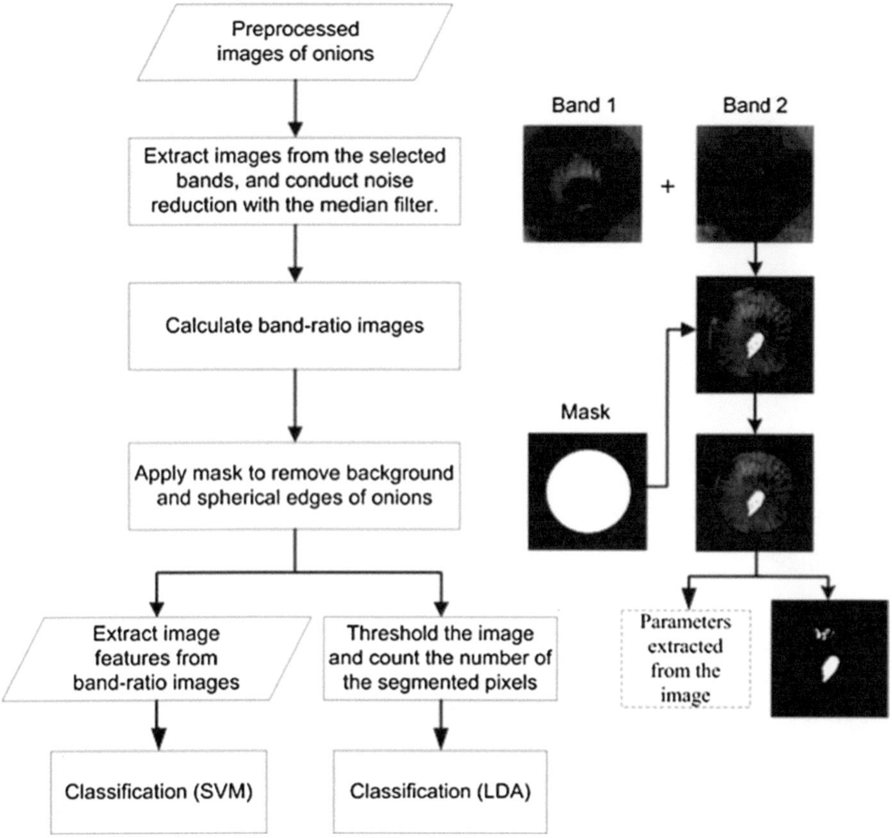

Fig. 14.6 Flowchart of the imaging processing procedure for the feature selection and the detections of sour skin-infected onions. Reproduced with permission from Wang et al. (2012); ©2012 Elsevier

1420 nm are quite close. Thus, there is no need to include both images at 1400 and 1420 nm in classifications.

14.4.1.4 Image Processing for Features Extraction

After thc key wavebands were determined, image processing techniques were applied to extract features for classifications (Fig. 14.6). Onion images at two selected wavelengths (1070 and 1400 nm) were extracted from their hyperspectral images, and then combined to a single grayscale image by applying the logarithmic conversion of the ratio (log-ratio): $I_R = log_{10}\frac{I_{1070}}{I_{1400}}$. To avoid infinite values in ratio images, the pixels with zero values in the image at 1400 nm were excluded from computation, and the 3×3 median filter was applied to pre-process both images

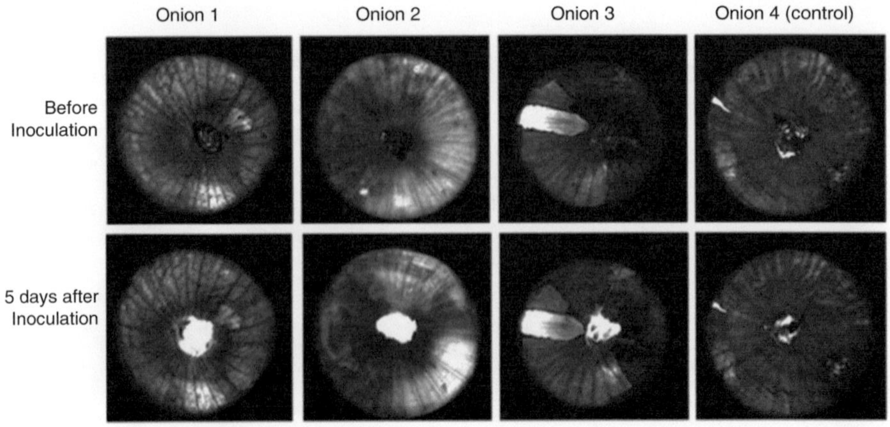

Fig. 14.7 Log-ratio images of four onion samples. Onions 1–3 were inoculated with *B.cepacia* and onion 4 was inoculated with sterilized tap water. The images at the *top row* were extracted from the HSI images of onions before inoculation, and the ones at the *bottom row* was obtained from the HSI images taken on 5 days after inoculation. Reproduced with permission from Wang et al. (2012); ©2012 Elsevier

before log-ratio conversions. Compare to regular band ratio methods, the advantage of using log-ratio conversion is that it can transform the residual multiplicative speckle noise to an additive noise component (Chen 2007).

Figure 14.7 illustrates the log-ratio images of three onion samples inoculated with *B. cepacia* and one control sample inoculated by tap water. For the onion samples inoculated with *B. cepacia*, the neck area on their log-ratio images on 5 DAI was much brighter than the body area, while this difference was not observed in the images before inoculation. The log-ratio images of the control sample (onion 4 in Fig. 14.7) before inoculation and on 5 DAI, in contrast, are quite close. Therefore, the bright neck areas (high ratio values) in the log-ratio images of sour skin-infected onions were associated with the sour skin symptoms appearing on the neck area of the onions.

The bright region on the bulb area of the onions are also indicative of the infection of sour skin. However, the contrast of sour skin-infected tissue to healthy tissue in body area was not as apparent as that in the neck area. This could be explained by the physical structure of the onion bulb. When an onion is infected by sour skin, the fluid released from rot onion tissues flows in the spaces between internal scales of the onion. Compared to dry skins/flesh scales on body area, dry neck tissues of onions could more easily absorb and hold the fluid, which results in a higher light absorbance at 1400 nm (a strong light absorption band of water). In sum, the presence of the sour skin changed the reflectance intensity with different rates at 1070 and 1400 nm, which are detected by the log-ratio images. Therefore, these log-ratio images are useful to distinguish the sour skin onions from healthy ones.

As illustrated in Fig. 14.6, two types of image features were further extracted based on the log-ratio image of onions: one directly segregated the diseased areas of onions and used the total pixel number as the feature, and the other extracted the statistical and textural image features from the log-ratio images. The first approach basically estimated the size of areas that show sour skin symptom. As for the second approach, three statistical or textural image features were used: max, contrast, and homogeneity, which were selected from 12 common image features (max, min, range, median, mean, standard deviation, skewness, kurtosis, entropy, contrast, energy, and homogeneity) by applying stepwise discriminant analysis. The feature "max" describes the largest difference in intensity between the onion images at the wavelengths of 1070 and 1400 nm. For a sour skin-infected onion, this parameter is related to the most serious sour skin infection area that gives the maximal value in log-ratio images. The feature "contrast" measures the local gray level variations among pixels and the last parameter "homogeneity" measures the closeness of local gray values. The "contrast" and "homogeneity" are related to both the size and the level of the infection of overall sour skin infection area on the log-ratio images. For example, if an onion has a larger infection area and higher infection level than others, its log-ratio image should have a high "contrast" and low "homogeneity" values. In sum, the combination of "max", "contrast", and 'homogeneity" considered both the size of the infected area and the infection level of sour skin in onions.

14.4.1.5 Classification

Two types of supervised classifiers were developed to classify healthy and sour skin-infected onions. The first type of classifier used Fisher's linear discriminant analysis (LDA) to make discriminations, based on the pixel number of the segregated diseased areas in onion log-ratio images. The second type of classifier was developed on support vector machine (SVM) based on the three image features extracted from the onion log-ratio images. The optimal configurations of the LDA and SVM classifiers were searched and evaluated using the tenfold cross validation, in which 126 log-ratio images were used for training and 14 log-ratio images were used for classification in each fold. The classification results are listed in Table 14.1

Table 14.1 The classification results of the LDA and the SVM classifiers

		LDA			SVM		
		Healthy	Sour skin	Accuracy	Healthy	Sour skin	Accuracy
Actual	Healthy	68	2	97.14 %	57	13	81.43 %
	Sour skin	26	44	62.86 %	5	65	92.86 %
	Sum	94	46	80 %	62	78	87.14 %

Reproduced with permission from Wang et al. (2012); ©2012 Elsevier
LDA linear discriminant analysis based on the pixel number after thresholding (threshold = 0.45), *SVM* support vector machine using three image parameters (max, contrast, and homogeneity)

For the approach of using LDA, the best classifier using the global threshold 0.45 achieved a classification rates of 80 %, with a relatively high false negative (26 out of 70) and low false positive (2 out of 70). The mis-classifications were mainly caused by the pixels of the wet flesh (no dry skin) of healthy onion, which has high log-ratio values due to high water content. For instance, on the onion 3 in Fig. 14.7, there is an small onion bulb area where the dry skin fell off. The exposed flesh scale, due to its high moisture content ($>80\,\%$), also showed higher log-ratio values similar to the sour skin-infected tissues. Due to these false positive areas, the global threshold value determined by the LDA was relatively high. As a result, the onion samples with relatively small infection area were mis-classified as healthy onions.

The optimal SVM classifier (RBF kernel, $\gamma = 1.5$) using three image features achieved a higher classification accuracy of 87.14 %. It performed a little better than the LDA classifier because that the SVM classifier using three image features considered both the size and the infection level of onion sour skin infections, whereas the LDA method can only count on the total size of sour skin infected areas. Thus, the SVM classifier achieved a better balance between false negative (5 out of 70) and false positive (13 out of 70). Similar to the LDA classifier, the relatively high false positive of the SVM classifier was mostly caused by those healthy onions with wet flesh scales on the bulb areas. Therefore, to further improve the performance of the proposed algorithm, the wet flesh scales on onions should be identified and excluded before classification.

In sum, this application illustrates an example of using the LCTF-based HSI to detect sour skin-infected onions. The log-ratio image utilizing two optimal wavelengths (1070 and 1400 nm) was proven to be effective in magnifying the spectral differences between healthy and sour skin-infected onions. Moreover, the application shows the flexibility of the LCTF-based spectral imaging system, which could be used as the hyperspectral imaging system to develop the inspection approach, and then be used as the multispectral imaging system for verifying the developed method.

14.4.2 Onion Internal Quality Prediction

The second application example is provided to demonstrate onion internal quality prediction using the LCTF-based HSI system.

A total of 308 onion samples harvested in three states in the United States in 2011 were used to develop calibration models in this study. All onion samples were stored in a cold storage room ($2 \pm 1\,^{\circ}$C and relative humidity 70 %) in the Vidalia Onion Lab of the University of Georgia until experiments were conducted. The onions were moved to a room with an ambient temperature of 23.9 °C about 2 h before data acquisition. All onions were cleaned by removing the surface dirt and dry leaves before imaging. Firmness reference was measured manually using the Magness-Taylor testing platform (FT 30, Wagner Instruments, Greenwich, Connecticut, USA) with a 11 mm in diameter probe. The squeezed onion juice was tested with a refractometer (range of 0–30°Brix) for the soluble solids content. Onion samples were placed in aluminum cups and were heated in an oven

(Model 845, Precision Scientific Company, Winchester, IL, USA) with a temperature of 75 °C for 12 h. The dry matter content was calculated as the ratio of the onion weight after and before the oven drying.

The LCTF-based NIR hyperspectral imaging system was developed by the Bio-Sensing and Instrumentation Lab at the University of Georgia, which was the same as used in the first case study (Sect. 14.4.1). Diffuse reflectance mode was used for image acquisition. Onions were sampled at two points (180° apart) when the root-neck axis was placed horizontally. A total of 616 hyperspectral images were saved for 308 onion samples. Dark spectra images with lens covered and white spectra images of the standard white board (Spectralon, Labsphere, Inc., North Sutton, New Hampshire, USA) were collected for every 10 onions. Then, the flat-field correction was conducted for each of the spectra image of onions (Fig. 14.8).

The spectra of onions were extracted from five regions of interests (ROI) in the onion image (Fig. 14.9). Each ROI was a 10 × 10 pixel square, corresponding to

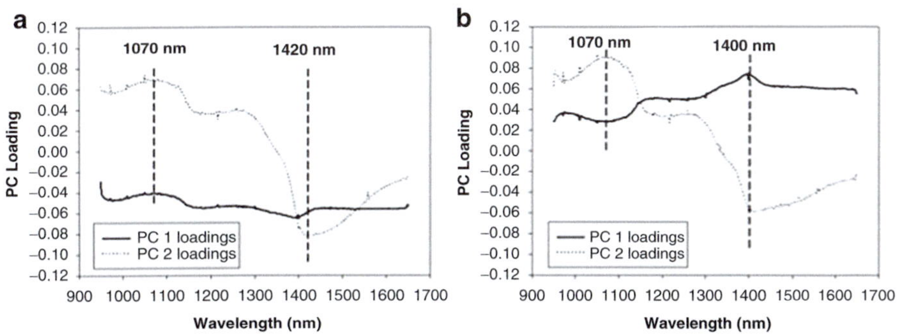

Fig. 14.8 The loading values of the first two principal components (PC 1 and PC 2) for the PCA on the spectra of the bulb tissues (**a**) and the spectra of the neck tissues (**b**). Reproduced with permission from Wang et al. (2012); ©2012 Elsevier

Fig. 14.9 ROI selection on a reflectance image. The numbers indicate the distance between two ROIs (in pixel)

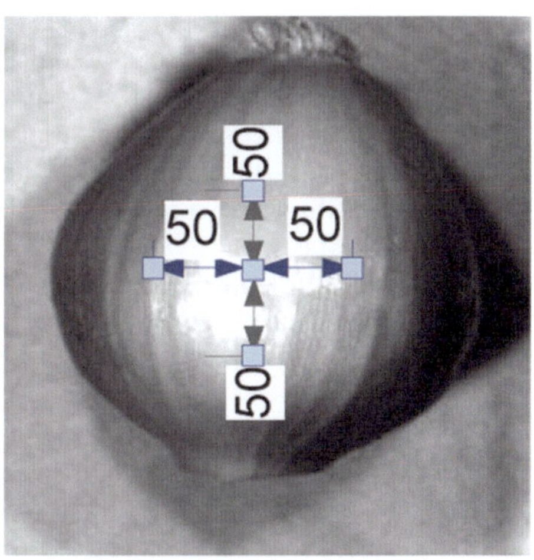

100 individual spectra. In order to reduce the amount of data and enhance the signal to noise ratio, the 100 spectra in each ROI were averaged as one individual spectrum for later analysis.

Partial least square regression (PLSR) technique was used to utilize the spectral data for onion internal quality prediction. The extracted spectra were preprocessed using wavelet smoothing and multiplicative scatter correction. The Interactive Dynamic Language (IDL) programming language (ITT Visual the Information Solutions, Boulder, CO, USA) was used to pre-process the spectral images. MATLAB 2009b (the Math Works Inc., in Natick,, Massachusetts, USA) was used for PLSR model development.

The performance of the calibration and validation model for onion internal quality prediction was presented in Table 14.2. The prediction of SSC and DM were comparable ($R^2 = 0.81$–0.83), both of which were substantially better than firmness ($R^2 = 0.46$). The RPD (ratio of SEP to SD) is an indicator of robustness of the model. The RPD value of the validation for both SSC and DM was greater than 2, suggesting that both modes could achieve a reasonable prediction accuracy (Nicolaï et al. 2007). Several studies have reported that SSC can be reliably predicted using NIR spectroscopy for several fruits such as kiwifruits (SEP = 0.80°Brix) (Schaare and Fraser 2000), apples (SEP = 0.28–0.56°Brix) (Park et al. 2003), and sweet cherries (SEP = 0.65–0.71°Brix) (Lu 2001). Although the SSC prediction presented in this study was not as good as those for other fruits in the literature, it was better than that in a similar study done on onions (SEP = 3.41°Brix) (Birth et al. 1985).

Given the high correlation between the DM and the SSC, it is not surprising that the performance of the calibration models for DM prediction was comparable to that of the SSC prediction. In Birth et al.'s study (Birth et al. 1985), the SEP of DM prediction was between 0.79 and 1.73 %, which was better than the result obtained in this study. Provided that the spectral imaging system typically has lower SNR than the spectrometer, the result obtained from this study was respectable.

It seemed that firmness could not be well predicted using the NIR spectral imaging system with reflectance mode. This could be due to two reasons: (1) firmness is determined by cellular structure of the tissue, which is more of a physical property than a chemical property, whereas the strength of the NIR spectroscopy lies in measuring chemical properties, such as sugar and water content; (2) the reference measurement of firmness could be prone to error. As Lu et al. (2000) acknowledged, there was a poor correlation between apple firmness and the maximum force of Magness-Taylor (MT) firmness measurement (a penetration test). In addition, the multi-layer structure of the onion may also pose a challenge for accurate firmness measurement.

Table 14.2 Calibration results of the PLSR model for onion internal quality prediction using LCTF-based HSI system

		Calibration				Validation			
Quality	Factors	R^2	SEC	RPD	SD	R^2	SEP	RPD	SD
Firmness	12	0.73	8.19	1.91	15.66	0.46	12.00	1.36	16.29
SSC	14	0.94	1.39	4.14	5.74	0.83	2.54	2.42	6.15
DM	16	0.96	1.34	4.78	6.41	0.81	2.91	2.30	6.71

SEC standard error of calibration, *RPD* residual predictive deviation, *SD* standard deviation, *SEP* standard error of prediction

14.5 Conclusion

The hyperspectral imaging system based on the liquid crystal tunable filter has its unique advantages over the line scan systems. First, the filter based HSI system does not have moving parts, which can be easily implemented in certain applications. Second, an existing CCD or NIR camera system could be readily retrofitted into a hyperspectral imaging system by adding the LCTF component. Therefore, LCTF-based spectral imaging provides an alternative and competitive nondestructive solution compared to other hyperspectral imaging systems. As demonstrated in two cases of this chapter, the HSI system based on the LCTF showed great promise to detect invisible bacterial diseases and to predict internal qualities of onions. We expect that the LCTF-based HSI systems will have more applications in vegetable quality evaluation in the future. There are many aspects that LCTF HSI systems can be further improved, such as having a faster tuning speed, larger working aperture, and broader wavelength range.

Acknowledgements Authors would like to thank Dr. Haihua Wang for his work on some of the data presented in this chapter.

References

Archibald D, Thai C, Dowell F (1999) Development of short-wavelength near-infrared spectral imaging for grain color classification. Proc SPIE 3543:189–198

Ariana D, Lu R, Guyer DE (2006) Near-infrared hyperspectral reflectance imaging for detection of bruises on pickling cucumbers. Comput Electron Agric 53:60–70

Birth G, Dull G, Magee J, Chan H, Cavaletto C (1984) An optical method for estimating papaya maturity. J ASHS 109(1):62–66

Birth G, Dull G, Renfroe W, Kays S (1985) Nondestructive spectrophotometric determination of dry matter in onions. J ASHS 110(2):297–303

Boyhan GE, Torrance RL (2002) Vidalia onions–sweet onion production in southeastern georgia. HortTechnology 12(2):196–202

van de Broek W, Wienke D, Melssen W, de Crom C, Buydens L (1995) Identification of plastics among nonplastics in mixed waste by remote sensing near-infrared imaging spectroscopy. 1. Image improvement and analysis by singular value decomposition. Anal Chem 67 (20):3753–3759

Burden D (2008) Onion profile. http://www.agmrc.org. Retrieved on 6 Aug 2008

Burkholder WH (1950) Sour skin, a bacterial rot of onion bulbs. Phytopathology 40(1):115

Ceponis M, Cappellini R, Lightner G (1986) Disorders in onion shipments to the New York market, 1972–1984. Plant Dis 70(10):988–991

Chen C (2007) Image processing for remote sensing. Taylor & Francis, Boca Raton

Cho SI, Krutz GW, Gibson HG, Haghighi K (1990) Magnet console design of an NMR-based sensor to detect ripeness of fruit. Trans ASAE 33(4):1043–1050

Cogdill RP, Hurburgh Jr, CR, Rippke GR (2004) Single-kernel maize analysis by near-infrared hyperspectral imaging. Trans ASAE 47(1):311–320

Cubero S, Aleixos N, Moltó E, Gómez-Sanchis J, Blasco J (2011) Advances in machine vision applications for automatic inspection and quality evaluation of fruits and vegetables. Food Bioprocess Technol 4:487–504

Dull G, Birth G, Leffler R (1989) Use of near infrared analysis for the nondestructive measurement of dry matter in potatoes. Am Potato J 66:215–225

Dull GG, Birth GS, Smittle DA, Leffler RG (1989) Near infrared analysis of soluble solids in intact cantaloupe. J Food Sci 54(2):393–395

ElMasry G, Wang N, ElSayed A, Ngadi M (2007) Hyperspectral imaging for nondestructive determination of some quality attributes for strawberry. J Food Eng 81(1):98–107

ElMasry G, Wang N, Vigneault C, Qiao J, ElSayed A (2008) Early detection of apple bruises on different background colors using hyperspectral imaging. LWT - Food Sci Technol 41:337–345

Evans M, Thai C, Grant J (1998) Development of a spectral imaging system based on a liquid crystal tunable filter. Trans ASAE 41(6):1845–1852

Ezzati M, Riboli E (2012) Can noncommunicable diseases be prevented? Lessons from studies of populations and individuals. Science 337:1482–1487

Fernandes A, Oliveira P, Moura J, Oliveira A, Falco V, Correia M, Melo-Pinto P (2011) Determination of anthocyanin concentration in whole grape skins using hyperspectral imaging and adaptive boosting neural networks. J Food Eng 105(2):216–226

Florkowski WJ, Shewfelt RL, Brueckner B, Prussia SE (2009) Postharvest handling a system approach, 2nd edn. Academic, New York

Food and Agriculture Organization of the United Nations (2012) World onion production 2010. http://faostat.fao.org/site/567/default.aspx. Retrieved 10 Nov 2012

Gat N (2000) Imaging spectroscopy using tunable filters: a review. Proc SPIE 4056:50–64

Gómez-Sanchis J, Gomez-Chova L, Aleixos N, Camps-Valls G, Montesinos-Herrero C, Molt E, Blasco J (2008) Hyperspectral system for early detection of rottenness caused by *Penicillium digitatum* in mandarins. J Food Eng 89(1):80–86

Holmes A, Govan J, Goldstein R (1998) Agricultural use of Burkholderia (Pseudomonas) cepacia: a threat to human health? Emerg Infect Dis 4(2):221–227

Jiang L, Zhu B, Rao X, Berney G, Tao Y (2007) Discrimination of black walnut shell and pulp in hyperspectral fluorescence imagery using gaussian kernel function approach. J Food Eng 81:108–117

Kazemi S, Wang N, Ngadi M, Prasher SO (2005) Evaluation of frying oil quality using VIS/NIR hyperspectral analysis. Agric Eng Int VII

Kim MS, Chen YR, Mehl PM (2001) Hyperspectral reflectance and fluorescence imaging system for food quality and safety. Trans ASAE 44(3):720–729

Kim MS, Chao K, Chan DE, Jun W, Lefcourt AM, Delwiche SR, Kang S, Lee K (2011) Line-scan hyperspectral imaging platform for agro-food safety and quality evaluationl system enhancement and characterization. Trans ASABE 54(2):703–711

Lawrence KC, Park B, Windham WR, Mao C (2003) Calibration of a pushbroom hyperspectral imaging system for agricultural inspection. Trans ASAE 46(2):513–521

Lawrence KC, Park B, Windham G, Thai CN (2007) Evaluation of LED and tungsten-halogen lighting for fecal contaminant detection. Appl Eng Agric 23(6): 811–818

Li C, Gitaitis R, Tollner B, Sumner P, MacLean D (2009) Onion sour skin detection using a gas sensor array and support vector machine. Sens Instrumen Food Qual 3(4):193–202

Liao K, Paulsen M, Reid J (1994) Real-time detection of colour and surface defects of maize kernels using machine vision. J Agric Eng 59:263–271

Lorente D, Aleixos N, Gómez-Sanchis J, Cubero S, García-Navarrete O, Blasco J (2012) Recent advances and applications of hyperspectral imaging for fruit and vegetable quality assessment. Food and Bioprocess Technol 5(4):1121–1142

Lu R (2001) Predicting firmness and sugar content of sweet cherries using near-infrared diffuse reflectance spectroscopy. Trans ASAE 44(5):1265–1271

Lu R (2003) Imaging spectroscopy for assessing internal quality of apple fruit. In: ASABE annual international meeting, Las Vegas. Paper Number 036012

Lu R (2004) Multispectral imaging for predicting firmness and soluble solids content of apple fruit. Postharvest Biol Technol 31(2):147–157

Lu R, Chen YR (1998) Hyperspectral imaging for safety inspection of food and agricultural products. Proc SPIE 3544: 121–133
Lu R, Peng Y (2006) Hyperspectral scattering for assessing peach fruit firmness. Biosyst Eng 93:161–171
Lu R, Guyer D, Beaudry R (2000) Determination of firmness and sugar content of apples using near-infrared diffuse reflectance. J Texture Stud 31(6):615–630
Mahesh S, Manickavasagan A, Jayas DS, Paliwal J, White NDG (2008) Feasibility of near-infrared hyperspectral imaging to differentiate canadian wheat classes. Biosyst Eng 101 (1):50–57
National Onion Association (2008) About onions: bulb onion production. http://www.onions-usa.org/about/season.asp. Retrieved 2 Aug 2008
Nicolaï BM, Beullens K, Bobelyn E, Peirs A, Saeys W, Theron KI, Lammertyn J (2007) Nondestructive measurement of fruit and vegetable quality by means of NIR spectroscopy: a review. Postharvest Biol Technol 46(2):99–118
Osborne B, Fearn T (1986) Near infrared spectroscopy in food analysis. Longman Scientific and Technical, Harlow
Park B, Lawrence KC, Windham WR, Buhr RJ (2001) Hyperspectral imaging for detecting fecal and ingesta contamination on poultry carcasses. In: ASAE annual international meeting, Sacramento. Paper Number 013130
Park B, Lawrence K, Windham W, Buhr R (2002) Hyperspectral imaging for detecting fecal and ingesta contamination on poultry carcasses. Trans ASAE 45(6):2017–2026
Park B, Abbott J, Lee K, Choi C, Choi K (2003) Near-infrared diffuse reflectance for quantitative and qualitative measurement of soluble solids and firmness of delicious and gala apples. Trans ASAE 46(6):1721–1732
Park B, Yoon SC, Lee S, Sundaram J, Windham WR, Lawrence KC (2012) Acousto-optic tunable filter hyperspectral microscope imaging method for characterizing spectra from foodborne pathogens. Trans ASABE 55(5): 1997–2006
Qin J, Lu R (2005) Detection of pits in tart cherries by hyperspectral transmission imaging. Trans ASAE 48(5):1963–1967
Rajkumar P, Wang N, Elmasry G, Raghavan G, Gariepy Y (2012) Studies on banana fruit quality and maturity stages using hyperspectral imaging. J Food Eng 108:194–200
Reich G (2005) Near-infrared spectroscopy and imaging: basic principles and pharmaceutical applications. Adv Drug Deliv Rev 57(8):1109–1143
Schaare P, Fraser D (2000) Comparison of reflectance, interactance and transmission modes of visible-near infrared spectroscopy for measuring internal properties of kiwifruit (actinidia chinensis). Postharvest Biol Technol 20(2):175–184
Schwartz HF, Mohan SK (2008) Compendium of onion and garlic diseases and pests, 2nd edn. The American Phytopathological Society, St. Paul
Shahin MA, Tollner EW, Gitaitis RD, Sumner DR, Maw BW (2002) Classification of sweet onions based on internal defects using image processing and neural network techniques. Trans ASAE 45(5):1613–1618
Shewfelt RL, Prussia SE (1993) Postharvest handling a system approach. Academic, San Diego
Singh CB, Jayas DS, Paliwal J, White NDG (2009) Detection of insect-damaged wheat kernels using near-infrared hyperspectral imaging. J Stored Prod Res 45(3):151–158
Slaughter D, Barrett D, Boersig M (1996) Nondestructive determination of soluble solids in tomatoes using near infrared spectroscopy. J Food Sci 61(4):695–697
Slaughter D, Crisosto CH (1998) Nondestructive internal quality assessment of kiwifruit using near-infrared spectroscopey. Semin Food Anal 3:131–140
Slaughter D, Thompson J, Tan E (2003) Nondestructive determination of total and soluble solids in fresh prune using near infrared spectroscopy. Postharvest Biol Technol 28(3):437–444
Tao Y (1998) Closed loop search method for on-line automatic calibration of multi-camera inspection systems. Trans ASAE 41(5):1549–1555

USDA-NASS (2008) 2004–2005 statistical highlight of US agriculture: crops. http://www.usda.gov/nass/pubs/stathigh//2005/cropindex.htm. Retrieved 23 July 2008

USDA-NASS (2012) U.S. Onion Statistics (94013). http://usda.mannlib.cornell.edu/MannUsda/viewDocumentInfo.do?documentID=1396. Retrieved 3 Jan 2012

Wang W, Li C, Tollner EW, Gitaitis R, Rains G (2012) A liquid crystal tunable filter based shortwave infrared spectral imaging system: design and integration. Comput Electron Agric 80:126–134

Wang W, Li C, Tollner EW, Rains GC, Gitaitis RD (2012) A liquid crystal tunable filter based shortwave infrared spectral imaging system: calibration and characterization. Comput Electron Agric 80:135–144

Wang W, Li C, Tollner EW, Rains GC (2012) Development of software for spectral imaging data acquisition using LabVIEW. Comput Electron Agric 84:68–75

Wang W, Li C, Tollner EW, Gitaitis RD, Rains GC (2012) Shortwave infrared hyperspectral imaging for detecting sour skin (*Burkholderia cepacia*)-infected onions. J Food Eng 109 (1):38–48

Williams P, Norris K (2001) Near-infrared technology in the agricultural and food industries. American Association of Cereal Chemists, St. Pual

Williams P, Geladi P, Fox G, Manley M (2009) Maize kernel hardness classification by near infrared (NIR) hyperspectral imaging and multivariate data analysis. Anal Chim Acta 653 (2):121–130

Yoon SC, Park B, Lawrence KC, Windham WR, Heitschmidt GW (2011) Line-scan hyperspectral imaging system for real-time inspection of poultry carcasses with fecal material and ingesta. Comput Electron Agric 79(2):159–168

Chapter 15
AOTF Hyperspectral Imaging for Foodborne Pathogen Detection

Bosoon Park

15.1 Introduction

Hyperspectral imaging is an imaging technique that combines conventional imaging with spectrophotometry and radiometry. The technique is capable of providing absolute radiometric measurements over a contiguous spectral range for every pixel of an image from target objects. Hyperspectral imaging, which was first developed for earth remote sensing (Melgani and Bruzzone 2004), is now being utilized in medical diagnosis (Lawlor et al. 2002; Carrasco et al. 2003; Sorg et al. 2005; Dicker et al. 2006; Liu et al. 2007a) and biological (Burger and Geladi 2006), agricultural (Gowen et al. 2007), and industrial (Tatzer et al. 2005) applications. Since hyperspectral images contain physical and geometric observations of size, orientation, shape, color, and texture in the visible spectral range, in addition to chemical/molecular information such as water, fat, proteins, and other hydrogen–bonded constituents in near-infrared spectral regions, hyperspectral imaging has been extensively researched for food safety inspection (Kim et al. 2001; Vargas et al. 2005; Yao et al. 2006; Park et al. 2006, 2007a, b; Kong et al. 2006; Liu et al. 2007b; Jun et al. 2009; Peng et al. 2011; Feng and Sun 2012), quality evaluation (Mehl et al. 2004; Nagata et al. 2006) and food processing (Park et al. 2002). Most recently, hyperspectral imaging techniques have been developed for detecting or identifying foodborne pathogens from poultry and meat (Windham et al. 2012; Yoon et al. 2009, 2011, 2013).

Food safety is an important issue for public health; about 3,562 outbreaks of food commodities including poultry, egg, beef, pork, leafy greens, fruits, nuts, and dairy

B. Park (✉)
U.S. Department of Agriculture, Agricultural Research Service, Athens, GA, USA
e-mail: bosoon.park@ars.usda.gov

B. Park, R. Lu (eds.), *Hyperspectral Imaging Technology in Food and Agriculture*,
Food Engineering Series, DOI 10.1007/978-1-4939-2836-1_15

occurred during the past decade in the U.S. (CDC 2012), causing illness to a large number of people. Germs and related foods responsible for most foodborne illnesses are *Salmonella* in eggs, poultry, meat, produce; *Campylobacter* in poultry; *E. coli* O157 in ground beef, leafy greens and raw milk; *Listeria* in deli meats, produce; *Vibrio* in raw oysters; *Norovirus* in sandwiches, salads; and *Toxoplasma* in meats.

In 2011, about 48 million incidences of foodborne illness occurred, resulting in 128,000 hospitalizations and 3,000 deaths in the United States. The cost of foodborne illness in the U.S. is estimated to be approximately $77.7 billion a year (Scharff 2012). Among the serious outbreaks from foodborne pathogens, *Salmonella* had the most infections and incidence cases (15.1 %) followed by *Campylobacter* (13 %) (CDC 2010). Challenges for food safety continue to arise in unpredictable ways due to changes in food production and supply, changes in the environment causing food contamination, rising numbers of multi-state outbreaks, new and emerging germs, toxins and antibiotic resistance, and new and different contaminated foods such as prepackaged raw materials.

Current detection methods for foodborne pathogens include ISO method 6579 (ISO 6579 2002), direct fluorescence antibody detection (Munson et al. 1976), immunodetection such as enzyme-linked immunosorbant assay (ELISA) (Tian et al. 1996), and polymerase chain reaction (PCR) (Correa et al. 2006). One of the most widely applied methods of subtyping is pulsed-field gel electrophoresis (PFGE), a technique in which fragments of the bacterial chromosome generated by digestion with a restriction enzyme selected to cut the DNA into 20–25 pieces are separated by electrophoresis (Gerner-Smidt et al. 2006; Swaminathan et al. 2006; Terajima et al. 2006). However, all these methods have limitations as field testing tools due to the time-consuming, cumbersome procedures for obtaining results and sensitivity concerns. Hence traditional culture-based methods still remain the most reliable and accurate “gold standard” techniques for pathogen detection (Velusamy et al. 2010). This method involves the culturing of an inoculum to amplify the microbial cell numbers followed by plating on a selective or differential media to generate colonies that can be detected based on their distinct colony morphologies. Culture-based methods are very sensitive with good specificity and relatively inexpensive; and they also can give both colony count estimations and qualitative information of the microorganisms present in food samples. However, culture-based methods are labor intensive and take at least 2–3 days for the microorganisms to multiply to visible colonies for a presumptive positive result. Another challenge for culturing methods is that unwanted background microflora grow together with target microorganisms on agar media and often look similar. Hence, highly skilled personnel are required to pick up presumptive-positive colonies by trial and error. However, this limitation can be improved if it is coupled with optical detection methods such as hyperspectral imaging, because they are more sensitive, accurate and rapid for detecting foodborne pathogens.

The research group at the USDA Agricultural Research Service (ARS) facility in Athens, Georgia has developed a macro-scale hyperspectral imaging technique to identify shiga toxin-producing *Escherichia coli* (STEC) serogroups by acquiring both spatial and spectral information from colonies of each STEC serogroup on Rainbow agar plate (Windham et al. 2012; Yoon et al. 2013). The spectral fingerprints of bacteria obtained by a hyperspectral imaging method can be used for detection and identification of pathogens grown on agar media. In particular, a visible/near-infrared hyperspectral imaging technique with multivariate classification models was developed to differentiate colonies of non-O157 STEC as well as *Campylobacter* (Yoon et al. 2009). They found that hyperspectral imaging technique has the potential for rapidly identifying colonies of non-O157 STEC serogroups on Rainbow agar plates inoculated with mixed cultures. Spatial and spectral data analysis demonstrated that differences in the appearance of the non-O157 STEC serogroup colonies are mainly due to the differences in absorption bands and color tones from each colony. Color is the major feature exploited in the classification model for STEC detection with 97 % classification accuracy (Yoon et al. 2013). However, this macro-scale hyperspectral imaging method requires incubation process at least 24 h. for fully grown bacterial colony on agar plates for measurement. With micro-colonies grown in agar media, ARS scientists developed a micro-scale optical method to identify foodborne pathogens with acousto-optic tunable filter (AOTF)-based hyperspectral microscopic imaging technology. In this chapter, we describe a new optical method of detecting and identifying foodborne pathogenic bacteria with hyperspectral microscopic imaging technique and classification methods.

15.2 Hyperspectral Microscope Imaging Technology and Hyperspectral Imaging Platform

A non-invasive optical method with hyperspectral microscope imaging is promising for real-time, in-situ foodborne pathogen detection with less colony biomass or microcolony by minimizing incubation time (Park et al. 2012a, b). In order to understand optical properties of foodborne pathogenic bacteria, a hyperspectral microscope imaging (HMI) system that provides both spatial and spectral information of bacterial samples at the cell level can be an effective tool (Park et al. 2011a). Hyperspectral microscopy has been studied for biological and medical applications (Huebschman et al. 2002; Zimmerman et al. 2003; De Beule et al. 2007; Ibraheem et al. 2006; Vermaas et al. 2008; Gehm and Brady 2009), yet no successful results have been obtained for bacterial live cell identification.

There are several different platforms for hyperspectral imaging, including pushbroom, acousto-optic tunable filter (AOTF), filter wheel, and liquid crystal tunable filter (LCTF). In selecting a platform, we need to consider several parameters, including imaging technique, spectral and spatial resolution, data capturing mode or speed, transmittance intensity, switching speed of the filter, and so on. A pushbroom-based platform commonly uses a dispersive and grating-prism-grating scanning method.

The pushbroom technique has a high spectral resolution and relatively high transmission and variable switching speed for image acquisition. In contrast, AOTF uses solid-state non-linear crystal to generate hyperspectral images. It has a dynamically variable spectral resolution, and a variable sequential band-pass width with a random access ability. A LCTF-based platform, however, has a fixed and priori defined spectral resolution and fixed sequential band-pass with a random access image acquisition. This platform has relatively low transmission. Similar to the AOTF platform, a switching speed is not fast. Thus, the selection of hyperspectral imaging platform is fully dependent on applications. The ARS research group developed two hyperspectral imaging platforms including pushbroom (Windham et al. 2012; Yoon et al. 2013) and AOTF (Park et al. 2012a, b) for the detection of fecal contaminants (Yoon et al. 2011) and foodborne pathogens (Park et al. 2012a, b; Windham et al. 2012; Yoon et al. 2013). They demonstrated the two different hyperspectral imaging platforms, i.e., the pushbroom platform for bacterial colony on agar plates (Windham et al. 2012; Yoon et al. 2013) and AOTF for microscopic imaging for live bacterial cells from micro-colonies on agar plates (Park et al. 2012a, b), respectively.

15.3 AOTF Hyperspectral Microscope Imaging System

Figure 15.1 shows hyperspectral microscope imaging (HMI) system (ChromoDynamics HSi-400, Lakewood, NJ) for spectral image acquisition from foodborne bacterial samples on the glass slides.

The HMI system consists of a Nikon upright microscope (Eclipse e80i, Lewisville, TX), acousto-optic tunable filters (AOTF) (HSi-400, Gooch & Housego, Orlando, FL), a high performance cooled electron-multiplying charge coupled device (EMCCD) 16-bit camera (iXon, Andor Technology, Belfast, Northern Ireland), and dark-field illumination lighting sources (CytoViva 150 Unit, 24W Metal Halide, CytoViva, Auburn, AL). The AOTF used for HMI research has a high-speed, high-throughput, random-access solid-state optical filter with an adjustable optical pass-band and exceptionally high rejected light levels. AOTF delivers diffraction limited image quality with a variable bandwidth resolution as low as 2 nm in the spectral range from 450 to 800 nm. An AOTF-based hyperspectral microscope is scanning spectrophotometers with no moving parts, capable of high speed of scan with random access to any number of wavelengths selected prior to scanning samples.

15.3.1 Principle of Acousto-Optic Tunable Filter

For AOTF modules, a piezoelectric material is attached to one end of the crystal under excitation from an external radio frequency (RF) signal, which produces an acoustic wave that propagates through the crystal. The acoustic wave produces a periodic variation of the refractive index of the crystal in a frequency determined by the RF signal. The interaction of the electromagnetic wave and the acoustic wave

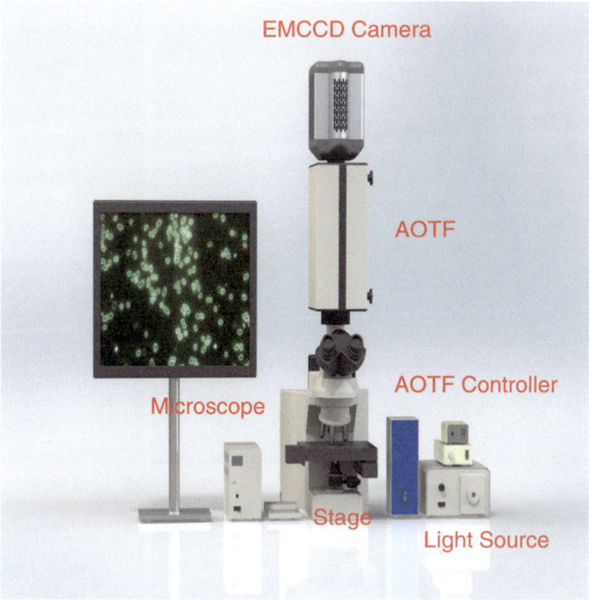

Fig. 15.1 A schematic of AOTF hyperspectral microscope imaging system

causes the crystal to refract selectively a narrow wavelength band. The relationship between diffracted radiation wavelength (λ) and frequency of the acoustic wave (F_a) is given by $\lambda = \frac{\Delta v\ \alpha\ v_a}{F_a}$, where Δv is the difference in the refractive indexes of the bi-refringent crystal, v_a is the velocity of the acoustic wave, and α is an AOTF design parameter.

15.3.2 Microscope Light Sources

Two different lighting sources, but not limited to, such as metal-halide and tungsten-halogen, are available for hyperspectral microscope. A metal-halide lamp is an electric light that produces light by an electric arc through a gaseous mixture of vaporized mercury and metal halides. Metal-halide lamps have high luminous efficacy of around 75–100 lm/W that is about twice that of mercury vapor lights and 3–5 times that of incandescent lights, and produce an intense white light. The output spectrum of a typical metal-halide lamp shows peaks at 385, 422, 497, 540, 564, 583 nm (highest), 630, and 674 nm.

Whereas, a tungsten-halogen lamp is operated at a higher temperature than a standard gas-filled lamp of similar power and operating life, producing light of a higher luminous efficacy and color temperature. A halogen lamp produces a continuous spectrum of light from near ultraviolet to deep into the infrared. Since the lamp filament can operate at a higher temperature than a non-halogen lamp, the spectrum is shifted toward blue, producing light with a higher effective color temperature. Figure 15.2 illustrates hyperspectral microscopic images of

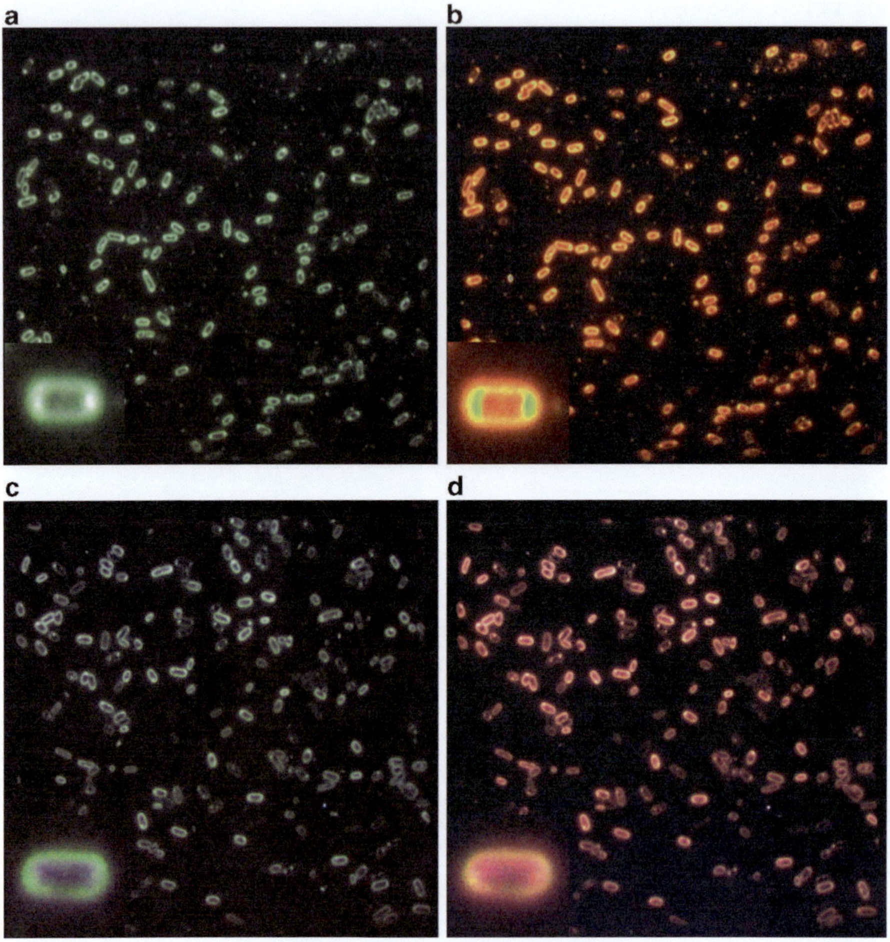

Fig. 15.2 Illustration of *Salmonella* Enteritidis (**a**, **b**) and *Salmonella* Typhimurium (**c**, **d**) scanned by a hyperspectral microscope imaging system with two different light sources; metal-halide (**a**, **c**) and tungsten-halogen (**b**, **d**). *Note*: each image contains sub-image from a single cell in the left corner

Salmonella Enteritidis and Typhimurium scanned with metal-halide and tungsten-halogen lighting sources. Spectral patterns from *S*. Enteritidis and *S*. Typhimurium are different depending on lighting excited to bacteria samples. Figure 15.3 illustrates spectral difference between metal-halide and tungsten-halogen as lighting sources. Also, different peaks of scattering intensity from the cells of *S*. Enteritidis and *S*. Typhimurium are observed at various electromagnetic spectral bands. Thus, selection of lighting source is important for HMI applications, especially for pathogenic bacteria identification and characterization, because the spectral characteristics change with lighting sources.

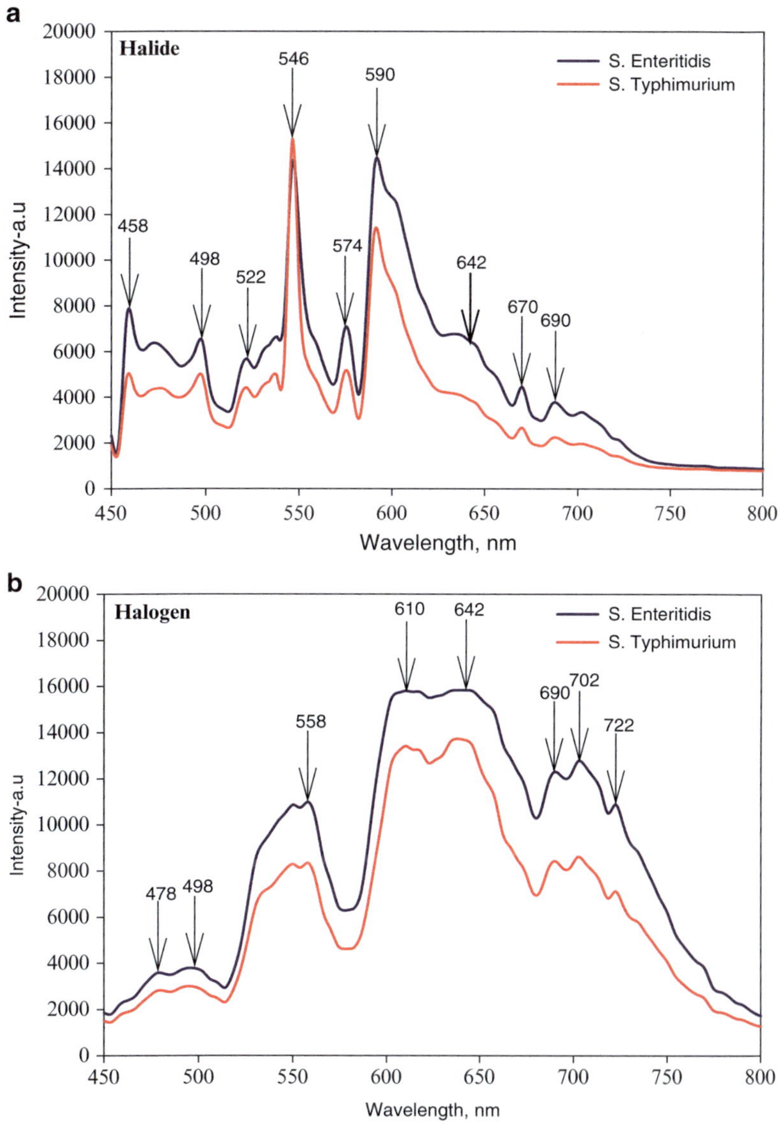

Fig. 15.3 Illustration of uncorrected, raw spectra from *S*. Enteritidis and *S*. Typhimurium with (**a**) metal-halide and (**b**) tungsten-halogen

15.3.3 Bright-Field and Dark-Field Illumination

Among the bright-field, dark-field, and fluorescence microscopic methods, dark-field describes an illumination technique used to enhance the contrast for unstained samples. Optical microscopy with dark-field illumination is useful for biological

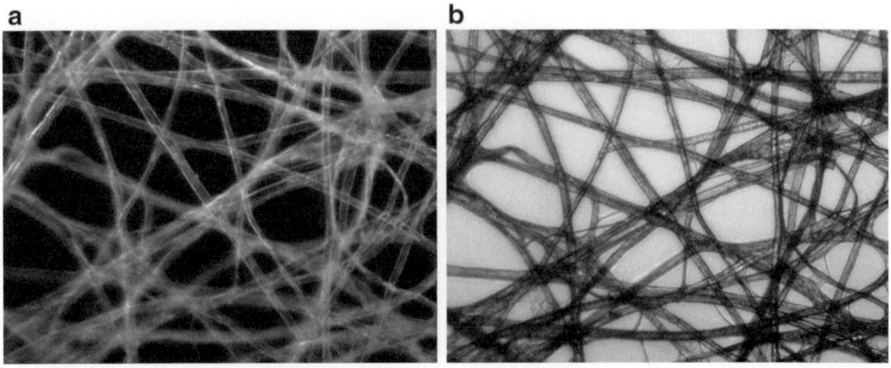

Fig. 15.4 Illustration of sample from (**a**) dark-field illumination; (**b**) bright-field illumination (courtesy of Google image®)

sample measurement, because it is effective in terms of image collection from live and unstained foodborne bacterial cell organisms. Also, the quality of images obtained from this method using scattering phenomena is better for bacteria detection, while the main limitation of dark field microscopy is its low light levels. However, this limitation can be mitigated by the flexible gain and integration time control capability with an electron multiplying charge-coupled device (EMCCD) detector of the hyperspectral microscope imaging system. A dark-field illumination technique produces a dark background with bright objects that can show bacterial cells distinctively. In contrast with bright-field illumination that measures absorbance of target objects, dark-field illumination measures scattered intensity from the bacterial cells. For the upright microscope, light enters the microscope to illuminate the bacterial cell. A specially designed disc blocks some light from the light source (metal-halide or tungsten halogen) and light leaves an outer ring of illumination. The condenser lens focuses the light towards the cells to allow it to enter the cell organisms. At this stage, most energy is directly transmitted, while some is scattered from the cell. The scattered light enters the objective lens, while the directly transmitted light that misses the lens is not collected due to a direct illumination block. Thus, only the scattered light produces the images, while the directly transmitted light is omitted as shown in Fig. 15.4.

15.3.4 Region of Interest of Gram-Negative Images

Figure 15.5 illustrates a hyperspectral microscope image with a region of interest (ROI) from *Salmonella* Enteritidis serotype. To compare spectral signatures between inner and outer cell region from the cell, two scattered imge ROIs, one from inner cell region and the other from outer cell region are acquired from *S.* Enteritidis bacterial cells.

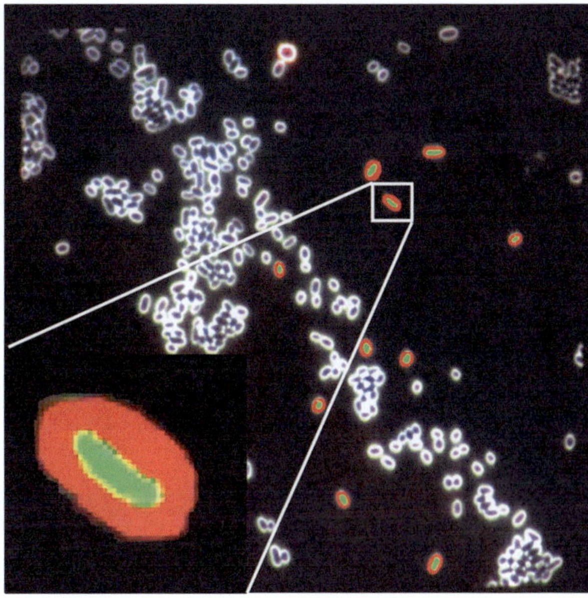

Fig. 15.5 Hyperspectral microscope image with a region of interest (ROI) (**a**) inner cell region (*green*) and (**b**) outer cell region (*red*) from gram-negative (*Salmonella* Enteritidis) bacterial cells

15.3.5 Region of Interest of Gram-Positive Images

Figure 15.6 illustrates a hyperspectral microscope image with a ROI from *Staphylococcus aureus* species. In contrast to the gram-negative bacteria, these species have round shapes. Similar to the *Salmonella*, two ROIs, one from inner region (green) and the other from outer region (red) of the cell, are acquired from *Staphylococcus aureus* bacterial cells to compare spectral signatures between the inner and outer cells of gram-positive bacteria.

15.4 Preparation of Bacteria Cell Cultures

For the hyperspectral microscopic image acquisition, foodborne pathogenic bacteria cell cultures are prepared by inoculating pure isolates form poultry carcass rinses into trypticase soy broth (TSB) tubes and incubated at 35 °C for 18–24 h. The culture of bacteria species grown overnight are centrifuged at 5000 rpm for 10 min. The bacterial pellet is re-suspended in water. From the culture of serotypes or serogroups of bacteria species, tenfold serial dilutions are prepared in 0.1 % peptone water and 10^{-6} final dilutions are plated onto selective agar media such as brilliant green sulfa (BGS) agar plates in duplicate for *Salmonella* (gram-negative) and *Staphylococcus* (gram-positive). From the plates incubated at 35 °C for 24 h, one colony is picked from BGS plate of each *Salmonella* serotype and re-suspended in 10 μL of water. For hyperspectral microscope imaging, 3 μL of

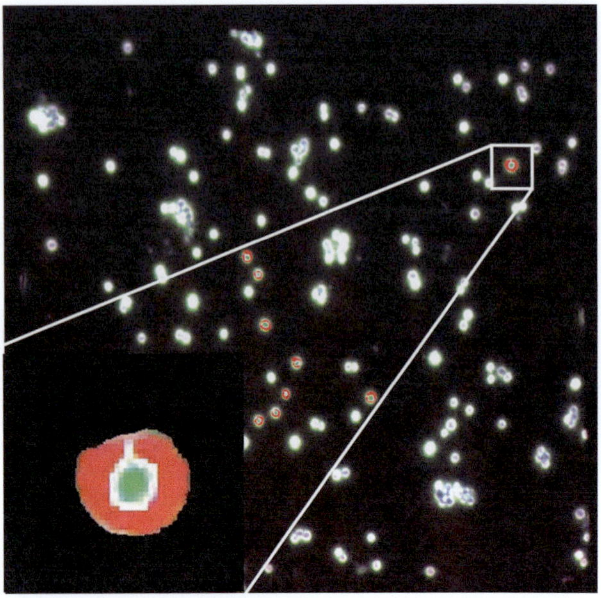

Fig. 15.6 Hyperspectral microscope image with a region of interest (ROI) (**a**) inner cell region (*green*) and (**b**) outer cell region (*red*) from gram-positive (*Staphylococcus aureus*) bacterial cells

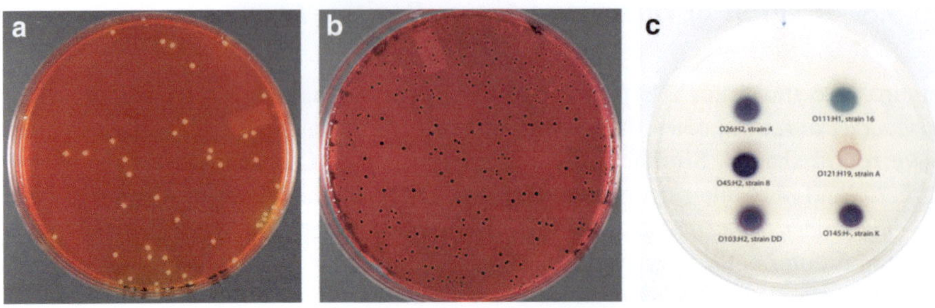

Fig. 15.7 Illustration of bacteria growth (**a**) *Salmonella* Enteritidis, (**b**) *Salmonella* Typhimurium on XLT4 (Xylose Lysine Tergitol-4) agar media, and (**c**) STEC on RBA (*rainbow agar*)

bacterial suspension from the samples are spread on microscopic glass slide in the area of approximately 20 × 20 mm. The slide is dried for 10 min in the biosafety cabinet (Nuaire, Labgard Class II, Type A2 BSC, Plymouth, MN) followed by adding 0.8 μL water on the center for a cover slip prior to scanning the sample. Figure 15.7 illustrates bacterial cultures grown on selective agar media XLT4 (Xylose Lysine Tergitol-4) for *Salmonella* and rainbow agar (RBA) for *E. coli*.

15.5 Bacterial Sample Fixation for Microscope

To prepare bacterial samples for HMI scan, 200 μL of 1× PBS (Phosphate-buffered saline) (filter sterile) in 1.5 mL centrifuge tube is added to the culture. A few colonies are scrapped and re-suspended in PBS buffer followed by vortex mixing for 3–5 s for homogeneous mixture. For formalin fixed samples, 6 μL of formalin (2 % final concentration) in 100 μL bacterial suspension is added and waited for 1 h before making a slide for scan. For live and wet mount slide preparation, 10 μL of PBS is applied on the top of the bacterial suspension and a cover slip is placed on the slide. Samples are then scanned under the hyperspectral microscope immediately with oil drops on top of the cover slip. For dry slides, 10 μL of bacterial suspension is applied on the slide followed by drying for 15 min in the biosafety cabinet.

15.6 Immobilization of Live Bacterial Cells

Figure 15.8 illustrates bacterial live cells with different treatments of fixation. Since hyperspectral image acquisition at the wavelengths between 450 and 800 nm with a 4-nm increment requires longer time than regular microscope imaging, the challenge of the HMI method is to completely immobilize live cells during image acquisition; otherwise, hyperspectral data or hypercube are unable to represent spectral characteristics of live bacterial cells. Thus, immobilization of live bacterial cell is the most crucial aspect to acquire high quality microscopic images.

For cell immobilization on a glass slide, three types of cell movements can be observed: (1) motility of bacterial cell, (2) Brownian motion (Li et al. 2008) caused by bombardment of the bacteria from water molecule that causes significant movement even if cells are immotile, and (3) bacteria movement with the flow of liquid media. For quality hyperspectral image acquisition, five different methods are examined for cell fixation on the glass slide during the scan as follows.

Formalin Fixation: Bacterial cells can be fixed with formalin (2 % concentration). Usually, this method shows no effect on the movement of cells; however, Brownian motion is still causing significant movement of cells even if the cells are immotile.

Adhesive Trapping with Poly (L-lysine) Coated Glass Slide: Poly (L-lysine) is a cationic polymer (positively charged) being used for attaching and immobilizing cells onto the glass substrates. As bacterial cells are negatively charged, cells are able to adhere to positively charged slide. However, this method allows wiggling as well as Brownian motion (Fig. 15.8a).

Agar Coated Slide: Glass slides are coated with 2 % agar which absorbs the water of the cell. With this method cells can be immobilized, yet are not clear, and the background is saturated under the light source of the imaging system.

Dry Method: For this process, approximately 3–10 μL of bacterial suspension is applied on the slide. Cells are dried in the biosafety cabinet for 15 min followed by adding 1 μL of oil on the sample before the sample is covered with a cover slip. One drop of oil is applied onto the cover slip prior to scanning. With this method

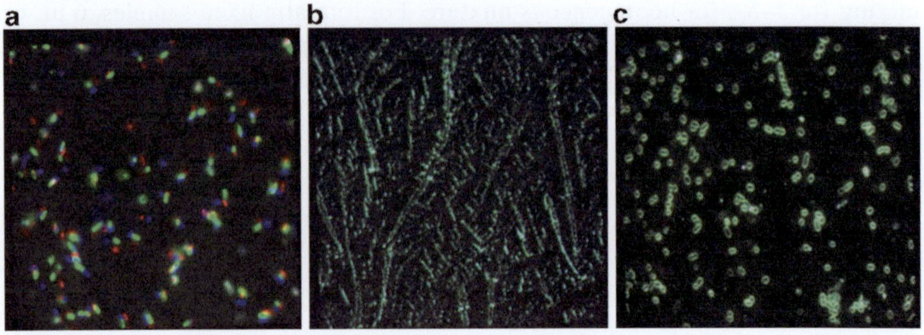

Fig. 15.8 Illustration of bacterial live cells with different treatments of fixation; (**a**) cell moved, (**b**) buffer crystallized, (**c**) cell fixed

only small bright dots instead of shape of cells under the microscope are observed, yet crystallization effect due to the PBS buffer is observed (Fig. 15.8b).

Modified Dry Method: In addition to sample slides prepared by dry method, approximately 1 μL of sterilized deionized (DI) water is applied under the cover slip instead of oil, and then the cover slip is pressed firmly for 30 s to confirm no air bubble remains between sample slide and a cover slip. Although sometimes Brownian motion is observed, most bacteria cells are thoroughly immobilized during the scan for 45 s at least. Hyperspectral microscope images can be successfully acquired without any bacterial mobility during the scan with this method (Fig. 15.8c).

15.7 Gram-Negative and Gram-Positive Bacteria

Bacteria are traditionally divided into two main groups, gram-negative and gram-positive based on their gram-stain retention. This classification system is ambiguous, as it can refer to three distinct aspects including staining result, cell-envelope organization and taxonomic group. Compared to gram-positive bacteria, gram-negative bacteria are more resistant against antibodies because of their impenetrable cell wall from additional outer cell region. For the hyperspectral microscope imaging, *Salmonella* and *Staphylococcus* are demonstrated as gram-negative and gram-positive samples. Figure 15.9 illustrates hyperspectral microscopic composite images based on the wavelengths of 648 nm (red), 550 nm (green), and 436 nm (blue) from *Staphylococcus aureus* (Fig. 15.9a) and *Salmonella* Enteritidis (Fig. 15.9b). These images have different morphology with round shape for gram-positive and ellipse shape for gram-negative bacterial cells.

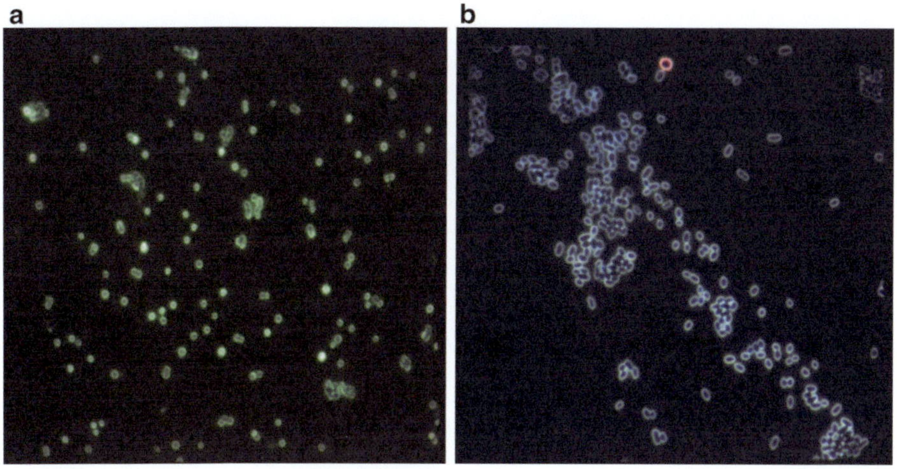

Fig. 15.9 Illustration of hyperspectral composite images from (**a**) gram-positive (*Staphylococcus aureus*) and (**b**) gram-negative (*Salmonella* Enteritidis)

15.8 Hyperspectral Microscope Image Acquisition and Analysis

Figure 15.10 illustrates the procedure of hyperspectral microscope image acquisition and analysis from live bacterial cells. Hyperspectral images from five gram-negative *Salmonella* serotypes (Enteritidis, Typhimurium, Heidelberg, Kentucky, and Infantis) and five gram-positive *Staphylococcus* species (aureus, haemolyticus, hyicus, sciuri, and simulans) are acquired with an AOTF-based hyperspectral microscope imaging (HMI) system. Visible/NIR hyperspectral microscope images are collected with a TIFF format at the wavelength ranges from 450 to 800 nm with 2 nm bandwidth, 4 nm spectral intervals with a scanning exposure time of 250 ms and a gain of 9 selected from parameter optimization for quality image acquisition (Park et al. 2012a, b). All images are acquired with a dark-field illumination equipped with metal-halide lighting in spectral sweep mode for collecting 89 contiguous spectral images (Park et al. 2012b). The images acquired with TIFF format are converted to hyperspectral image format or hypercube with HSiAnalysis™ software (Gooch & Housego, Orlando, FL) for further processing and analysis.

Figure 15.11 illustrates the overview of hyperspectral microscope image acquisition scheme with an AOTF platform. Three selective spectral images at 550, 590, 670 nm and their corresponding spectra from *E. coli* are illustrated. The intensities of spectral images at 550, 590 and 670 nm are higher than their neighboring spectral bands. This information can be used for identification of specific bacterial serotypes as well as species with further multivariate data analysis using their spectral signatures.

Fig. 15.10 A flow diagram of procedure for hyperspectral microscope image acquisition and analysis from live bacterial cells

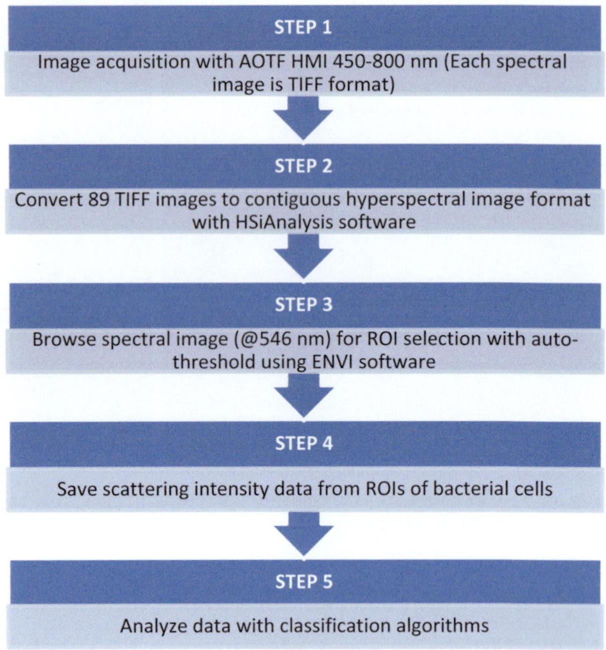

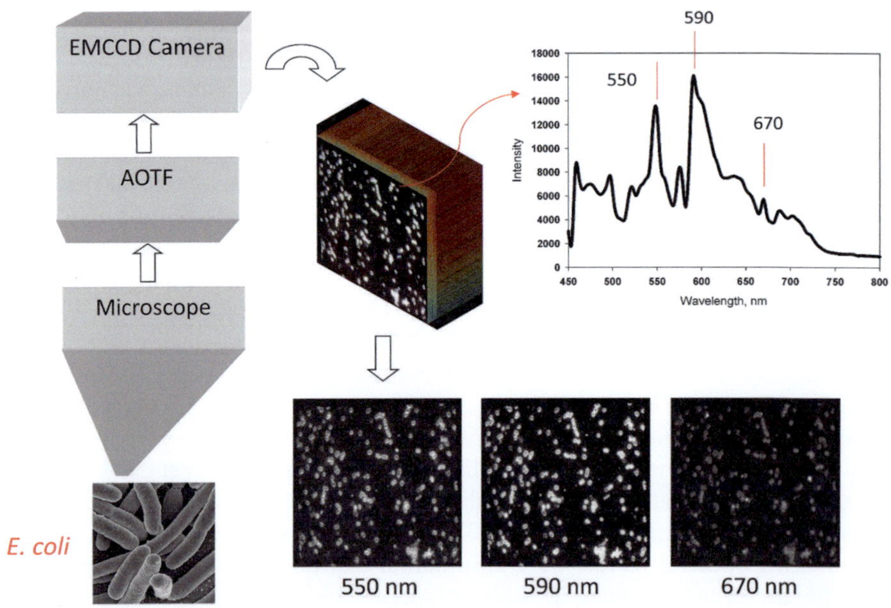

Fig. 15.11 Schematic of AOTF hyperspectral microscope image acquisition

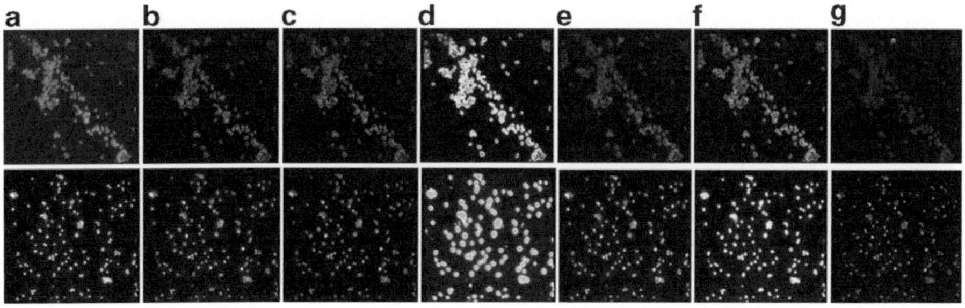

Fig. 15.12 Image comparison of *Salmonella* Enteritidis (*top*) and *Staphylococcus aureus* (*bottom*) at wavelength of (**a**) 458, (**b**) 498, (**c**) 522, (**d**) 546, (**e**) 574, (**f**) 590, and (**g**) 670 nm

Selected spectral images at the wavelength of 458, 498, 522, 546, 574, 590, and 670 nm that show scattering peak intensities from *Salmonella* Enteritidis and *Staphylococcus aureus* are illustrated in Fig. 15.12. Higher scattering intensities at 546 nm are observed than at other wavelengths for both bacterial samples.

After image conversion from TIFF to hypercube, one spectral image of 546 nm (Fig. 15.12d) is selected to create a region-of-interest (ROI) from bacterial cells with ENVI (Exelis Visual Information Solutions, Inc., Boulder, CO) software (version 4.8). The scattering intensity data from the ROI of each cell are saved for further analysis for classification model development.

15.9 Classification Methods

Open-source R software (version 3.0.1) is used for developing classification models with five different algorithms such as Mahalanobis distance (MD), k-nearest neighbor (kNN), linear discriminant analysis (LDA), quadratic discriminant analysis (QDA), and support vector machine (SVM) to identify different species and serotypes using their spectral signatures collected by the HMI system. In addition, Matlab (Mathworks, Natick, MA) software is also used for analyzing spectral characteristics from bacteria to develop classification models to identify bacterial species and serotypes with their spectral signatures.

15.9.1 Mahalanobis Distance

Mahalanobis distance (MD) (De Maesschalck et al. 2000) is commonly used for distance measure with multivariate chemometric techniques in the principal component (PC) space. MD is used for several different purposes such as the detection of outliers, the selection of calibration samples from a large set of measurement, and the observation of difference between two data sets. In pattern recognition, MD is applied to the clustering techniques such as the k-nearest neighbor method (kNN) (Vandeginste et al. 1998), in discrimination techniques such as linear and quadratic analysis (LDA and QDA) (Wu et al. 1996). The MD measures the distance using variance-covariance of two points as $MD_i = \sqrt{(x_i - \overline{x})C_x^{-1}(x_i - \overline{x})^T}$, where C_x is the variance-covariance matrix of two variables, x_1 and x_2. The ellipse represents the equal MDs towards the center point of data. In order to discriminate two classes, a threshold value is determined using a prior knowledge.

15.9.2 *k-Nearest Neighbor*

k-nearest neighbor (kNN) (Vandeginste et al. 1998) is one of the simplest methods for pattern classification. It is a non-parametric method for classifying objects based on the closest training data in a feature space. For high-dimensional data sets (i.e. with more than ten number of dimensions), dimension reduction is usually performed prior to applying the kNN algorithm. The k-nearest neighbor assigns a given unlabeled data based on their closeness with the k closest labeled data in the training data set. Integer k means the number of samples to measure the distance from unknown data. The kNN measures the distance using Euclidean distance from unknown data to the closed samples.

15.9.3 *Linear Discriminant Analysis*

Linear discriminant analysis (LDA) (Dixon and Brereton 2009) is a method to find a linear combination of features that separates two or more classes of objects. Similar to principal component analysis (PCA) and factor analysis, LDA is looking for linear combinations of variables that best explain the data, so that LDA explicitly attempts to model the difference between the classes of data. LDA seeks for reducing dimensionality of data matrix while preserving as much of the class discriminating information as possible. In order to find maximum distance between two classes, LDA utilizes the Mahalanobis distance (Park et al. 2007a).

15.9.4 *Quadratic Discriminant Analysis*

Quadratic discriminant analysis (QDA) is a more general version of the linear classifier, so that it is used in statistical classification to separate two or more classes of objects by a quadric surface. Similar to LDA, QDA assumes the measurements from each class are normally distributed. Unlike LDA, however, there is no assumption that the covariance of each class is identical in QDA. QDA is based on the Mahalanobis distance to measure the discriminating information between two classes as LDA (Dixon and Brereton 2009). In contrast, even though the distributions are significantly non-Gaussian, in the hyperbolic space, the QDA projections may preserve complex structures in the data for classification.

15.9.5 Support Vector Machine

Support vector machine (SVM) (Furey et al. 2000) is a supervised learning algorithm with associated learning algorithms that analyze data and recognize patterns used for classification. SVM can efficiently perform a non-linear classification in high-dimensional feature spaces, and constructs a hyperplane in a high-dimensional space that can be used for classification. Intuitively, a good separation is achieved by the hyperplane that has the largest distance to the nearest training data point of any classes. SVM solves binary classification problems by creating an optimal hyperplane that maximizes the distance between the boundary points of two classes. It can be used to solve both linear and non-linear problems. The hyperplane is defined as

$$f(x) = x^T w + b_0 = 0,$$

where, x is training data set, w is normal to the hyperplane, $x^T w$ is inner product between x and w. $1/\|w\|$ is the perpendicular distance from the hyperplane to the boundary of each class.

To evaluate classification models developed, cross validation needs to be conducted. For instant, the entire data can be divided into 30 data groups for the cross validation purpose. For every single cross validation, 30 % of all data are used for calibration, and the other 70 % of the data are used for validation. This process can be iterated ten times to complete the validation.

15.10 Spectral Data Collection from Bacterial Cells

In order to collect spectral image data from bacterial cells for further processing and analysis, an auto-threshold method is used. Figure 15.13 illustrates a hyperspectral microscope image, along with a ROI, collected from *Staphylococcus simulans* species with the selected threshold method. It is important to select an optimum threshold value for data collection from the cells. For instant, the ROIs for *Staphylococcus simulans* (Fig. 15.13b) are selected with the threshold values of minimum 9,000 and maximum 20,000 from the spectral image at 546 nm. The threshold values vary with species and serotypes of bacteria. For *Salmonella* serotypes, the minimum threshold values of 4,000 for Typhimurium, 5,000 for Heidelberg and Infantis, and 6,000 for both Enteritidis

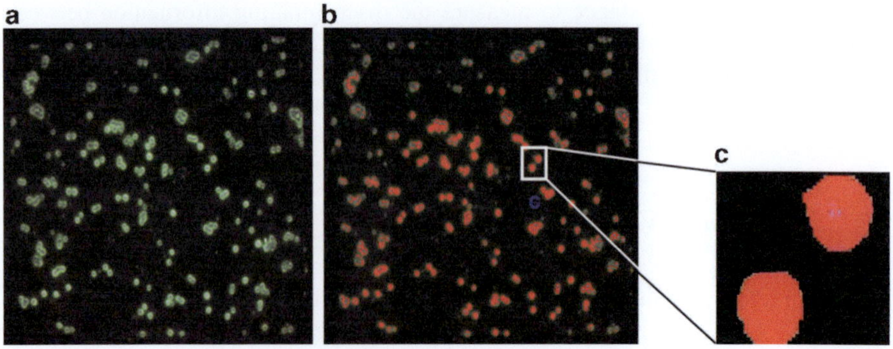

Fig. 15.13 Hyperspectral microscope image with a region of interest (ROI) collected from *Staphylococcus simulans* using an auto-threshold method (min: 9,000 and max: 20,000); (**a**) composite cell image; (**b**) ROI for spectral data collection; (**c**) magnified cells with ROI

and Kentucky are selected. However, maximum value is fixed as 15,000 for collecting spectral data from *Salmonella*. Whereas, for *Staphylococcus* species, different minimum threshold values are selected at 5,000 for *haemolyticus*, 6,000 for *sureus*, 7,000 for hyicus, 8,000 for *sciuri*, and 9,000 for *simulans* species. In the case of *Staphylococcus*, the maximum threshold is selected as 20,000 for all species.

15.10.1 Spectral Characteristics of Salmonella

Figure 15.14 illustrates mean spectra and standard deviation (SD) of scattering intensity from *Salmonella* Enteritidis serotype. As seen in the figures, the spectral signatures from both outer cells (Fig. 15.14a) and inner cells (Fig. 15.14b) are similar to each other, and spectral peaks are observed at the wavelengths of 462, 498, 522, 546, 574, 598, 642, 670, and 690 nm, respectively. However, the variability of each spectrum depends on the scattering wavelength of *S*. Enteritidis bacterial cells.

Based on the spectral data from *Salmonella* Enteritidis serotype, less spectral variation from inner cells is observed than from outer cells. Specifically, spectral variation for both inner and outer cells of *Salmonella* is much less at near-infrared wavelengths than at the visible wavelength range. The largest variation for inner cells is observed at 598 nm, followed by 546 nm. Also, the largest variation for outer cells is observed at 598 nm followed by the wavelength of 642 nm. The range of spectral variation is between 24 and 29 % for inner cells; whereas, the spectral variation for outer cells is slightly higher than for inner cells. This higher scattering variability from outer cells could be caused by the additional cell region at the outer cells of gram-negative bacteria.

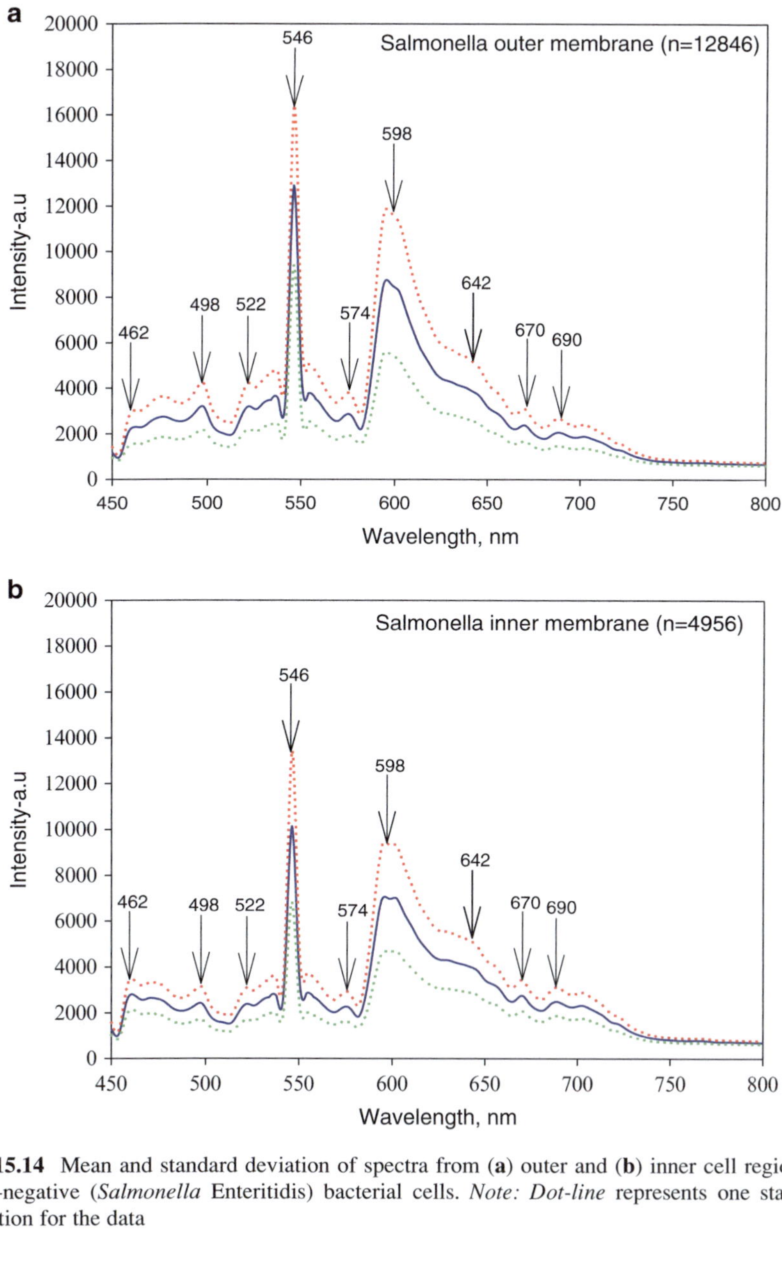

Fig. 15.14 Mean and standard deviation of spectra from (**a**) outer and (**b**) inner cell regions of gram-negative (*Salmonella* Enteritidis) bacterial cells. *Note: Dot-line* represents one standard deviation for the data

15.10.2 Spectral Characteristics of Staphylococcus

Figure 15.15 illustrates mean spectra and standard deviation (SD) of scattering intensity from *Staphylococcus aureus* species. From the figures, the spectral patterns of both inner and outer cells are similar to each other. The spectral peaks for

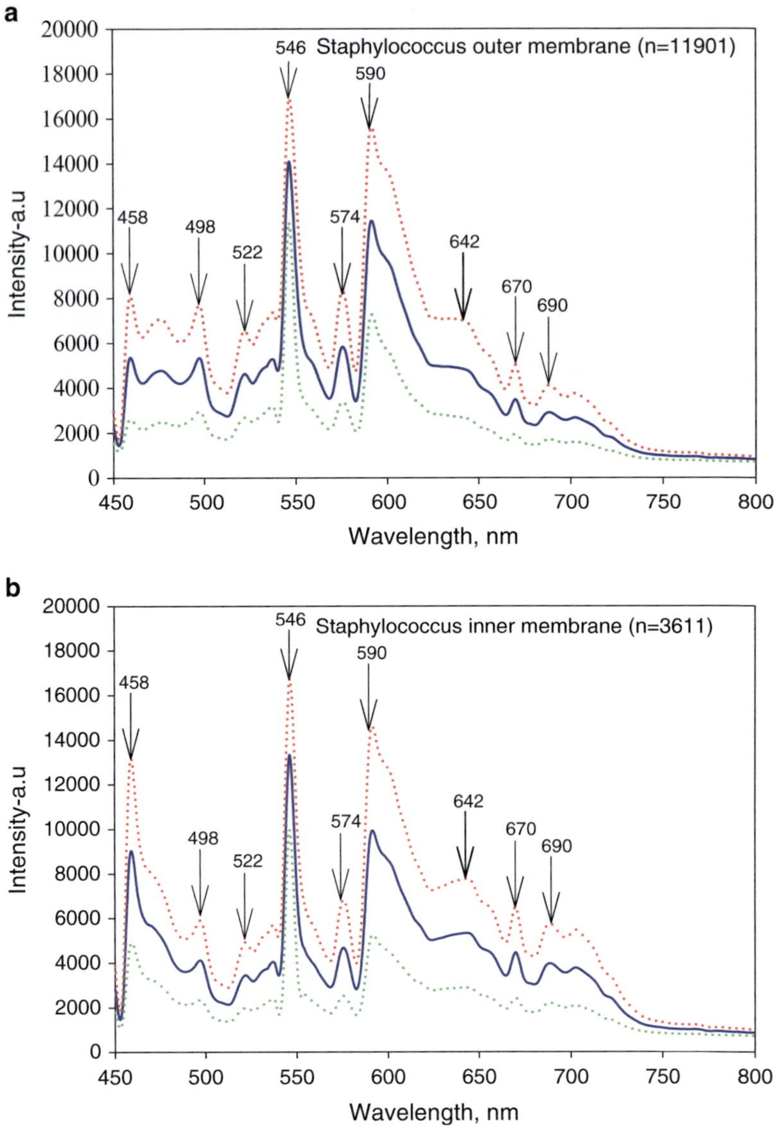

Fig. 15.15 Mean and standard deviation of spectra from (**a**) outer and (**b**) inner cell regions of gram-positive (*Staphylococcus aureus*) bacterial cells. *Note*: *Dot-line* represents one standard deviation for the data

both cases are observed at the wavelengths of 458, 498, 522, 546, 574, 590, 646, 670, and 690 nm, respectively. However, the SD of each spectrum varies with the wavelength of the *Staphylococcus aureus* cells.

In contrast to gram-negative bacteria (*Salmonella*), higher spectral variation from both inner and outer cells is observed than for gram-positive bacteria

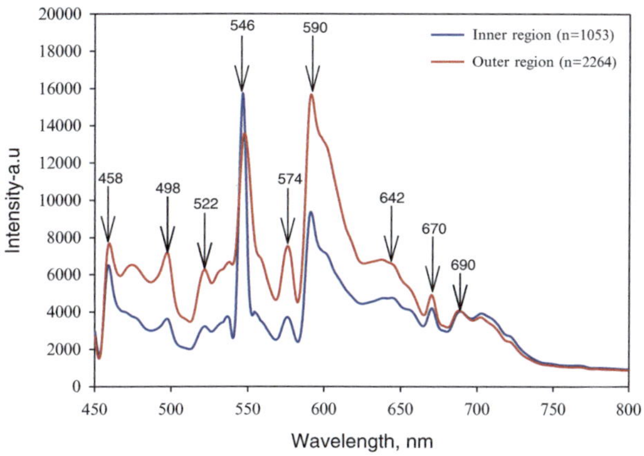

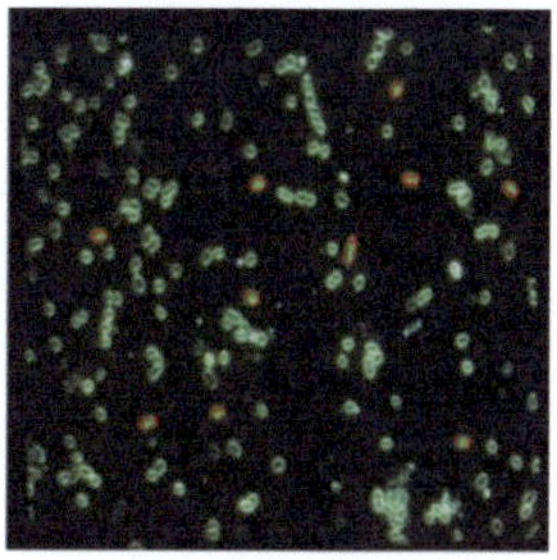

Fig. 15.16 Mean spectra of STEC serotype O121 for the inner and outer regions of the cell: (**a**) O121 image with ROIs of inner (*red*) and outer (*blue*) cells

(*Staphylococcus*). Based on the spectral data from *Staphylococcus aureus* species, higher spectral variation from inner cells is observed than from outer cells. More specifically, spectral variation for both inner and outer cells of *Staphylococcus* is much less at 546 nm, which represents the strongest excitation band for the hyperspectral microscopic images generated by the metal-halide lighting source. The highest variation of inner cells is observed at 590 nm followed by 458 nm, 646 nm, and 670 nm, respectively. However, the highest variation of outer cells is observed at 458 nm followed by 646 nm. For gram-positive bacteria, the range of spectral variation at other scattering peaks are between 40 and 45 % for inner cells; whereas, the spectral variation (between 33 and 44 %) for outer cells is broader than for inner cells. The higher scattering variability from the outer cells could be due to the fact that they are more sensitive to dark-field illuminator than inner cells of gram-positive bacteria.

15.10.3 Spectral Characteristics of Escherichia coli

Figure 15.16 illustrates average spectra of shiga toxin-producing E. coli (STEC), particularly O121 serogroup from inner and outer cells. The scattering intensity for outer cells is brighter than for inner cells, except at 546 nm and beyond 700 nm. The brightness at 546 nm might be due to stronger scattering from inner cells or possibly the stronger back-illumination of the metal halide lighting source. The cause for the peak at 546 nm can be determined using a different light source such as quartz halogen. Similar spectral patterns are observed between inner and outer cells, except for the wavelength region of 450–500 nm.

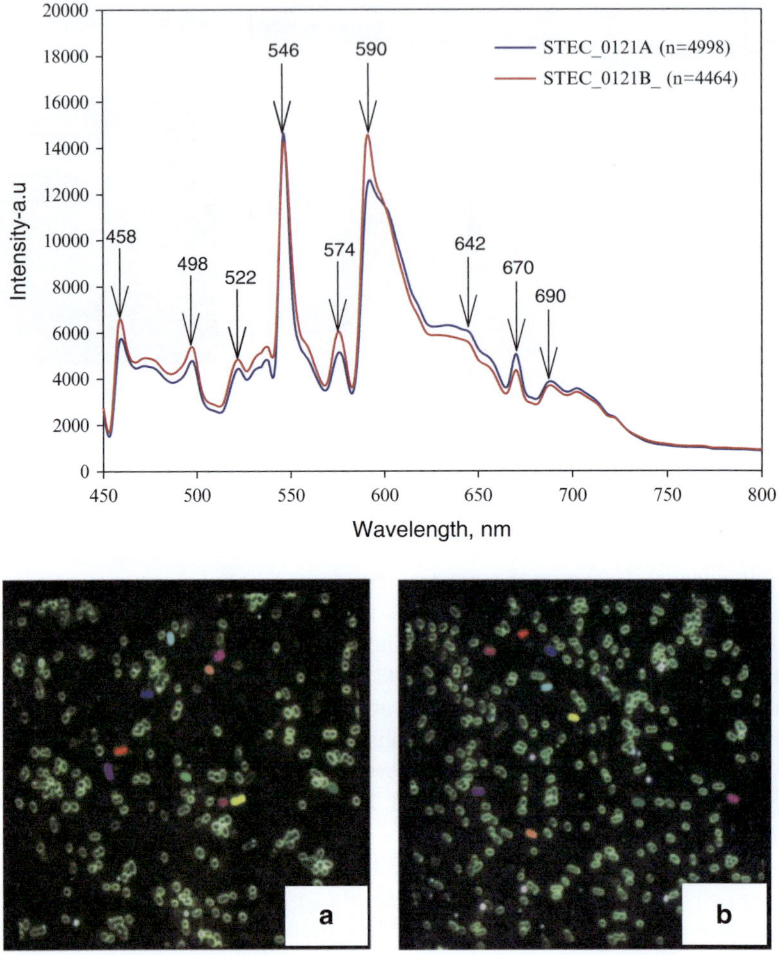

Fig. 15.17 Mean spectra and corresponding regions of interest (ROIs) from the replicate samples of STEC_O121A and STEC_O121B serogroups

In order to analyze spectral variability of each cell within and across the sample treatments, the spectra of ROIs are obtained from STEC O121 serogroup cells of each treatment A and B (Fig. 15.17). The size of ROIs of each cell varies from 345 to 666 pixels in treatment "A" samples; whereas, the sizes of cells in treatment "B" are more uniform varying between 408 and 490 pixels. Regardless of the size of cells, no significant differences in the spectral signatures between samples collected from different agar plates are observed, which means the spectral signatures of STEC O121 are replicable and independent of cell size.

15.10.4 Spectral Characteristics of STEC Serogroups

Figure 15.18 illustrates the mean spectra and corresponding hyperspectral images collected from six different shiga toxin-producing *E. coli* (STEC) serotypes of O26, O45, O103, O111, O121, and O145 with an exposure time of 250 ms and gain of 9 for the HMI system. The ROIs of each serotype are acquired with a threshold from the hypercube. The spectra of each serogroup are generated from as many cells as possibly to minimize the variation among the cells. The spectra are averaged for O26 (14,652 pixels), O45 (13,755 pixels), O103 (16,797 pixels), O111 (11,955 pixels), O121 (12,253 pixels), and O145 (10,409 pixels), respectively. Overall, all spectra have similar patterns with the scattering intensity peaks at 458, 498, 522, 546, 574, 590, 670, and 690 nm. Spectral intensity from the cells of O45 serogroup is higher than for other serogroups, while the spectral intensity from O121 serogroup is lowest at the wavelengths between 450 and 500 nm. However, the intensities of six serogroups between 540 and 560 nm are similar. Distinct peaks at 546 and 574 nm are observed from the spectra. The identical peaks at 546 nm regardless of bacteria serotypes can be explained by the strong excitation of the metal-halide lighting source used for the experiment. The spectral

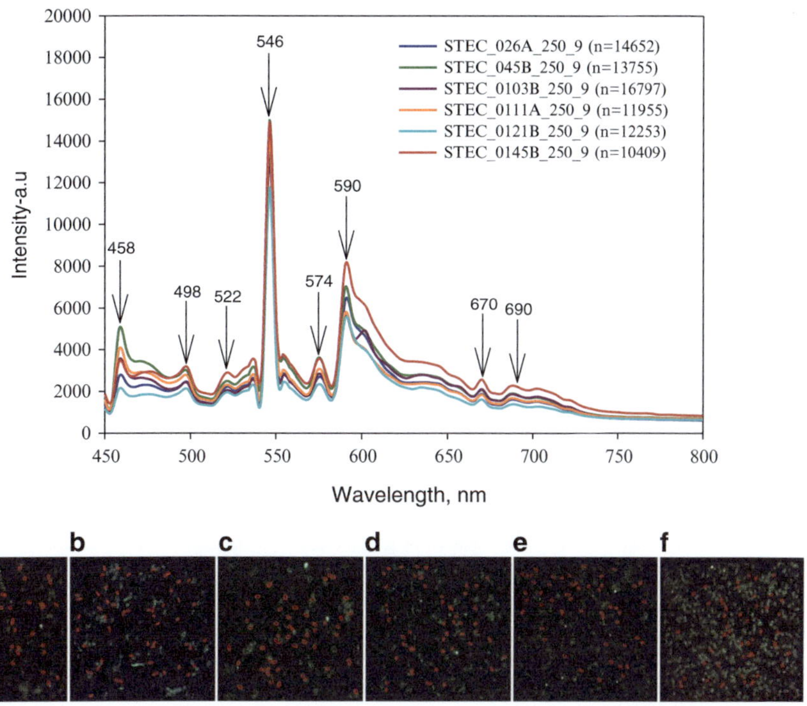

Fig. 15.18 Mean spectra of six different STEC serogroups and corresponding images with ROIs selected by a thresholding method: (**a**) O26A, (**b**) O45B, (**c**) O103B, (**d**) O111A, (**e**) O121B, and (**f**) O145B

characteristics of this pattern can again be confirmed if another lighting source such as quartz halogen is utilized.

The spectra for the O145 serogroup have much higher scattering intensity than for any other serogroups above 590 nm. The spectral patterns, however, are similar to each other with the peaks at 670 and 690 nm, respectively. At these wavelengths, the intensity from the O121 serogroup is the lowest. No significant patterns, however, are observed from all serogroups beyond 750 nm.

Since the growth pattern of cells for each serogroup is not uniform during the incubation process, the size and number of cells are possibly different when the samples are immobilized on the slides for hyperspectral image scanning as shown in Fig. 15.18a–f. The aggregation of cells is also observed from the serogroups of O45 (Fig. 15.18b), O111 (Fig. 15.18d), and O145 (Fig. 15.18f).

15.10.5 Comparison of Spectral Signatures Between **Salmonella** *and* **Staphylococcus**

Figure 15.19 illustrates the comparison of spectra between inner and outer cell regions of *Staphylococcus* and *Salmonella* bacteria. As seen in Fig. 15.19, scattering intensity from the inner cells of *Staphylococcus* is higher than any other scattering intensity at 458, 642, 670, and 690 nm, respectively, yet the highest scattering intensity peaks are observed at 496, 546, 574 and 590 nm for the outer cells of *Staphylococcus*.

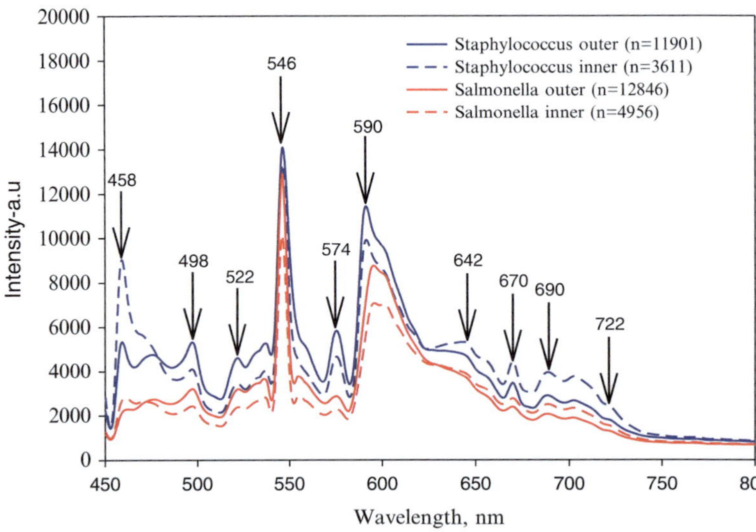

Fig. 15.19 Comparison of spectra between the inner and outer cell regions of *Staphylococcus* and *Salmonella* bacteria

The variation of scattering intensity from *Salmonella* is lower than that from *Staphylococcus*. The intensity variation of *Salmonella* at near-infrared is much lower than at visible wavelengths. The highest variation is observed at 594 nm followed by 498 and 522 nm. Whereas, the lowest variation of scattering intensity of *Staphylococcus* is observed at 546 nm which is close to the highest excitation intensity of the metal-halide lighting source and the highest intensity variation is observed at the wavelength of 458 nm followed by 498 nm, 646 nm, and 670 nm, respectively.

15.11 Classification of Bacteria Species and Serotypes

15.11.1 Accuracy for Identification of Gram-Negative and Gram-Positive Bacteria

For differentiation between gram-negative (*Salmonella*) and gram-positive (*Staphylococcus*), all classification methods including kNN, LDA, QDA, SVM, and MD demonstrate high accuracy (above 99 %) to classify *Salmonella* and *Staphylococcus*. These classification models have low overall errors (less than 0.45 %). Specifically, *Salmonella* bacteria are perfectly classified from *Staphylococcus* using LDA and MD models; whereas, *Staphylococcus* bacteria are identified with 100 % classification accuracy using QDA and SVM. The highest classification accuracy is obtained from SVM (99.99 %), followed by MD (99.98 %), kNN (99.98 %), LDA (99.98 %), and QDA (99.75 %), with the corresponding Kappa coefficients of 0.9998, 0.9995, 0.9995, 0.9995, and 0.9937, respectively. Thus, all models demonstrate an outstanding performance for the classification of gram-negative and gram-positive foodborne bacteria.

15.11.2 Classification of Salmonella *Serotypes*

Figure 15.20 illustrates spectral differences for the five *Salmonella* serotypes. Classification accuracies range between 73.6 % from MD model for *S*. Heidelberg and 97.6 % from SVM model for *S*. Typhimurium. The best classification model for *Salmonella* serotypes is SVM, which has 93.8 % accuracy. Among the five *Salmonella* serotypes, Typhimurium is better classified (93.1 % accuracy) than others (88.5 % or below). Table 15.1 summarizes the classification accuracy of each serotype with the five different models.

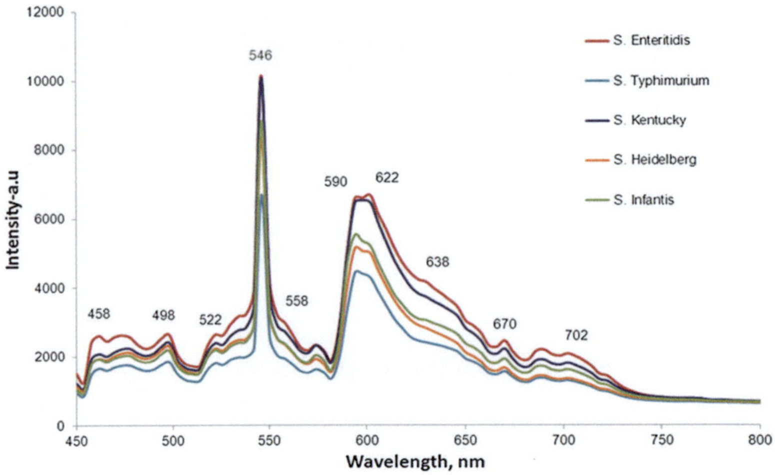

Fig. 15.20 Comparison of spectra from five different *Salmonella* serotypes

Table 15.1 Classification accuracies (in percent) of *Salmonella* serotypes with five classification algorithms

Serotypes	MD	kNN	LDA	QDA	SVM	Mean
S. Enteritidis	87.9	89.7	88.1	83.6	93.6	88.5
S. Typhimurium	91.6	92.5	92.5	91.7	**97.6**	**93.1**
S. Kentucky	80.7	85.9	78.2	84.7	90.7	84.0
S. Heidelberg	73.6	88.4	84.7	81.1	93.0	84.1
S. Infantis	86.4	88.0	87.3	82.8	94.2	87.7

i.e., Mahalanobis distance (MD), k-nearest neighbor (kNN), linear discriminant analysis (LDA), quadratic discriminant analysis (QDA), support vector machine (SVM)

15.11.3 Classification of Staphylococcus *Species*

Figure 15.21 illustrates the spectral differences for the five *Staphylococcus* species. All five classification algorithms perform better for *Staphylococcus* than for *Salmonella*. The classification accuracies range between 91.3 % for *S.* haemolyticus and 99.7 % for *S.* hyicus. The SVM model again achieves the best classification of 99.7 %. Among the five *Staphylococcus* species, hyicus is slightly better classified (97.0 % accuracy) than other species. Table 15.2 summarizes classification accuracies for each species of *Staphylococcus* with the five classification models.

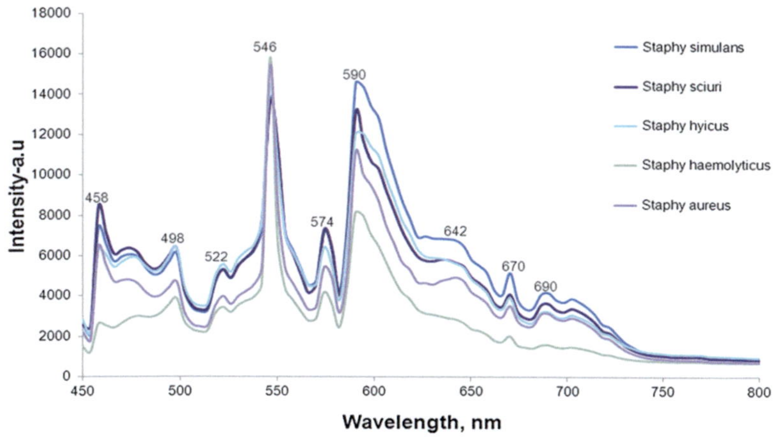

Fig. 15.21 Comparison of spectra from five different *Staphylococcus* species

Table 15.2 Classification accuracy (in percent) of *Staphylococcus* species with five classification algorithms

Serotypes	MD	kNN	LDA	QDA	SVM	Mean
S. aureus	94.9	96.5	95.5	91.6	99.5	95.6
S. haemolyticus	94.7	97.4	95.7	91.3	99.1	**95.6**
S. hyicus	95.1	95.3	98.7	96.5	**99.7**	**97.0**
S. sciuri	94.7	94.6	94.3	96.4	98.7	95.7
S. simulans	94.0	96.5	93.5	98.1	99.1	96.2

i.e., Mahalanobis distance (MD), k-nearest neighbor (kNN), linear discriminant analysis (LDA), quadratic discriminant analysis (QDA), support vector machine (SVM)

15.11.4 Classification of STEC Serogroups

The SVM algorithm performs better than the other four algorithms in the classification of STEC serotypes. The highest accuracy (92 %) is obtained from O45 serogroup followed by O26 (89 %), O145 (84 %), and O111 (72 %), respectively. Classification accuracies, however, are low from O103 (57 %) and O121 (16 %). In conjunction with spectral patterns (Fig. 15.18), the scattering intensity of O45 serogroup between 450 and 500 nm has major contributions to the classification accuracy. For STEC serogroups, using average spectral data obtained from ROIs instead of individual pixels from cell images would help improve classification performance.

A Sparse Kernel-based Ensemble Learning (SKEL) algorithm (Park et al. 2014) is another candidate for the classification of STEC serogroups. Similar to the SVM results, the highest accuracy (92 %) is obtained for O45 serogroup followed by O26 (87 %), O145 (84 %), and O111 (73 %), respectively. The classification accuracies, however, are relatively low for O103 (57 %) and O121 (16 %) serogroups. The low classification accuracies are possibly caused by the variability of spectral image data obtained from the image pixels. To improve classification performance, averaged spectral data obtained from ROIs of each serogroup should be further tested with the SKEL algorithm.

15.11.5 Graphical Description of Classification Performance of STEC Serogroups

Figure 15.22 illustrates the separability of the SVM algorithm for STEC bacteria from the hyperspectral microscope imagery using a graphical description for better understanding of poor classification of STEC O121 serogroup. In the plot, PCA is first applied on the training data set and the first two-dimensions (maximum variance) are extracted. All classes are projected onto these two dimensions. The data are then projected onto a higher dimensional space using Gaussian RBF (radial basis function) kernel where SVM is applied, and then a separating hyperplane (linear in higher dimensional space but non-linear in input space) is built to separate the particular sample. This separating hyperplane is plotted along with the data on two-dimensional spaces. In the figure, the plot illustrates good separability of STEC O45 serogroup (Fig. 15.22a) from other serogroups. According to the separating hyperplane, the separability of STEC O121 serogroup (Fig. 15.22b) from other serogroups is very poor, which confirms the low classification accuracy of this serogroup.

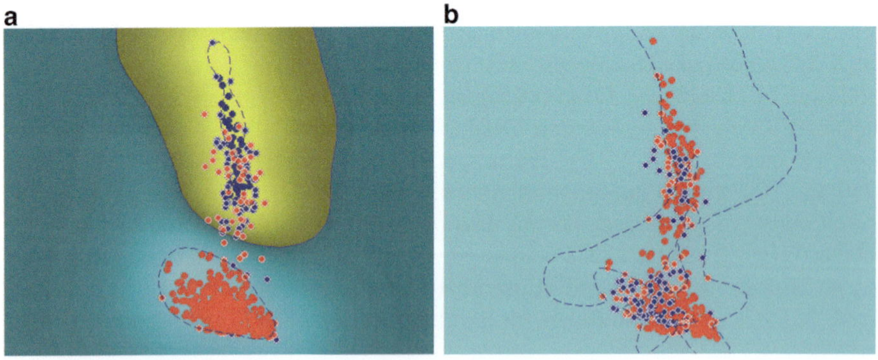

Fig. 15.22 Separability of the support vector machine (SVM) algorithm for (**a**) STEC O45 (*red*) and (**b**) STEC O121 serogroup (*red*) on two-dimensional hyperplane

15.12 Concluding Remarks

To acquire spectral information from live bacterial cells using hyperspectral microscope imaging (HMI) technique, it is important and also challenging to immobilize live cells on glass slides until the scanning has been completed. If live cells move during scanning, all hypercube data would be invalid. Since AOTF-based HMI is able to scan whole cells at different bands, hyperspectral images can be obtained, depending on the AOTF scanning wavelength range and the sensitivity of the detector used. The speed of acquiring hyperspectral images depends upon number of scans and exposure time of the imaging system. With HMI technology, we are able to acquire quality hyperspectral images from foodborne bacterial cells, using a protocol for live cell immobilization on a glass slide. The scattering intensity of hyperspectral images obtained with dark-field illumination for visible and near-infrared spectral ranges can be used for differentiating bacteria species and serotypes. Both spatial and spectral information can be further analyzed to identify and classify the signatures of foodborne pathogenic bacteria with appropriate classification algorithms. However, more research needs to be conducted to understand spectral variability from bacterial cells, optimize parameter selection of the imaging system including light sources. Also, additional algorithms need to be evaluated for hyperspectral image data analysis to improve classification accuracy. For the identification or classification of unknown foodborne pathogen samples, the ground truth regions of interest can be selected as "spectrally pure fingerprints" for bacteria serotypes as well as species. These fingerprints can be compiled into a spectral library for further identification of unknown samples from food matrices.

Mention of trade names or commercial products in this article is solely for the purpose of providing specific information and does not imply recommendation or endorsement by the U.S. Department of Agriculture.

References

Burger J, Geladi P (2006) Hyperspectral NIR imaging for calibration and prediction: a comparison between image and spectrometer data for studying organic and biological samples. Analyst 131:1152–1160

Carrasco O, Gomez RB, Chainani A, Roper WE (2003) Hyperspectral imaging applied to medical diagnoses and food safety. In: Proceedings of SPIE 5097, geo-spatial and temporal image and data exploitation III 215. doi:10.1117/12.502589

CDC (2010) FoodNet surveillance report for 2008

Center for Disease Control and Prevention (CDC) (2012) National Center for Emerging and Zoonotic Infectious Diseases, Division of Foodborne, Waterborne and Environmental Diseases

Correa AA, Toso J, Albarnaz JD, Simoes CMO, Barardi CRM (2006) Detection of *Salmonella* typhimurium in oysters by PCR and molecular hybridization. J Food Qual 29(5):458–469

De Beule P, Owen DM, Manning HB, Talbot CB, Requejo-Isidro J (2007) Rapid hyperspectral fluorescence lifetime imaging. Microsc Res Tech 70:481–484

De Maesschalck R, Jouan-Rimbaud D, Massart DL (2000) The Mahalanobis distance. Chemom Intell Lab Syst 50:1–18

Dicker DT, Lerner J, Van Belle P, Herlyn M, Elder DE (2006) Differentiation of normal skin and melanoma using high resolution hyperspectral imaging. Cancer Biol Ther 5(8):1003–1008

Dixon SJ, Brereton RG (2009) Comparison of performance of five common classifiers represented as boundary methods: Euclidean distance to centroids, linear discriminant analysis, quadratic discriminant analysis, learning vector quantization and support vector machines, as dependent on data structure. Chemom Intell Lab Syst 95:1–17

Feng YZ, Sun DW (2012) Application of hyperspectral imaging in food safety inspection and control: a review. Crit Rev Food Sci Nutr 52(11):1039–1058

Furey TS, Cristianini N, Duffy N, Bednarski DW, Schummer M, Haussler D (2000) Support vector machine classification and validation of cancer tissue samples using microarray expression data. Bioinfomatics 16:906–914

Gehm ME, Brady DJ (2009) High-throughput hyperspectral microscopy. In: Proceedings of SPIE 6090

Gerner-Smidt P, Hise K, Kincaid J, Hunter S, Rolando S, Hyytia-Trees E, Ribot EM, Swaminathan B (2006) Pulsenet Taskforce. PulseNet USA: a five-year update. Foodborne Pathog Dis 3:9–19

Gowen AA, O'Donnell CP, Cullen PJ, Downey G, Frias JM (2007) Hyperspectral imaging—an emerging process analytical tool for food quality and safety control. Trends Food Sci Technol 18(12):590–598

Huebschman ML, Schultz RA, Garner HR (2002) Characteristics and capabilities of the hyperspectral imaging microscope. IEEE Eng Med Biol Mag 21(4):104–117

Ibraheem I, Leitner R, Mairer H, Cerroni L, Smolle J (2006) Hyperspectral analysis of stained histological preparations for the detection of melanoma. In: Proceedings of 3rd international spectral imaging workshop, Graz, Austria, 13 May 2001, pp 24–32

ISO 6579 (2002) Microbiology of food and animal feeding stuffs—horizontal method for the detection of Salmonella spp. Geneva, Switzerland

Jun W, Kim MS, Lee K, Millner P, Chao K (2009) Assessment of bacterial biofilm on stainless steel by hyperspectral fluorescence imaging. J Food Meas Charact 3(1):41–48

Kim MS, Chen YR, Mehl PM (2001) Hyperspectral reflectance and fluorescence imaging system for food quality and safety. Trans ASABE 44(3):721–729

Kong SG, Martin ME, Vo-Dinh T (2006) Hyperspectral fluorescence imaging for mouse skin tumor detection. ETRI J 28(6):770–776

Lawlor J, Fletcher-Holmes DW, Harvey AR, McNaught AI (2002) In vivo hyperspectral imaging of human retina and optic disc. Invest Ophthalmol Vis Sci 43:4350

Li G, Tam LK, Tan JX (2008) Amplified effect of Brownian motion in bacterial near-surface swimming. Proc Natl Acad Sci U S A 105(47):18355–18359

Liu Z, Li Q, Yan J, Tang Q (2007a) A novel hyperspectral medical sensor for tongue diagnosis. Sens Rev 27(1):57–60

Liu Y, Chen YR, Kim MS, Chan DE, Lefcourt AM (2007b) Development of simple algorithms for the detection of fecal contaminants on apples from visible/near infrared hyperspectral reflectance imaging. J Food Eng 81(2):412–418

Mehl PM, Chen YR, Kim MS, Chan DE (2004) Development of hyperspectral imaging technique for the detection of apple surface defects and contaminations. J Food Eng 61(1):67–81

Melgani F, Bruzzone L (2004) Classification of hyperspectral remote sensing images with support vector machines. IEEE Geosci Remote Sens 42(8):1778–1790

Munson TE, Schrade JP, Bisciello NB Jr, Fantasia LD, Hartung WH, O'Connor JJ (1976) Evaluation of an automated fluorescent antibody procedure for detection of salmonella in foods and feeds. Appl Environ Microbiol 31(4):514–521

Nagata M, Tallada JG, Kobayahi T (2006) Bruise detection using NIR hyperspectral imaging for strawberry (Fragaria * ananassa Duch.). Environ Control Biol 44(2):133–142

Park B, Lawrrence KC, Windham WR, Smith DP, Feldner PW (2002) Hyperspectral imaging for food processing automation. In: Proceedings of SPIE 4816, imaging spectrometry VIII 308. doi:10.1117/12.447917

Park B, Lawrence KC, Windham WR, Smith DP (2006) Performance of hyperspectral imaging system for poultry surface fecal contaminant detection. J Food Eng 75(3):340–348

Park B, Windham WR, Lawrence KC, Smith DP (2007a) Contaminant classification of poultry hyperspectral imagery using a spectral angle mapper algorithm. Biosyst Eng 96(3):323–333

Park B, Yoon SC, Lawrence KC, Windham WR (2007b) Fisher Linear discriminant analysis for improving fecal detection accuracy with hyperspectral images. Trans ASABE 50:2275–2283

Park B, Lee S, Yoon SC, Sundaram J, Windham WR, Hinton Jr A, Lawrence KC (2011a) AOTF hyperspectral microscope imaging for foodborne pathogenic bacteria detection. In: Proceedings of SPIE defense, security and sensing, vol 8027, pp 1–11

Park B, Yoon SC, Lee S, Sundaram J, Windham WR, Hinton A Jr, Lawrence KC (2012a) Acousto-optic tunable filter hyperspectral microscope imaging for identifying foodborne pathogens. Trans ASABE 55:1997–2006

Park B, Windham WR, Ladely AR, Gurram P, Kwon H, Yoon SC, Lawrence KC, Narrang N, Cray WC (2012b) Classification of Shiga toxin-producing Escherichia coli (STEC) serotypes with hyperspectral microscope imagery. In: Proceedings of SPIE defense, security and sensing 83690L, pp 1–13

Park B, Windham WR, Ladely SR, Gurram P, Kwon H, Yoon SC, Lawrence KC, Narang N, Cray WC (2014) Detection of non-O157 Shiga toxin-producing Escherichia coli (STEC) serogroups with hyperspectral microscope imaging technology. Trans ASABE 57:973–986

Peng Y, Zhang J, Wang W, Li Y, Wu J, Huang H, Gao X, Jiang W (2011) Potential prediction of the microbial spoilage of beef using spatially resolved hyperspectral scattering profiles. J Food Eng 102(2):163–169

Scharff R (2012) Economic burden from health losses due to foodborne illness in the United States. J Food Prot 75:123–131

Sorg BS, Moeller BJ, Donovan O, Cao Y, Dewhirst MW (2005) Hyperspectral imaging of hemoglobin saturation in tumor microvasculature and tumor hypoxia development. J Biomed Opt 5694(1):74–81

Swaminathan B, Gerner-Smidt P, Ng LK, Lukinmaa S, Kam KM, Rolando S, Gutierrez EP, Binsztein N (2006) Building PulseNet International: an interconnected system of laboratory networks to facilitate timely public health recognition and response to foodborne disease outbreaks and emerging foodborne diseases. Foodborne Pathog Dis 3:36–50

Tatzer P, Wolf M, Panner T (2005) Industrial application for inline material sorting using hyperspectral imaging in the NIR range. Real-Time Imaging 11(2):99–107

Terajima J, Izumiya H, Iyoda S, Mitobe J, Miura M, Watanabe H (2006) Effectiveness of pulsed-field gel electrophoresis for the early detection of diffuse outbreaks due to Shiga toxin producing *Escherichia coli* in Japan. Foodborne Pathog Dis 3:68–73

Tian H, Miyamoto T, Okabe T, Kuramitsu Y, Honjoh KI, Hatano S (1996) Rapid detection of *Salmonella* spp. In foods by combination of a new selective enrichment and a sandwich ELISA using two monoclonal antibodies against Dulcitol1-phosphate dehydrogenase. J Food Prot 59 (11):1158–1163

Vandeginste BMG, Massart DL, Buydens LMC, De Jong S, Lewi PJ, Smeyers-Verbeke J (1998) Handbook of chemometrics and qualimetrics: part B. Elsevier, Amsterdam

Vargas AM, Kim MS, Tao Y, Lefcourt AM, Chen YR, Luo Y, Song Y, Buchanan R (2005) Detection of fecal contamination on cantaloupes using hyperspectral fluorescence imagery. J Food Sci 70(8):471–476

Velusamy V, Arshak K, Korostynska O, Oliwa K, Adley C (2010) An overview of foodborne pathogen detection: in the perspective of biosensors. Biotechol Adv 28:232–254

Vermaas WFJ, Timlin JA, Jones T, Sinclair MB, Nieman LT (2008) *In vivo* hyperspecral confocal fluorescence imaging to determine pigment localization and distribution in cyanobacterial cells. Proc Natl Acad Sci 105(10):4050–4055

Windham WR, Yoon SC, Ladely SR, Heitschmidt GW, Lawrence KC, Park B, Narang N, Cray WC (2012) The effect of regions of interest and spectral pre-processing on the detection of non-O157 Shiga toxin-producing *Escherichia coli* serogroups on agar media by hyperspectral imaging. J Near Infrared Spectrosc 20:547–558

Wu W, Mallet Y, Walczak B, Penninckx W, Massart DL, Heurding S, Erini F (1996) Comparison of regularised discriminant analysis, linear discriminant analysis and quadratic discriminant analysis, applied to NIR data. Anal Chim Acta 329:257–265

Yao H, Hruska Z, Brown RL, Cleveland TE (2006) Hyperspectral bright greenish-yellow fluorescence (BGYF) imaging of aflatoxin contaminated corn kernels. In: Proceedings of SPIE 6381, optics for natural resources, agriculture, and foods 63810B. doi:10.1117/12.686217

Yoon SC, Lawrence KC, Siragusa G, Line J, Park B, Feldner P (2009) Hyperspectral reflectance imaging for detecting a foodborne pathogen: *Campylobacter*. Trans ASABE 52:651–662

Yoon SC, Park B, Lawrence KC, Windham WR, Heitschmidt GW (2011) Line-scan hyperspectral imaging system for real-time inspection of poultry carcasses with fecal material and ingesta. Comput Electron Agric 79:159–168

Yoon SC, Windham WR, Ladely SR, Heitschmidt GW, Lawrence KC, Park B, Narang N, Cray WC (2013) Hyperspectral imaging for differentiating colonies of non-O157 Shiga-toxin producing Escherichia coli (STEC) serogroups on spread plates of pure cultures. J Near Infrared Spectrosc 21:81–95

Zimmerman T, Reitdorf J, Pepperkok R (2003) Spectral imaging and its applications in live cell microscopy. FEBS Lett 546(1):87–92

Index

B. Park, R. Lu (eds.), *Hyperspectral Imaging Technology in Food and Agriculture*, Food Engineering Series, DOI 10.1007/978-1-4939-2836-1

D

M

N

O

Q

R